PROBLEM-SOLVING CASES IN
MICROSOFT® ACCESS™ AND EXCEL®

PROBLEM-SOLVING CASES IN MICROSOFT® ACCESS™ AND EXCEL®

Annual Eleventh Edition

Ellen F. Monk

Joseph A. Brady

Gerard S. Cook

COURSE TECHNOLOGY
CENGAGE Learning®

Australia • Brazil • Japan • Korea • Mexico • Singapore • Spain • United Kingdom • United States

COURSE TECHNOLOGY
CENGAGE Learning

**Problem-Solving Cases in Microsoft®
Access™ and Excel®, Annual 11th Edition**
Ellen F. Monk, Joseph A. Brady,
Gerard S. Cook

Publisher: Joe Sabatino

Senior Acquisitions Editor: Charles
McCormick, Jr.

Senior Product Manager: Kate Mason

Development Editor: Dan Seiter

Editorial Assistant: Anne Merrill

Sr. Brand Manager: Robin LeFevre

Market Development Manager:
Jon Monahan

Marketing Coordinator: Mike Saver

Design Direction, Production Management,
and Composition: PreMediaGlobal

Media Editor: Chris Valentine

Cover Images: Image #: 71633476 (tablet)
Copyright: © mitya73/Shutterstock
Image #: 71717314 (color clouds)
Copyright: © qushe/Shutterstock

Manufacturing Planner: Julio Esperas

For product information and technology assistance, contact us at
Cengage Learning Customer & Sales Support, 1-800-354-9706.
For permission to use material from this text or product,
submit all requests online at **www.cengage.com/permissions**.
Further permissions questions can be emailed to
permissionrequest@cengage.com.

Some of the product names and company names used in this book have been used for identification purposes only and may be trademarks or registered trademarks of their respective manufacturers and sellers.

Library of Congress Control Number: 2012953791

ISBN-13: 978-1-133-62837-8

ISBN-10: 1-133-62837-0

Course Technology
20 Channel Center Street
Boston, MA 02210
USA

Screenshots for this book were created using Microsoft Access and Excel®, and were used with permission from Microsoft.

Microsoft and the Office logo are either registered trademarks or trademarks of Microsoft Corporation in the United States and/or other countries. Course Technology, a part of Cengage Learning, is an independent entity from the Microsoft Corporation, and is not affiliated with Microsoft in any manner.

The programs in this book are for instructional purposes only. They have been tested with care, but are not guaranteed for any particular intent beyond educational purposes. The author and the publisher do not offer any warranties or representations, nor do they accept any liabilities with respect to the programs.

Course Technology, a part of Cengage Learning, reserves the right to revise this publication and make changes from time to time in its content without notice.

Cengage Learning is a leading provider of customized learning solutions with office locations around the globe, including Singapore, the United Kingdom, Australia, Mexico, Brazil, and Japan. Locate your local office at: **www.cengage.com/global**.

Cengage Learning products are represented in Canada by Nelson Education, Ltd.

To learn more about Course Technology, visit **www.cengage.com/coursetechnology**.

Purchase any of our products at your local college store or at our preferred online store, **www.cengagebrain.com**.

Printed in the United States of America
1 2 3 4 5 6 7 16 15 14 13 12

To our development editor, Dan Seiter,
who has made our books the best they can be.

BRIEF CONTENTS

Part 3: Decision Support Cases Using Microsoft Excel Solver

Part 4: Decision Support Case Using Basic Excel Functionality

Part 5: Integration Cases Using Access and Excel

Part 6: Advanced Skills Using Excel

Part 7: Presentation Skills

For two decades, we have taught MIS courses at the university level. From the start, we wanted to use good computer-based case studies for the database and decision-support portions of our courses.

At first, we could not find a casebook that met our needs! This surprised us because we thought our requirements were not unreasonable. First, we wanted cases that asked students to think about real-world business situations. Second, we wanted cases that provided students with hands-on experience, using the kind of software that they had learned to use in their computer literacy courses—and that they would later use in business. Third, we wanted cases that would strengthen students' ability to analyze a problem, examine alternative solutions, and implement a solution using software. Undeterred by the lack of casebooks, we wrote our own, and Course Technology, part of Cengage Learning, published it.

This is the eleventh casebook we have written for Course Technology. The cases are all new, and the tutorials are updated.

As with our prior casebooks, we include tutorials that prepare students for the cases, which are challenging but doable. The cases are organized to help students think about the logic of each case's business problem and then about how to use the software to solve the business problem. The cases fit well in an undergraduate MIS course, an MBA information systems course, or a computer science course devoted to business-oriented programming.

BOOK ORGANIZATION

The book is organized into seven parts:

- Database cases using Access
- Decision support cases using the Excel Scenario Manager
- Decision support cases using the Excel Solver
- A decision support case using basic Excel functionality
- Integration cases using Access and Excel
- Advanced Excel skills
- Presentation skills

Part 1 begins with two tutorials that prepare students for the Access case studies. Parts 2 and 3 each begin with a tutorial that prepares students for the Excel case studies. All four tutorials provide students with hands-on practice in using the software's more advanced features—the kind of support that other books about Access and Excel do not provide. Part 4 asks students to use Excel's basic functionality for decision support. Part 5 challenges students to use both Access and Excel to find a solution to a business problem. Part 6 is a tutorial that teaches advanced skills students might need to complete some of the Excel cases. Part 7 is a tutorial that hones students' skills in creating and delivering an oral presentation to business managers. The next sections explore these parts of the book in more depth.

Part 1: Database Cases Using Access

This section begins with two tutorials and then presents five case studies.

Tutorial A: Database Design

This tutorial helps students understand how to set up tables to create a database, without requiring students to learn formal analysis and design methods, such as data normalization.

Tutorial B: Microsoft Access

The second tutorial teaches students the more advanced features of Access queries and reports—features that students will need to know to complete the cases.

Cases 1–5

Five database cases follow Tutorials A and B. The students must use the Access database in each case to create forms, queries, and reports that help management. The first case is an easier "warm-up" case. The next four cases require more effort to design the database and implement the results.

Part 2: Decision Support Cases Using Excel Scenario Manager

This section has one tutorial and two decision support cases that require the use of the Excel Scenario Manager.

Tutorial C: Building a Decision Support System in Excel

This section begins with a tutorial that uses Excel to explain decision support and fundamental concepts of spreadsheet design. The case emphasizes the use of Scenario Manager to organize the output of multiple "what-if" scenarios.

Cases 6–7

Students can perform these two cases with or without Scenario Manager, although it is nicely suited to both cases. In each case, students must use Excel to model two or more solutions to a problem. Students then use the model outputs to identify and document the preferred solution in a memorandum. The instructor might also require students to summarize their solutions in oral presentations.

Part 3: Decision Support Cases Using Microsoft Excel Solver

This section has one tutorial and two decision support cases that require the use of Excel Solver.

Tutorial D: Building a Decision Support System Using Microsoft Excel Solver

This section begins with a tutorial for using Excel Solver, a powerful decision support tool for solving optimization problems.

Cases 8–9

Once again, students use Excel and the Solver tool in each case to analyze alternatives and identify and document the preferred solution.

Part 4: Decision Support Case Using Basic Excel Functionality
Case 10

The book continues with a case that uses basic Excel functionality. (In other words, the case does not require Scenario Manager or the Solver.) Excel is used to test students' analytical skills in "what-if" analyses.

Part 5: Integration Cases Using Access and Excel
Cases 11 and 12

These cases integrate Access and Excel. The cases show students how to share data between Access and Excel to solve problems.

Part 6: Advanced Skills Using Excel

This part contains one tutorial that focuses on using advanced techniques in Excel.

Tutorial E: Guidance for Excel Cases

Some cases require the use of Excel techniques that are not discussed in other tutorials or cases in this casebook. For example, techniques for using data tables and pivot tables are explained in Tutorial E rather than in the cases themselves. Tutorial E also includes a section on creating macros.

Part 7: Presentation Skills

Tutorial F: Giving an Oral Presentation

Each case includes an optional assignment that lets students practice making a presentation to management to summarize the results of their case analysis. This tutorial gives advice for creating oral presentations. It also includes technical information on charting, a technique that is useful in case analyses or as support for presentations. This tutorial will help students to organize their recommendations, to present their solutions both in words and graphics, and to answer questions from the audience. For larger classes, instructors may want to have students work in teams to create and deliver their presentations, which would model the team approach used by many corporations.

INDIVIDUAL CASE DESIGN

The format of the cases uses the following template:

- Each case begins with a *Preview* and an overview of the tasks.
- The next section, *Preparation*, tells students what they need to do or know to complete the case successfully. Again, the tutorials also prepare students for the cases.
- The third section, *Background*, provides the business context that frames the case. The background of each case models situations that require the kinds of thinking and analysis that students will need in the business world.
- The *Assignment* sections are generally organized to help students develop their analyses.
- The last section, *Deliverables*, lists the finished materials that students must hand in: printouts, a memorandum, a presentation, and files. The list is similar to the deliverables that a business manager might demand.

USING THE CASES

We have successfully used cases like these in our undergraduate MIS courses. We usually begin the semester with Access database instruction. We assign the Access database tutorials and then a case to each student. Then, to teach students how to use the Excel decision support system, we do the same thing: we assign a tutorial and then a case.

Some instructors have asked for access to extra cases, especially in the second semester of a school year. For example, they assigned the integration case in the fall, and they need another one for the spring. To meet this need, we have set up an online "Hall of Fame" that features some of our favorite cases from prior editions. These password-protected cases are available to instructors on the Cengage Learning Web site. Go to *www.cengage.com/coursetechnology* and search for this textbook by title, author, or ISBN. Note that the cases are in Microsoft Office 2010 format.

TECHNICAL INFORMATION

This textbook was tested for quality assurance using the Windows 7 operating system, Microsoft Access 2010, and Microsoft Excel 2010.

Data Files and Solution Files

We have created "starter" data files for the Excel cases, so students need not spend time typing in the spreadsheet skeleton. Cases 11 and 12 also require students to load an Access database file. All these files are on the Cengage Learning Web site, which is available both to students and instructors. Instructors should go to *www.cengage.com* and search for this textbook by title, author, or ISBN. Students will find the files at *www.cengagebrain.com*. You are granted a license to copy the data files to any computer or computer network used by people who have purchased this textbook.

Solutions to the material in the text are available to instructors at *login.cengage.com/sso*. Search for this textbook by title, author, or ISBN. The solutions are password protected.

ACKNOWLEDGMENTS

We would like to give many thanks to the team at Cengage Learning, including our Development Editor, Dan Seiter; Senior Product Manager, Kate Hennessy Mason; and our Content Project Manager, Divya Divakaran. As always, we acknowledge our students' diligent work.

PART 1

DATABASE CASES USING ACCESS

DATABASE DESIGN

This tutorial has three sections. The first section briefly reviews basic database terminology. The second section teaches database design. The third section features a database design problem for practice.

REVIEW OF TERMINOLOGY

You will begin by reviewing some basic terms that will be used throughout this textbook. In Access, a **database** is a group of related objects that are saved in one file. An Access **object** can be a table, form, query, or report. You can identify an Access database file by its suffix, .accdb.

A **table** consists of data that is arrayed in rows and columns. A **row** of data is called a **record**. A **column** of data is called a **field**. Thus, a record is a set of related fields. The fields in a table should be related to one another in some way. For example, a company might want to keep its employee data together by creating a database table called Employee. That table would contain data fields about employees, such as their names and addresses. It would not have data fields about the company's customers; that data would go in a Customer table.

A field's values have a **data type** that is declared when the table is defined. Thus, when data is entered into the database, the software knows how to interpret each entry. Data types in Access include the following:

- Text for words
- Integer for whole numbers
- Double for numbers that have a decimal value
- Currency for numbers that represent dollars and cents
- Yes/No for variables that have only two values (such as 1/0, on/off, yes/no, and true/false)
- Date/Time for variables that are dates or times

Each database table should have a **primary key** field—a field in which each record has a *unique* value. For example, in an Employee table, a field called Employee Identification Number (EIN) could serve as a primary key. (This assumes that each employee is given a number when hired, and that these numbers are not reused later.) Sometimes a table does not have a single field whose values are all different. In that case, two or more fields are combined into a **compound primary key**. The combination of the fields' values is unique.

Database tables should be logically related to one another. For example, suppose a company has an Employee table with fields for EIN, Name, Address, and Telephone Number. For payroll purposes, the company has an Hours Worked table with a field that summarizes Labor Hours for individual employees. The relationship between the Employee table and Hours Worked table needs to be established in the database so you can determine the number of hours worked by any employee. To create this relationship, you include the primary key field from the Employee table (EIN) as a field in the Hours Worked table. In the Hours Worked table, the EIN field is then called a **foreign key** because it's from a "foreign" table.

In Access, data can be entered directly into a table or it can be entered into a form, which then inserts the data into a table. A **form** is a database object that is created from an existing table to make the process of entering data more user-friendly.

A **query** is the database equivalent of a question that is posed about data in a table (or tables). For example, suppose a manager wants to know the names of employees who have worked for the company for more than five years. A query could be designed to search the Employee table for the information. The query would be run, and its output would answer the question.

Queries can be designed to search multiple tables at a time. For this to work, the tables must be connected by a **join** operation, which links tables on the values in a field that they have in common. The common field acts as a "hinge" for the joined tables; when the query is run, the query generator treats the joined tables as one large table.

In Access, queries that answer a question are called *select queries* because they select relevant data from the database records. Queries also can be designed to change data in records, add a record to the end of a table, or delete entire records from a table. These queries are called **update**, **append**, and **delete** queries, respectively.

Access has a **report** generator that can be used to format a table's data or a query's output.

DATABASE DESIGN

Designing a database involves determining which tables belong in the database and then creating the fields that belong in each table. This section begins with an introduction to key database design concepts, then discusses design rules you should use when building a database. First, the following key concepts are defined:

- Entities
- Relationships
- Attributes

Database Design Concepts

Computer scientists have highly formalized ways of documenting a database's logic. Learning their notations and mechanics can be time-consuming and difficult. In fact, doing so usually takes a good portion of a systems analysis and design course. This tutorial will teach you database design by emphasizing practical business knowledge; the approach should enable you to design serviceable databases quickly. Your instructor may add more formal techniques.

A database models the logic of an organization's operation, so your first task is to understand the operation. You can talk to managers and workers, make your own observations, and look at business documents such as sales records. Your goal is to identify the business's "entities" (sometimes called *objects*). An **entity** is a thing or event that the database will contain. Every entity has characteristics, called **attributes**, and one or more **relationships** to other entities. Take a closer look.

Entities

As previously mentioned, an entity is a tangible thing or an event. The reason for identifying entities is that *an entity eventually becomes a table in the database*. Entities that are things are easy to identify. For example, consider a video store. The database for the video store would probably need to contain the names of DVDs and the names of customers who rent them, so you would have one entity named Video and another named Customer.

In contrast, entities that are events can be more difficult to identify, probably because they are more conceptual. However, events are real, and they are important. In the video store example, one event would be Video Rental and another event would be Hours Worked by employees.

In general, your analysis of an organization's operations is made easier when you realize that organizations usually have physical entities such as these:

- Employees
- Customers
- Inventory (products or services)
- Suppliers

Thus, the database for most organizations would have a table for each of these entities. Your analysis also can be made easier by knowing that organizations engage in transactions internally (within the company) and externally (with the outside world). Such transactions are explained in an introductory accounting course, but most people understand them from events that occur in daily life. Consider the following examples:

- Organizations generate revenue from sales or interest earned. Revenue-generating transactions include event entities called Sales and Interest Earned.
- Organizations incur expenses from paying hourly employees and purchasing materials from suppliers. Hours Worked and Purchases are event entities in the databases of most organizations.

Thus, identifying entities is a matter of observing what happens in an organization. Your powers of observation are aided by knowing what entities exist in the databases of most organizations.

Relationships

As an analyst building a database, you should consider the relationship of each entity to the other entities you have identified. For example, a college database might contain entities for Student, Course, and Section to contain data about each. A relationship between Student and Section could be expressed as "Students enroll in sections."

An analyst also must consider the **cardinality** of any relationship. Cardinality can be one-to-one, one-to-many, or many-to-many:

- In a one-to-one relationship, one instance of the first entity is related to just one instance of the second entity.
- In a one-to-many relationship, one instance of the first entity is related to many instances of the second entity, but each instance of the second entity is related to only one instance of the first.
- In a many-to-many relationship, one instance of the first entity is related to many instances of the second entity, and one instance of the second entity is related to many instances of the first.

For a more concrete understanding of cardinality, consider again the college database with the Student, Course, and Section entities. The university catalog shows that a course such as Accounting 101 can have more than one section: 01, 02, 03, 04, and so on. Thus, you can observe the following relationships:

- The relationship between the entities Course and Section is one-to-many. Each course has many sections, but each section is associated with just one course.
- The relationship between Student and Section is many-to-many. Each student can be in more than one section, because each student can take more than one course. Also, each section has more than one student.

Thinking about relationships and their cardinalities may seem tedious to you. However, as you work through the cases in this text, you will see that this type of analysis can be valuable in designing databases. In the case of many-to-many relationships, you should determine the tables a given database needs; in the case of one-to-many relationships, you should decide which fields the tables need to share.

Attributes

An attribute is a characteristic of an entity. You identify attributes of an entity because *attributes become a table's fields*. If an entity can be thought of as a noun, an attribute can be considered an adjective that describes the noun. Continuing with the college database example, consider the Student entity. Students have names, so Last Name would be an attribute of the Student entity and therefore a field in the Student table. First Name would be another attribute, as well as Address, Phone Number, and other descriptive fields.

Sometimes it can be difficult to tell the difference between an attribute and an entity, but one good way is to ask whether more than one attribute is possible for each entity. If more than one instance is possible, but you do not know the number in advance, you are working with an entity. For example, assume that a student could have a maximum of two addresses—one for home and one for college. You could specify attributes Address 1 and Address 2. Next, consider that you might not know the number of student addresses in advance, meaning that all addresses have to be recorded. In that case, you would not know how many fields to set aside in the Student table for addresses. Therefore, you would need a separate Student Addresses table (entity) that would show any number of addresses for a given student.

Database Design Rules

As described previously, your first task in database design is to understand the logic of the business situation. Once you understand this logic, you are ready to build the database. To create a context for learning about database design, look at a hypothetical business operation and its database needs.

Example: The Talent Agency

Suppose you have been asked to build a database for a talent agency that books musical bands into nightclubs. The agent needs a database to keep track of the agency's transactions and to answer day-to-day questions. For example, a club manager often wants to know which bands are available on a certain date at a certain time, or wants to know the agent's fee for a certain band. The agent may want to see a list of all band members and the instrument each person plays, or a list of all bands that have three members.

Suppose that you have talked to the agent and have observed the agency's business operation. You conclude that your database needs to reflect the following facts:

1. A booking is an event in which a certain band plays in a particular club on a particular date, starting and ending at certain times, and performing for a specific fee. A band can play more than once a day. The Heartbreakers, for example, could play at the East End Cafe in the afternoon and then at the West End Cafe on the same night. For each booking, the club pays the talent agent. The agent keeps a five percent fee and then gives the remainder of the payment to the band.

2. Each band has at least two members and an unlimited maximum number of members. The agent notes a telephone number of just one band member, which is used as the band's contact number. No two bands have the same name or telephone number.

3. Band member names are not unique. For example, two bands could each have a member named Sally Smith.

4. The agent keeps track of just one instrument that each band member plays. For the purpose of this database, "vocals" are considered an instrument.

5. Each band has a desired fee. For example, the Lightmetal band might want $700 per booking, and would expect the agent to try to get at least that amount.

6. Each nightclub has a name, an address, and a contact person. The contact person has a telephone number that the agent uses to call the club. No two clubs have the same name, contact person, or telephone number. Each club has a target fee. The contact person will try to get the agent to accept that fee for a band's appearance.

7. Some clubs feed the band members for free; others do not.

Before continuing with this tutorial, you might try to design the agency's database on your own. Ask yourself: What are the entities? Recall that business databases usually have Customer, Employee, and Inventory entities, as well as an entity for the event that generates revenue transactions. Each entity becomes a table in the database. What are the relationships among the entities? For each entity, what are its attributes? For each table, what is the primary key?

Six Database Design Rules

Assume that you have gathered information about the business situation in the talent agency example. Now you want to identify the tables required for the database and the fields needed in each table. Observe the following six rules:

Rule 1: You do not need a table for the business. The database represents the entire business. Thus, in the example, Agent and Agency are not entities.

Rule 2: Identify the entities in the business description. Look for typical things and events that will become tables in the database. In the talent agency example, you should be able to observe the following entities:

- *Things*: The product (inventory for sale) is Band. The customer is Club.
- *Events*: The revenue-generating transaction is Bookings.

You might ask yourself: Is there an Employee entity? Isn't Instrument an entity? Those issues will be discussed as the rules are explained.

Rule 3: Look for relationships among the entities. Look for one-to-many relationships between entities. The relationship between those entities must be established in the tables, using a foreign key. For details, see the following discussion in Rule 4 about the relationship between Band and Band Member.

Look for many-to-many relationships between entities. Each of these relationships requires a third entity that associates the two entities in the relationship. Recall the many-to-many relationship from the college database scenario that involved Student and Section entities. To display the enrollment of specific students in specific sections, a third table would be required. The mechanics of creating such a table are described in Rule 4 during the discussion of the relationship between Band and Club.

Rule 4: Look for attributes of each entity and designate a primary key. As previously mentioned, you should think of the entities in your database as nouns. You should then create a list of adjectives that describe those nouns. These adjectives are the attributes that will become the table's fields. After you have identified fields for each table, you should check to see whether a field has unique values. If such a field exists, designate it as the primary key field; otherwise, designate a compound primary key.

In the talent agency example, the attributes, or fields, of the Band entity are Band Name, Band Phone Number, and Desired Fee, as shown in Figure A-1. Assume that no two bands have the same name, so the primary key field can be Band Name. The data type of each field is shown.

BAND	
Field Name	Data Type
Band Name (primary key)	Text
Band Phone Number	Text
Desired Fee	Currency

Source: © Cengage Learning 2014

FIGURE A-1 The Band table and its fields

Two Band records are shown in Figure A-2.

Band Name (primary key)	Band Phone Number	Desired Fee
Heartbreakers	981 831 1765	$800
Lightmetal	981 831 2000	$700

Source: © Cengage Learning 2014

FIGURE A-2 Records in the Band table

If two bands might have the same name, Band Name would not be a good primary key, so a different unique identifier would be needed. Such situations are common. Most businesses have many types of inventory, and duplicate names are possible. The typical solution is to assign a number to each product to use as the primary key field. For example, a college could have more than one faculty member with the same name, so each faculty member would be assigned an employee identification number. Similarly, banks assign a personal identification number (PIN) for each depositor. Each automobile produced by a car manufacturer gets a unique Vehicle Identification Number (VIN). Most businesses assign a number to each sale, called an invoice number. (The next time you go to a grocery store, note the number on your receipt. It will be different from the number on the next customer's receipt.)

At this point, you might be wondering why Band Member would not be an attribute of Band. The answer is that, although you must record each band member, you do not know in advance how many members are in each band. Therefore, you do not know how many fields to allocate to the Band table for members. (Another way to think about band members is that they are the agency's employees, in effect. Databases for organizations usually have an Employee entity.) You should create a Band Member table with the attributes Member ID Number, Member Name, Band Name, Instrument, and Phone. A Member ID Number field is needed because member names may not be unique. The table and its fields are shown in Figure A-3.

BAND MEMBER	
Field Name	Data Type
Member ID Number (primary key)	Text
Member Name	Text
Band Name (foreign key)	Text
Instrument	Text
Phone	Text

Source: © Cengage Learning 2014

FIGURE A-3 The Band Member table and its fields

Note in Figure A-3 that the phone number is classified as a Text data type because the field values will not be used in an arithmetic computation. The benefit is that Text data type values take up fewer bytes than Numerical or Currency data type values; therefore, the file uses less storage space. You should also use the Text data type for number values such as zip codes.

Five records in the Band Member table are shown in Figure A-4.

Member ID Number	Member Name	Band Name	Instrument	Phone
0001	Pete Goff	Heartbreakers	Guitar	981 444 1111
0002	Joe Goff	Heartbreakers	Vocals	981 444 1234
0003	Sue Smith	Heartbreakers	Keyboard	981 555 1199
0004	Joe Jackson	Lightmetal	Sax	981 888 1654
0005	Sue Hoopes	Lightmetal	Piano	981 888 1765

Source: © Cengage Learning 2014

FIGURE A-4 Records in the Band Member table

You can include Instrument as a field in the Band Member table because the agent records only one instrument for each band member. Thus, you can use the instrument as a way to describe a band member, much like the phone number is part of the description. Phone could not be the primary key because two members might share a telephone and because members might change their numbers, making database administration more difficult.

You might ask why Band Name is included in the Band Member table. The common-sense reason is that you did not include the Member Name in the Band table. You must relate bands and members somewhere, and the Band Member table is the place to do it.

To think about this relationship in another way, consider the cardinality of the relationship between Band and Band Member. It is a one-to-many relationship: one band has many members, but each member in the database plays in just one band. You establish such a relationship in the database by using the primary key field of one table as a foreign key in the other table. In Band Member, the foreign key Band Name is used to establish the relationship between the member and his or her band.

The attributes of the Club entity are Club Name, Address, Contact Name, Club Phone Number, Preferred Fee, and Feed Band?. The Club table can define the Club entity, as shown in Figure A-5.

CLUB	
Field Name	Data Type
Club Name (primary key)	Text
Address	Text
Contact Name	Text
Club Phone Number	Text
Preferred Fee	Currency
Feed Band?	Yes/No

Source: © Cengage Learning 2014

FIGURE A-5 The Club table and its fields

Two records in the Club table are shown in Figure A-6.

Club Name (primary key)	Address	Contact Name	Club Phone Number	Preferred Fee	Feed Band?
East End	1 Duce St.	Al Pots	981 444 8877	$600	Yes
West End	99 Duce St.	Val Dots	981 555 0011	$650	No

Source: © Cengage Learning 2014

FIGURE A-6   Records in the Club table

You might wonder why Bands Booked into Club (or a similar name) is not an attribute of the Club table. There are two reasons. First, you do not know in advance how many bookings a club will have, so the value cannot be an attribute. Second, Bookings is the agency's revenue-generating transaction, an event entity, and you need a table for that business transaction. Consider the booking transaction next.

You know that the talent agent books a certain band into a certain club for a specific fee on a certain date, starting and ending at a specific time. From that information, you can see that the attributes of the Bookings entity are Band Name, Club Name, Date, Start Time, End Time, and Fee. The Bookings table and its fields are shown in Figure A-7.

BOOKINGS	
Field Name	**Data Type**
Band Name	Text
Club Name	Text
Date	Date/Time
Start Time	Date/Time
End Time	Date/Time
Fee	Currency

Source: © Cengage Learning 2014

FIGURE A-7   The Bookings table and its fields—and no designation of a primary key

Some records in the Bookings table are shown in Figure A-8.

Band Name	Club Name	Date	Start Time	End Time	Fee
Heartbreakers	East End	11/21/13	21:30	23:30	$800
Heartbreakers	East End	11/22/13	21:00	23:30	$750
Heartbreakers	West End	11/28/13	19:00	21:00	$500
Lightmetal	East End	11/21/13	18:00	20:00	$700
Lightmetal	West End	11/22/13	19:00	21:00	$750

Source: © Cengage Learning 2014

FIGURE A-8   Records in the Bookings table

Note that no single field is guaranteed to have unique values, because each band is likely to be booked many times and each club might be used many times. Furthermore, each date and time can appear more than once. Thus, no one field can be the primary key.

If a table does not have a single primary key field, you can make a compound primary key whose field values will be unique when taken together. Because a band can be in only one place at a time, one possible solution is to create a compound key from the Band Name, Date, and Start Time fields. An alternative solution is to create a compound primary key from the Club Name, Date, and Start Time fields.

If you don't want a compound key, you could create a field called Booking Number. Each booking would then have its own unique number, similar to an invoice number.

You can also think about this event entity in a different way. Over time, a band plays in many clubs, and each club hires many bands. Thus, Band and Club have a many-to-many relationship, which signals the need for a table between the two entities. A Bookings table would associate the Band and Club tables. You implement an associative table by including the primary keys from the two tables that are associated. In this case, the primary keys from the Band and Club tables are included as foreign keys in the Bookings table.

Rule 5: Avoid data redundancy. You should not include extra (redundant) fields in a table. Redundant fields take up extra disk space and lead to data entry errors because the same value must be entered in multiple tables, increasing the chance of a keystroke error. In large databases, keeping track of multiple instances of the same data is nearly impossible, so contradictory data entries become a problem.

Consider this example: Why wouldn't Club Phone Number be included in the Bookings table as a field? After all, the agent might have to call about a last-minute booking change and could quickly look up the number in the Bookings table. Assume that the Bookings table includes Booking Number as the primary key and Club Phone Number as a field. Figure A-9 shows the Bookings table with the additional field.

BOOKINGS	
Field Name	**Data Type**
Booking Number (primary key)	Text
Band Name	Text
Club Name	Text
Club Phone Number	Text
Date	Date/Time
Start Time	Date/Time
End Time	Date/Time
Fee	Currency

Source: © Cengage Learning 2014

FIGURE A-9   The Bookings table with an unnecessary field—Club Phone Number

The fields Date, Start Time, End Time, and Fee logically depend on the Booking Number primary key—they help define the booking. Band Name and Club Name are foreign keys and are needed to establish the relationship between the Band, Club, and Bookings tables. But what about Club Phone Number? It is not defined by the Booking Number. It is defined by Club Name—*in other words, it is a function of the club, not of the booking.* Thus, the Club Phone Number field does not belong in the Bookings table. It is already in the Club table.

Perhaps you can see the practical data-entry problem of including Club Phone Number in Bookings. Suppose a club changed its contact phone number. The agent could easily change the number one time, in the Club table. However, the agent would need to remember which other tables contained the field and change the values there too. In a small database, this task might not be difficult, but in larger databases, having redundant fields in many tables makes such maintenance difficult, which means that redundant data is often incorrect.

You might object by saying, "What about all of those foreign keys? Aren't they redundant?" In a sense, they are. But they are needed to establish the relationship between one entity and another, as discussed previously.

Rule 6: Do not include a field if it can be calculated from other fields. A **calculated field** is made using the query generator. Thus, the agent's fee is not included in the Bookings table because it can be calculated by query (here, five percent multiplied by the booking fee).

PRACTICE DATABASE DESIGN PROBLEM

Imagine that your town library wants to keep track of its business in a database, and that you have been called in to build the database. You talk to the town librarian, review the old paper-based records, and watch people use the library for a few days. You learn the following about the library:

1. Any resident of the town can get a library card simply by asking for one. The library considers each cardholder a member of the library.

2. The librarian wants to be able to contact members by telephone and by mail. She calls members when books are overdue or when requested materials become available. She likes to mail a thank-you note to each patron on his or her anniversary of becoming a member of the library. Without a database, contacting members efficiently can be difficult; for example, multiple members can have the same name. Also, a parent and a child might have the same first and last name, live at the same address, and share a phone.

3. The librarian tries to keep track of each member's reading interests. When new books come in, the librarian alerts members whose interests match those books. For example, long-time member Sue Doaks is interested in reading Western novels, growing orchids, and baking bread. There must be some way to match her interests with available books. One complication is that, although the librarian wants to track all of a member's reading interests, she wants to classify each book as being in just one category of interest. For example, the classic gardening book *Orchids of France* would be classified as a book about orchids or a book about France, but not both.

4. The library stocks thousands of books. Each book has a title and any number of authors. Also, more than one book in the library might have the same title. Similarly, multiple authors might have the same name.

5. A writer could be the author of more than one book.

6. A book will be checked out repeatedly as time goes on. For example, *Orchids of France* could be checked out by one member in March, by another member in July, and by another member in September.

7. The library must be able to identify whether a book is checked out.

8. A member can check out any number of books in one visit. Also, a member might visit the library more than once a day to check out books.

9. All books that are checked out are due back in two weeks, with no exceptions. The librarian would like to have an automated way of generating an overdue book list each day so she can telephone offending members.

10. The library has a number of employees. Each employee has a job title. The librarian is paid a salary, but other employees are paid by the hour. Employees clock in and out each day. Assume that all employees work only one shift per day and that all are paid weekly. Pay is deposited directly into an employee's checking account—no checks are hand-delivered. The database needs to include the librarian and all other employees.

Design the library's database, following the rules set forth in this tutorial. Your instructor will specify the format of your work. Here are a few hints in the form of questions:

- A book can have more than one author. An author can write more than one book. How would you describe the relationship between books and authors?

- The library lends books for free, of course. If you were to think of checking out a book as a sales transaction for zero revenue, how would you handle the library's revenue-generating event?

- A member can borrow any number of books at one checkout. A book can be checked out more than once. How would you describe the relationship between checkouts and books?

MICROSOFT ACCESS

Microsoft Access is a relational database package that runs on the Microsoft Windows operating system. There are many different versions of Access; this tutorial was prepared using Access 2010.

Before using this tutorial, you should know the fundamentals of Access and know how to use Windows. This tutorial explains advanced Access skills you will need to complete database case studies. The tutorial concludes with a discussion of common Access problems and how to solve them.

To prevent losing your work, always observe proper file-saving and closing procedures. To exit Access, click the File tab and select Close Database, then click the File tab and select Exit. You can also simply select the Exit option to return to Windows. Always end your work with these steps. If you remove your USB key or other portable storage device when database forms and tables are shown on the screen, you will lose your work.

To begin this tutorial, you will create a new database called Employee.

AT THE KEYBOARD

Open a new database. Click the File tab, select New from the menu, and then click Blank database from the Available Templates list. Name the database Employee. Click the file folder next to the filename to browse for the folder where you want to save the file. Otherwise, your file will be saved automatically in the Documents folder. Click the Create button.

Your opening screen should resemble the screen shown in Figure B-1.

Source: Used with permission from Microsoft Corporation

FIGURE B-1   Entering data in Datasheet view

When you create a table, Access opens it in Datasheet view by default. Because you will use Design view to build your tables, close the new table by clicking the *X* in the upper-right corner of the table window that corresponds to Close Table I. You are now on the Home tab in the Database window of Access, as shown in Figure B-2. From this screen, you can create or change objects.

Source: Used with permission from Microsoft Corporation

FIGURE B-2   The Database window Home tab in Access

CREATING TABLES

Your database will contain data about employees, their wage rates, and the hours they worked.

Defining Tables

In the Database window, build three new tables using the following instructions.

AT THE KEYBOARD

Defining the Employee Table

This table contains permanent data about employees. To create the table, click the Create tab and then click Table Design in the Tables group. The table's fields are Last Name, First Name, Employee ID, Street Address, City, State, Zip, Date Hired, and US Citizen. The Employee ID field is the primary key field. Change the lengths of text fields from the default 255 spaces to more appropriate lengths; for example, the Last Name field might be 30 spaces, and the Zip field might be 10 spaces. Your completed definition should resemble the one shown in Figure B-3.

Field Name	Data Type	Description
Last Name	Text	
First Name	Text	
Employee ID	Text	
Street Address	Text	
City	Text	
State	Text	
Zip	Text	
Date Hired	Date/Time	
US Citizen	Yes/No	

Source: Used with permission from Microsoft Corporation

FIGURE B-3 Fields in the Employee table

When you finish, click the File tab, select Save Object As, and then enter a name for the table. In this example, the table is named Employee. Make sure to specify the name of the *table*, not the database itself. (In this example, it is a coincidence that the Employee table has the same name as its database file.) Close the table by clicking the Close button (X) that corresponds to the Employee table.

Defining the Wage Data Table

This table contains permanent data about employees and their wage rates. The table's fields are Employee ID, Wage Rate, and Salaried. The Employee ID field is the primary key field. Use the data types shown in Figure B-4. Your definition should resemble the one shown in Figure B-4.

Field Name	Data Type	Description
Employee ID	Text	
Wage Rate	Currency	
Salaried	Yes/No	

Source: Used with permission from Microsoft Corporation

FIGURE B-4 Fields in the Wage Data table

Click the File tab and then select Save Object As to save the table definition. Name the table Wage Data.

Defining the Hours Worked Table

The purpose of this table is to record the number of hours that employees work each week during the year. The table's three fields are Employee ID (which has a text data type), Week # (number–long integer), and Hours (number–double). The Employee ID and Week # are the compound keys.

In the following example, the employee with ID number 08965 worked 40 hours in Week 1 of the year and 52 hours in Week 2.

Employee ID	Week#	Hours
08965	1	40
08965	2	52

Note that no single field can be the primary key field because 08965 is an entry for each week. In other words, if this employee works each week of the year, 52 records will have the same Employee ID value at the end of the year. Thus, Employee ID values will not distinguish records. No other single field can distinguish these records either, because other employees will have worked during the same week number and some employees will have worked the same number of hours. For example, 40 hours—which corresponds to a full-time workweek—would be a common entry for many weeks.

All of this presents a problem because a table must have a primary key field in Access. The solution is to use a compound primary key; that is, use values from more than one field to create a combined field that will distinguish records. The best compound key to use for the current example consists of the Employee ID field and the Week # field, because as each person works each week, the week number changes. In other words, there is only *one* combination of Employee ID 08965 and Week # 1. Because those values *can occur in only one record*, the combination distinguishes that record from all others.

The first step of setting a compound key is to highlight the fields in the key. Those fields must appear one after the other in the table definition screen. (Plan ahead for that format.) As an alternative, you can highlight one field, hold down the Control key, and highlight the next field.

AT THE KEYBOARD

In the Hours Worked table, click the first field's left prefix area (known as the row selector), hold down the mouse button, and drag down to highlight the names of all fields in the compound primary key. Your screen should resemble the one shown in Figure B-5.

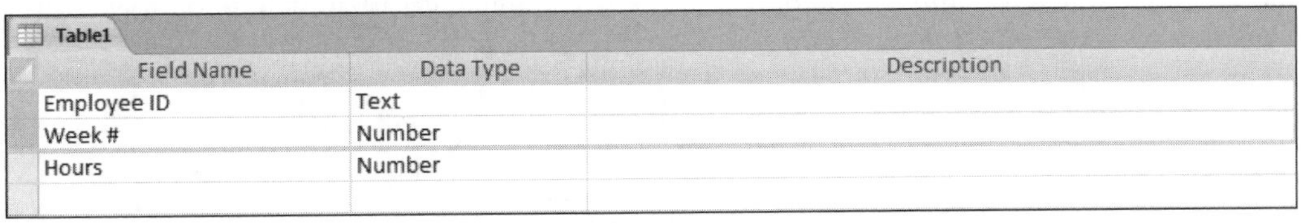

Table1		
Field Name	Data Type	Description
Employee ID	Text	
Week #	Number	
Hours	Number	

Source: Used with permission from Microsoft Corporation

FIGURE B-5   Selecting fields for the compound primary key for the Hours Worked table

Now click the Key icon. Your screen should resemble the one shown in Figure B-6.

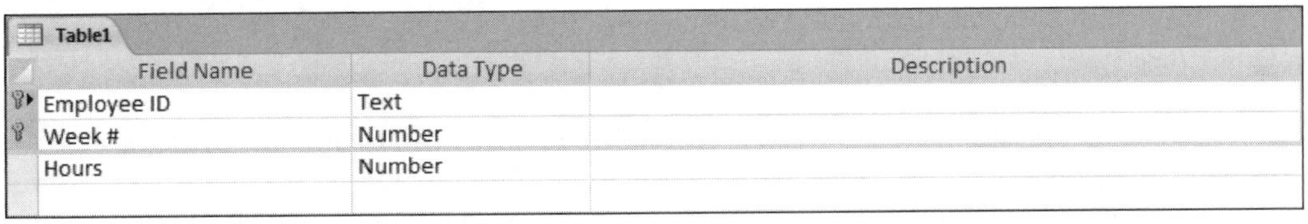

Table1		
Field Name	Data Type	Description
Employee ID	Text	
Week #	Number	
Hours	Number	

Source: Used with permission from Microsoft Corporation

FIGURE B-6   The compound primary key for the Hours Worked table

You have created the compound primary key and finished defining the table. Click the File tab and then select Save Object As to save the table as Hours Worked.

Adding Records to a Table

At this point, you have set up the skeletons of three tables. The tables have no data records yet. If you printed the tables now, you would only see column headings (the field names). The most direct way to enter data into a table is to double-click the table's name in the navigation pane at the left side of the screen and then type the data directly into the cells.

NOTE

To display and open the database objects, Access 2010 uses a navigation pane, which is on the left side of the Access window.

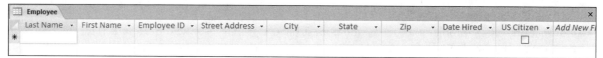
AT THE KEYBOARD

On the Home tab of the Database window, double-click the Employee table. Your data entry screen should resemble the one shown in Figure B-7.

Last Name ▾	First Name ▾	Employee ID ▾	Street Address ▾	City ▾	State ▾	Zip ▾	Date Hired ▾	US Citizen ▾	Add New Fi
*								☐	

Source: Used with permission from Microsoft Corporation

FIGURE B-7   The data entry screen for the Employee table

The Employee table has many fields, some of which may be off the screen to the right. Scroll to see obscured fields. (Scrolling happens automatically as you enter data.) Figure B-7 shows all of the fields on the screen.

Enter your data one field value at a time. Note that the first row is empty when you begin. Each time you finish entering a value, press Enter to move the cursor to the next cell. After you enter data in the last cell in a row, the cursor moves to the first cell of the next row *and* Access automatically saves the record. Thus, you do not need to click the File tab and then select Save Object As after entering data into a table.

When entering data in your table, you should enter dates in the following format: 6/15/10. Access automatically expands the entry to the proper format in output.

Also note that Yes/No variables are clicked (checked) for Yes; otherwise, the box is left blank for No. You can change the box from Yes to No by clicking it.

Enter the data shown in Figure B-8 into the Employee table. If you make errors in data entry, click the cell, backspace over the error, and type the correction.

Last Name ▾	First Name ▾	Employee ID ▾	Street Address ▾	City ▾	State ▾	Zip ▾	Date Hired ▾	US Citizen ▾	Click to Add ▾
Howard	Jane	11411	28 Sally Dr	Glasgow	DE	19702	8/1/2012	☑	
Smith	John	12345	30 Elm St	Newark	DE	19711	6/1/1996	☑	
Smith	Albert	14890	44 Duce St	Odessa	DE	19722	7/15/1987	☑	
Jones	Sue	22282	18 Spruce St	Newark	DE	19716	7/15/2004	☐	
Ruth	Billy	71460	1 Tater Dr	Baltimore	MD	20111	8/15/1999	☐	
Add	Your	Data	Here	Elkton	MD	21921		☑	
*								☐	

Source: Used with permission from Microsoft Corporation

FIGURE B-8   Data for the Employee table

Note that the sixth record is *your* data record. Assume that you live in Elkton, Maryland, were hired on today's date (enter the date), and are a U.S. citizen. Make up a fictitious Employee ID number. For purposes of this tutorial, the sixth record has been created using the name of one of this text's authors and the employee ID 09911.

After adding records to the Employee table, open the Wage Data table and enter the data shown in Figure B-9.

Employee ID ▾	Wage Rate ▾	Salaried ▾
11411	$10.00	☐
12345		☑
14890	$12.00	☐
22282		☑
71460		☑
Your Employee ID	$8.00	☐
*		☐

Source: Used with permission from Microsoft Corporation

FIGURE B-9   Data for the Wage Data table

In this table, you are again asked to create a new entry. For this record, enter your own employee ID. Also assume that you earn $8 an hour and are not salaried. Note that when an employee's Salaried box is not checked (in other words, Salaried = No), the implication is that the employee is paid by the hour. Because salaried employees are not paid by the hour, their hourly rate is 0.00.

When you finish creating the Wage Data table, open the Hours Worked table and enter the data shown in Figure B-10.

Hours Worked		
Employee ID ▾	Week # ▾	Hours ▾
11411	1	40
11411	2	50
12345	1	40
12345	2	40
14890	1	38
14890	2	40
22282	1	40
22282	2	40
71460	1	40
71460	2	40
Your Employee ID	1	60
Your Employee ID	2	55
*		

Source: Used with permission from Microsoft Corporation

FIGURE B-10   Data for the Hours Worked table

Notice that salaried employees are always given 40 hours. Nonsalaried employees (including you) might work any number of hours. For your record, enter your fictitious employee ID, 60 hours worked for Week 1, and 55 hours worked for Week 2.

CREATING QUERIES

Because you know how to create basic queries, this section explains the advanced queries you will create in the cases in this book.

Using Calculated Fields in Queries

A **calculated field** is an output field made up of *other* field values. A calculated field is *not* a field in a table; it is created in the query generator. The calculated field does not become part of the table—it is just part of the query output. The best way to understand this process is to work through an example.

AT THE KEYBOARD

Suppose you want to see the employee IDs and wage rates of hourly workers, and the new wage rates if all employees were given a 10 percent raise. To view that information, show the employee ID, the current wage rate, and the higher rate, which should be titled New Rate in the output. Figure B-11 shows how to set up the query.

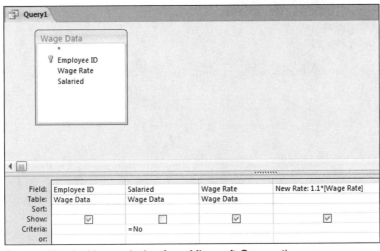

Source: Used with permission from Microsoft Corporation

FIGURE B-11   Query setup for the calculated field

To set up this query, you need to select hourly workers by using the Salaried field with Criteria = No. Note in Figure B-11 that the Show box for the field is not checked, so the Salaried field values will not appear in the query output.

Note the expression for the calculated field, which you can see in the far-right field cell:

New Rate: 1.1 * [Wage Rate]

The term *New Rate:* merely specifies the desired output heading. (Don't forget the colon.) The rest of the expression, 1.1 * [Wage Rate], multiplies the old wage rate by 110 percent, which results in the 10 percent raise.

In the expression, the field name Wage Rate must be enclosed in square brackets. Remember this rule: *Any time an Access expression refers to a field name, the expression must be enclosed in square brackets.*

If you run this query, your output should resemble that in Figure B-12.

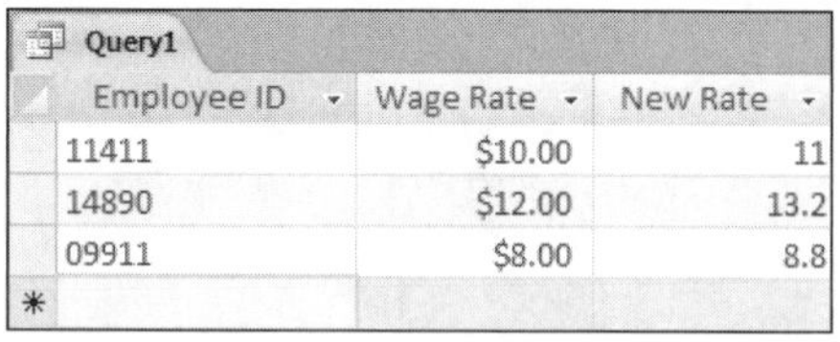

Employee ID	Wage Rate	New Rate
11411	$10.00	11
14890	$12.00	13.2
09911	$8.00	8.8

Source: Used with permission from Microsoft Corporation

FIGURE B-12   Output for a query with calculated field

Notice that the calculated field output is not shown in Currency format, but as a Double—a number with digits after the decimal point. To convert the output to Currency format, select the output column by clicking the line above the calculated field expression. The column darkens to indicate its selection. Your data entry screen should resemble the one shown in Figure B-13.

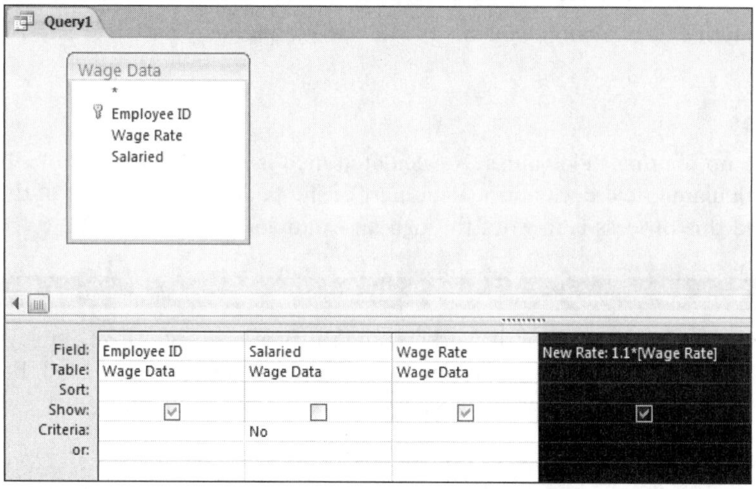

Source: Used with permission from Microsoft Corporation

FIGURE B-13   Activating a calculated field in query design

Then, on the Design tab, click Property Sheet in the Show/Hide group. The Field Properties window appears, as shown on the right in Figure B-14.

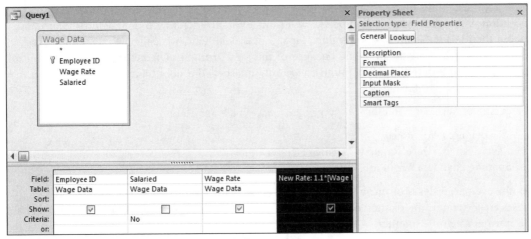

Source: Used with permission from Microsoft Corporation

FIGURE B-14   Field properties of a calculated field

Click Format and choose Currency, as shown in Figure B-15. Then click the *X* in the upper-right corner of the window to close it.

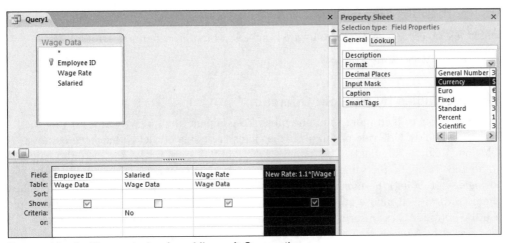

Source: Used with permission from Microsoft Corporation

FIGURE B-15   Currency format of a calculated field

When you run the query, the output should resemble that in Figure B-16.

Employee ID	Wage Rate	New Rate
11411	$10.00	$11.00
14890	$12.00	$13.20
09911	$8.00	$8.80

Source: Used with permission from Microsoft Corporation

FIGURE B-16   Query output with formatted calculated field

Next, you examine how to avoid errors when making calculated fields.

Avoiding Errors when Making Calculated Fields

Follow these guidelines to avoid making errors in calculated fields:

- Do not enter the expression in the *Criteria* cell as if the field definition were a filter. You are making a field, so enter the expression in the *Field* cell.

- Spell, capitalize, and space a field's name *exactly* as you did in the table definition. If the table definition differs from what you type, Access thinks you are defining a new field by that name. Access then prompts you to enter values for the new field, which it calls a Parameter Query field. This problem is easy to debug because of the tag *Parameter Query*. If Access asks you to enter values for a parameter, you almost certainly misspelled a field name in an expression in a calculated field or criterion.

For example, here are some errors you might make for Wage Rate:

Misspelling: (Wag Rate)
Case change: (wage Rate / WAGE RATE)
Spacing change: (WageRate / Wage Rate)

- Do not use parentheses or curly braces instead of the square brackets. Also, do not put parentheses inside square brackets. You *can*, however, use parentheses outside the square brackets in the normal algebraic manner.

For example, suppose that you want to multiply Hours by Wage Rate to get a field called Wages Owed. This is the correct expression:

Wages Owed: [Wage Rate] * [Hours]

The following expression also would be correct:

Wages Owed: ([Wage Rate] * [Hours])

But it would *not* be correct to omit the inside brackets, which is a common error:

Wages Owed: [Wage Rate * Hours]

"Relating" Two or More Tables by the Join Operation

Often, the data you need for a query is in more than one table. To complete the query, you must **join** the tables by linking the common fields. One rule of thumb is that joins are made on fields that have common *values*, and those fields often can be key fields. The names of the join fields are irrelevant; also, the names of the tables or fields to be joined may be the same, but it is not required for an effective join.

Make a join by bringing in (adding) the tables needed. Next, decide which fields you will join. Then click one field name and hold down the left mouse button while you drag the cursor over to the other field's name in its window. Release the button. Access inserts a line to signify the join. (If a relationship between two tables has been formed elsewhere, Access inserts the line automatically, and you do not have to perform the click-and-drag operation. Access often inserts join lines without the user forming relationships.)

You can join more than two tables. The common fields *need not* be the same in all tables; that is, you can daisy-chain them together.

A common join error is to add a table to the query and then fail to link it to another table. In that case, you will have a table floating in the top part of the QBE (query by example) screen. When you run the query, your output will show the same records over and over. The error is unmistakable because there is *so much* redundant output. The two rules are to add only the tables you need and to link all tables.

Next, you will work through an example of a query that needs a join.

AT THE KEYBOARD

Suppose you want to see the last names, employee IDs, wage rates, salary status, and citizenship only for U.S. citizens and hourly workers. Because the data is spread across two tables, Employee and Wage Data, you should add both tables and pull down the five fields you need. Then you should add the Criteria expressions. Set up your work to resemble that in Figure B-17. Make sure the tables are joined on the common field, Employee ID.

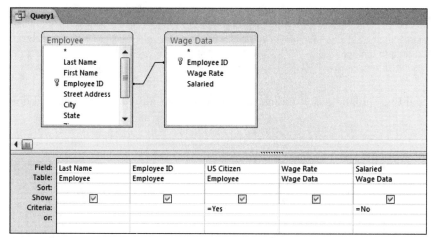

Source: Used with permission from Microsoft Corporation

FIGURE B-17   A query based on two joined tables

You should quickly review the criteria you will need to set up this join: If you want data for employees who are U.S. citizens *and* who are hourly workers, the Criteria expressions go in the *same* Criteria row. If you want data for employees who are U.S. citizens *or* who are hourly workers, one of the expressions goes in the second Criteria row (the one with the or: notation).

Now run the query. The output should resemble that in Figure B-18, with the exception of the name "Brady."

Last Name	Employee ID	US Citizen	Wage Rate	Salaried
Howard	11411	☑	$10.00	☐
Smith	14890	☑	$12.00	☐
Brady	09911	☑	$8.00	☐
*		☐		☐

Source: Used with permission from Microsoft Corporation

FIGURE B-18   Output of a query based on two joined tables

You do not need to print or save the query output, so return to Design view and close the query. Another practice query follows.

AT THE KEYBOARD

Suppose you want to see the wages owed to hourly employees for Week 2. You should show the last name, the employee ID, the salaried status, the week #, and the wages owed. Wages will have to be a calculated field ([Wage Rate] * [Hours]). The criteria are No for Salaried and 2 for the Week #. (This means that another "And" query is required.) Your query should be set up like the one in Figure B-19.

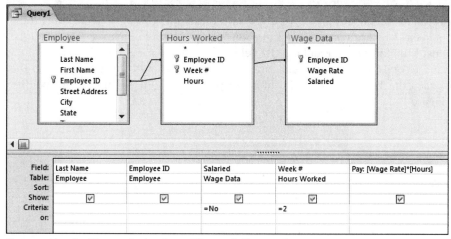

Source: Used with permission from Microsoft Corporation

FIGURE B-19   Query setup for wages owed to hourly employees for Week 2

NOTE

In the query in Figure B-19, the calculated field column was widened so you could see the whole expression. To widen a column, click the column boundary line and drag to the right.

Run the query. The output should be similar to that in Figure B-20, if you formatted your calculated field to Currency.

Last Name	Employee ID	Salaried	Week #	Pay
Howard	11411	☐	2	$500.00
Smith	14890	☐	2	$480.00
Brady	09911	☐	2	$440.00
*		☐		

Source: Used with permission from Microsoft Corporation

FIGURE B-20   Query output for wages owed to hourly employees for Week 2

Notice that it was not necessary to pull down the Wage Rate and Hours fields to make the query work. You do not need to save or print the query output, so return to Design view and close the query.

Summarizing Data from Multiple Records (Totals Queries)

You may want data that summarizes values from a field for several records (or possibly all records) in a table. For example, you might want to know the average hours that all employees worked in a week or the total (sum) of all of the hours worked. Furthermore, you might want data grouped or stratified in some way. For example, you might want to know the average hours worked, grouped by all U.S. citizens versus all non-U.S. citizens. Access calls such a query a **Totals query**. These queries include the following operations:

Sum	The total of a given field's values
Count	A count of the number of instances in a field—that is, the number of records. In the current example, you would count the number of employee IDs to get the number of employees.
Average	The average of a given field's values
Min	The minimum of a given field's values
Var	The variance of a given field's values
StDev	The standard deviation of a given field's values
Where	The field has criteria for the query output

AT THE KEYBOARD

Suppose you want to know how many employees are represented in the example database. First, bring the Employee table into the QBE screen. Because you will need to count the number of employee IDs, which is a Totals query operation, you must bring down the Employee ID field.

To tell Access that you want a Totals query, click the Design tab and then click the Totals button in the Show/Hide group. A new row called the Total row opens in the lower part of the QBE screen. At this point, the screen resembles that in Figure B-21.

Source: Used with permission from Microsoft Corporation
FIGURE B-21 Totals query setup

Note that the Total cell contains the words *Group By*. Until you specify a statistical operation, Access assumes that a field will be used for grouping (stratifying) data.

To count the number of employee IDs, click next to Group By to display an arrow. Click the arrow to reveal a drop-down menu, as shown in Figure B-22.

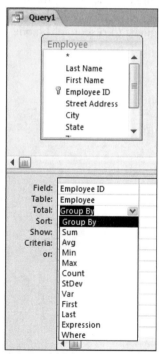

Source: Used with permission from Microsoft Corporation
FIGURE B-22 Choices for statistical operation in a Totals query

Select the Count operator. (You might need to scroll down the menu to see the operator you want.) Your screen should resemble the one shown in Figure B-23.

Source: Used with permission from Microsoft Corporation
FIGURE B-23   Count in a Totals query

Run the query. Your output should resemble that in Figure B-24.

Source: Used with permission from Microsoft Corporation
FIGURE B-24   Output of Count in a Totals query

Notice that Access created a pseudo-heading, "CountOfEmployee ID," by splicing together the statistical operation (Count), the word Of, and the name of the field (Employee ID). If you wanted a phrase such as "Count of Employees" as a heading, you would go to Design view and change the query to resemble the one shown in Figure B-25.

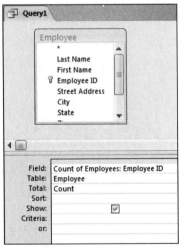

Source: Used with permission from Microsoft Corporation
FIGURE B-25   Heading change in a Totals query

When you run the query, the output should resemble that in Figure B-26.

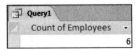

Source: Used with permission from Microsoft Corporation
FIGURE B-26   Output of heading change in a Totals query

You do not need to print or save the query output, so return to Design view and close the query.

AT THE KEYBOARD

As another example of a Totals query, suppose you want to know the average wage rate of employees, grouped by whether the employees are salaried. Figure B-27 shows how to set up your query.

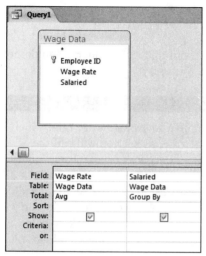

Source: Used with permission from Microsoft Corporation

FIGURE B-27   Query setup for average wage rate of employees

When you run the query, your output should resemble that in Figure B-28.

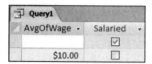

Source: Used with permission from Microsoft Corporation

FIGURE B-28   Output of query for average wage rate of employees

Recall the convention that salaried workers are assigned zero dollars an hour. Suppose you want to eliminate the output line for zero dollars an hour because only hourly-rate workers matter for the query. The query setup is shown in Figure B-29.

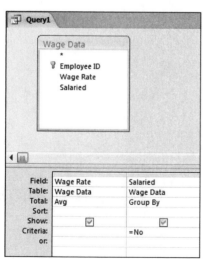

Source: Used with permission from Microsoft Corporation

FIGURE B-29   Query setup for nonsalaried workers only

When you run the query, you will get output for nonsalaried employees only, as shown in Figure B-30.

AvgOfWage ᵛ	Salaried ᵛ
$10.00	☐

Source: Used with permission from Microsoft Corporation

FIGURE B-30   Query output for nonsalaried workers only

Thus, it is possible to use Criteria in a Totals query, just as you would with a "regular" query. You do not need to print or save the query output, so return to Design view and close the query.

AT THE KEYBOARD

Assume that you want to see two pieces of information for hourly workers: (1) the average wage rate, which you will call Average Rate in the output; and (2) 110 percent of the average rate, which you will call the Increased Rate. To get this information, you can make a calculated field in a new query from a Totals query. In other words, you use one query as a basis for another query.

Create the first query; you already know how to perform certain tasks for this query. The revised heading for the average rate will be Average Rate, so type *Average Rate: Wage Rate* in the Field cell. Note that you want the average of this field. Also, the grouping will be by the Salaried field. (To get hourly workers only, enter *Criteria: No.*) Confirm that your query resembles that in Figure B-31, then save the query and close it.

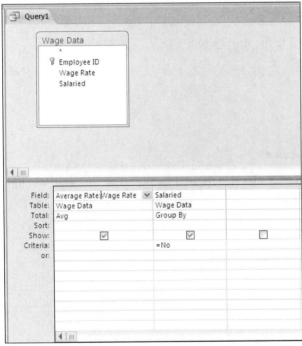

Source: Used with permission from Microsoft Corporation

FIGURE B-31   A totals query with average

Now begin a new query. However, instead of bringing in a table to the query design, select a query. To start a new query, click the Create tab and then click the Query Design button in the Queries group. The Show Table dialog box appears. Click the Queries tab instead of using the default Tables tab, and select the query you just saved as a basis for the new query. The most difficult part of this query is to construct the expression for the calculated field. Conceptually, it is as follows:

Increased Rate: 1.1 * [The current average]

You use the new field name in the new query as the current average, and you treat the new name like a new field:

Increased Rate: 1.1 * [Average Rate]

The query within a query is shown in Figure B-32.

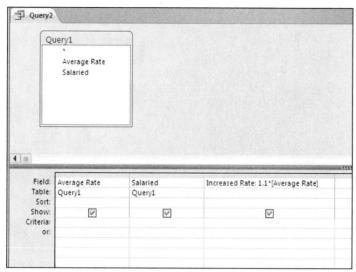

Source: Used with permission from Microsoft Corporation
FIGURE B-32   A query within a query

Figure B-33 shows the output of the new query. Note that the calculated field is formatted.

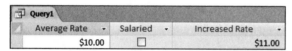

Source: Used with permission from Microsoft Corporation
FIGURE B-33   Output of an Expression in a Totals query

You do not need to print or save the query output, so return to Design view and close the query.

Using the Date() Function in Queries

Access has two important date function features:

- The built-in Date() function gives you today's date. You can use the function in query criteria or in a calculated field. The function "returns" the day on which the query is run; in other words, it inserts the value where the Date() function appears in an expression.
- Date arithmetic lets you subtract one date from another to obtain the difference—in number of days—between two calendar dates. For example, suppose you create the following expression:
  10/9/2012 – 10/4/2012
  Access would evaluate the expression as the integer 5 (9 minus 4 is 5).

As another example of how date arithmetic works, suppose you want to give each employee a one-dollar bonus for each day the employee has worked. You would need to calculate the number of days between the employee's date of hire and the day the query is run, and then multiply that number by $1.

You would find the number of elapsed days by using the following equation:

Date() – [Date Hired]

Also suppose that for each employee, you want to see the last name, employee ID, and bonus amount. You would set up the query as shown in Figure B-34.

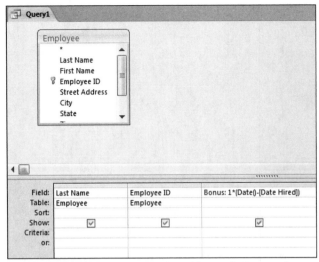

Source: Used with permission from Microsoft Corporation

FIGURE B-34   Date arithmetic in a query

Assume that you set the format of the Bonus field to Currency. The output will be similar to that in Figure B-35, although your Bonus data will be different because you used a different date.

Last Name	Employee ID	Bonus
Brady	09911	$0.00
Howard	11411	$137.00
Smith	12345	$4,581.00
Smith	14890	$7,825.00
Jones	22282	$1,615.00
Ruth	71460	$3,411.00

Source: Used with permission from Microsoft Corporation

FIGURE B-35   Output of query with date arithmetic

Using Time Arithmetic in Queries

Access also allows you to subtract the values of time fields to get an elapsed time. Assume that your database has a Job Assignments table showing the times that nonsalaried employees were at work during a day. The definition is shown in Figure B-36.

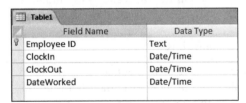

Field Name	Data Type
Employee ID	Text
ClockIn	Date/Time
ClockOut	Date/Time
DateWorked	Date/Time

Source: Used with permission from Microsoft Corporation

FIGURE B-36   Date/Time data definition in the Job Assignments table

Assume that the DateWorked field is formatted for Long Date and that the ClockIn and ClockOut fields are formatted for Medium Time. Also assume that for a particular day, nonsalaried workers were scheduled as shown in Figure B-37.

Employee ID	ClockIn	ClockOut	DateWorked	Click to Add
09911	8:30:00 AM	4:30:00 PM	Saturday, September 28, 2013	
11411	9:00:00 AM	3:00:00 PM	Saturday, September 28, 2013	
14890	7:00:00 AM	5:00:00 PM	Saturday, September 28, 2013	

Source: Used with permission from Microsoft Corporation

FIGURE B-37   Display of date and time in a table

You want a query showing the elapsed time that your employees were on the premises for the day. When you add the tables, your screen may show the links differently. Click and drag the Job Assignments, Employee, and Wage Data table icons to look like those in Figure B-38.

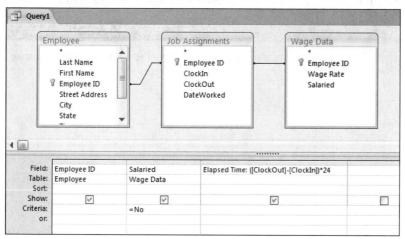

Source: Used with permission from Microsoft Corporation

FIGURE B-38   Query setup for time arithmetic

Figure B-39 shows the output, which looks correct. For example, employee 09911 was at work from 8:30 a.m. to 4:30 p.m., which is eight hours. But how does the odd expression that follows yield the correct answers?

Source: Used with permission from Microsoft Corporation

FIGURE B-39   Query output for time arithmetic

([ClockOut] – [ClockIn]) * 24

Why wouldn't the following expression work?

[ClockOut] – [ClockIn]

Here is the answer: In Access, subtracting one time from the other yields the *decimal* portion of a 24-hour day. Returning to the example, you can see that employee 09911 worked eight hours, which is one-third of a day, so the time arithmetic function yields .3333. That is why you must multiply by 24—to convert from decimals to an hourly basis. Hence, for employee 09911, the expression performs the following calculation: 1/3 × 24 = 8.

Note that parentheses are needed to force Access to do the subtraction *first*, before the multiplication. Without parentheses, multiplication takes precedence over subtraction. For example, consider the following expression:

[ClockOut] – [ClockIn] * 24

In this example, ClockIn would be multiplied by 24, the resulting value would be subtracted from ClockOut, and the output would be a nonsensical decimal number.

Deleting and Updating Queries

The queries presented in this tutorial so far have been Select queries. They select certain data from specific tables based on a given criterion. You also can create queries to update the original data in a database. Businesses use such queries often, and in real time. For example, when you order an item from a Web site, the company's database is updated to reflect your purchase through the deletion of that item from the company's inventory.

Consider an example. Suppose you want to give all nonsalaried workers a $0.50 per hour pay raise. Because you have only three nonsalaried workers, it would be easy to change the Wage Rate data in the table. However, if you had 3,000 nonsalaried employees, it would be much faster and more accurate to change the Wage Rate data by using an Update query that adds $0.50 to each nonsalaried employee's wage rate.

AT THE KEYBOARD

Now you will change each of the nonsalaried employees' pay via an Update query. Figure B-40 shows how to set up the query.

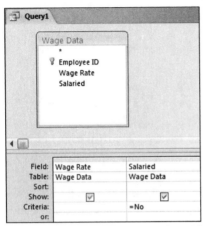

Source: Used with permission from Microsoft Corporation

FIGURE B-40 Query setup for an Update query

So far, this query is just a Select query. Click the Update button in the Query Type group, as shown in Figure B-41.

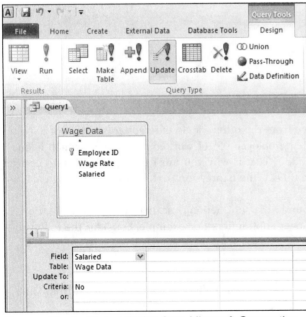

Source: Used with permission from Microsoft Corporation

FIGURE B-41 Selecting a query type

Notice that you now have another line on the QBE grid called Update To:, which is where you specify the change or update the data. Notice that you will update only the nonsalaried workers by using a filter under the Salaried field. Update the Wage Rate data to Wage Rate plus $0.50, as shown in Figure B-42. Note that the update involves the use of brackets [ ], as in a calculated field.

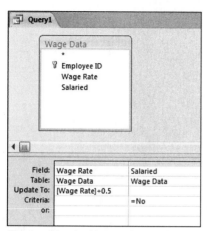

Source: Used with permission from Microsoft Corporation

FIGURE B-42   Updating the wage rate for nonsalaried workers

Now run the query by clicking the Run button in the Results group. If you cannot run the query because it is blocked by Disabled Mode, click the Database Tools tab, then click Message Bar in the Show/Hide group. Click the Options button, choose "enable this content," and then click OK. When you successfully run the query, the warning message in Figure B-43 appears.

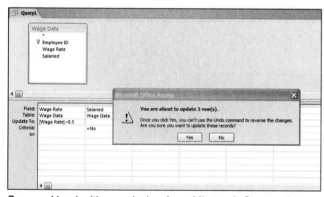

Source: Used with permission from Microsoft Corporation

FIGURE B-43   Update query warning

When you click Yes, the records are updated. Check the updated records by viewing the Wage Data table. Each nonsalaried wage rate should be increased by $0.50. You could add or subtract data from another table as well. If you do, remember to put the field name in square brackets.

Another type of query is the Delete query, which works like Update queries. For example, assume that your company has been purchased by the state of Delaware, which has a policy of employing only state residents. Thus, you must delete (or fire) all employees who are not exclusively Delaware residents. To do that, you would create a Select query. Using the Employee table, you would click the Delete button in the Query Type group, then bring down the State field and filter only those records that were not in Delaware (DE). Do not perform the operation, but note that if you did, the setup would look like the one in Figure B-44.

Source: Used with permission from Microsoft Corporation

FIGURE B-44   Deleting all employees who are not Delaware residents

Using Parameter Queries

A **Parameter query** is actually a type of Select query. For example, suppose your company has 5,000 employees and you want to query the database to find the same kind of information repeatedly, but about different employees each time. For example, you might want to know how many hours a particular employee has worked. You could run a query that you created and stored previously, but run it only for a particular employee.

AT THE KEYBOARD

Create a Select query with the format shown in Figure B-45.

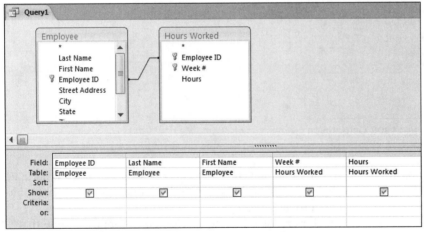

Source: Used with permission from Microsoft Corporation

FIGURE B-45   Design of a Parameter query beginning as a Select query

In the Criteria line of the QBE grid for the Employee ID field, type what is shown in Figure B-46.

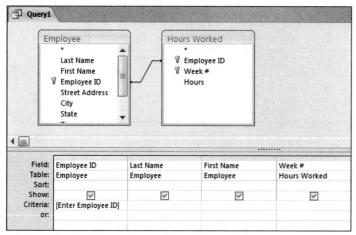

Source: Used with permission from Microsoft Corporation

FIGURE B-46   Design of a Parameter query, continued

Note that the Criteria line uses square brackets, as you would expect to see in a calculated field. Now run the query. You will be prompted for the employee's ID number, as shown in Figure B-47.

Source: Used with permission from Microsoft Corporation

FIGURE B-47   Enter Parameter Value dialog box

Enter your own employee ID. Your query output should resemble that in Figure B-48.

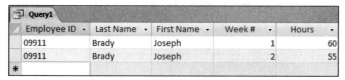

Source: Used with permission from Microsoft Corporation

FIGURE B-48   Output of a Parameter query

MAKING SEVEN PRACTICE QUERIES

This portion of the tutorial gives you additional practice in creating queries. Before making these queries, you must create the specified tables and enter the records shown in the "Creating Tables" section of this tutorial. The output shown for the practice queries is based on those inputs.

AT THE KEYBOARD

For each query that follows, you are given a problem statement and a "scratch area." You also are shown what the query output should look like. Set up each query in Access and then run the query. When you are satisfied with the results, save the query and continue with the next one. Note that you will work with the Employee, Hours Worked, and Wage Data tables.

1. Create a query that shows the employee ID, last name, state, and date hired for employees who live in Delaware *and* were hired after 12/31/99. Perform an ascending sort by employee ID. First click the Sort cell of the field, and then choose Ascending or Descending. Before creating your query, use the table shown in Figure B-49 to work out your QBE grid on paper.

Field					
Table					
Sort					
Show					
Criteria					
Or:					

Source: © Cengage Learning 2014

FIGURE B-49   QBE grid template

Your output should resemble that in Figure B-50.

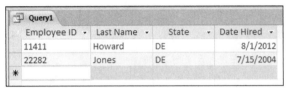

Source: Used with permission from Microsoft Corporation

FIGURE B-50   Number 1 query output

2. Create a query that shows the last name, first name, date hired, and state for employees who live in Delaware *or* were hired after 12/31/99. The primary sort (ascending) is on last name, and the secondary sort (ascending) is on first name. The Primary Sort field must be to the left of the Secondary Sort field in the query setup. Before creating your query, use the table shown in Figure B-51 to work out your QBE grid on paper.

Field					
Table					
Sort					
Show					
Criteria					
Or:					

Source: © Cengage Learning 2014

FIGURE B-51   QBE grid template

If your name was Joe Brady, your output would look like that in Figure B-52.

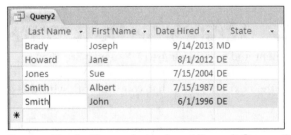

Source: Used with permission from Microsoft Corporation

FIGURE B-52   Number 2 query output

3. Create a query that sums the number of hours worked by U.S. citizens and the number of hours worked by non-U.S. citizens. In other words, create two sums, grouped on citizenship. The heading for total hours worked should be Total Hours Worked. Before creating your query, use the table shown in Figure B-53 to work out your QBE grid on paper.

Field					
Table					
Total					
Sort					
Show					
Criteria					
Or:					

Source: © Cengage Learning 2014

FIGURE B-53   QBE grid template

Your output should resemble that in Figure B-54.

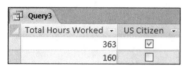

Query3	
Total Hours Worked ▾	US Citizen ▾
363	☑
160	☐

Source: Used with permission from Microsoft Corporation

FIGURE B-54   Number 3 query output

4. Create a query that shows the wages owed to hourly workers for Week 1. The heading for the wages owed should be Total Owed. The output headings should be Last Name, Employee ID, Week #, and Total Owed. Before creating your query, use the table shown in Figure B-55 to work out your QBE grid on paper.

Field					
Table					
Sort					
Show					
Criteria					
Or:					

Source: © Cengage Learning 2014

FIGURE B-55   QBE grid template

If your name was Joe Brady, your output would look like that in Figure B-56.

Query4			
Last Name ▾	Employee ID ▾	Week # ▾	Total Owed ▾
Howard	11411	1	$420.00
Smith	14890	1	$475.00
Brady	09911	1	$510.00
*			

Source: Used with permission from Microsoft Corporation

FIGURE B-56   Number 4 query output

5. Create a query that shows the last name, employee ID, hours worked, and overtime amount owed for hourly employees who earned overtime during Week 2. Overtime is paid at 1.5 times the normal hourly rate for all hours worked over 40. Note that the amount shown in the query should be just the overtime portion of the wages paid. Also, this is not a Totals query—amounts should be shown for individual workers. Before creating your query, use the table shown in Figure B-57 to work out your QBE grid on paper.

Field					
Table					
Sort					
Show					
Criteria					
Or:					

Source: © Cengage Learning 2014

FIGURE B-57   QBE grid template

If your name was Joe Brady, your output would look like that in Figure B-58.

Last Name	Employee ID	Hours	OT Pay
Howard	11411	50	$157.50
Brady	09911	55	$191.25

Source: Used with permission from Microsoft Corporation

FIGURE B-58   Number 5 query output

6. Create a Parameter query that shows the hours employees have worked. Have the Parameter query prompt for the week number. The output headings should be Last Name, First Name, Week #, and Hours. This query is for nonsalaried workers only. Before creating your query, use the table shown in Figure B-59 to work out your QBE grid on paper.

Field					
Table					
Sort					
Show					
Criteria					
Or:					

Source: © Cengage Learning 2014

FIGURE B-59   QBE grid template

Run the query and enter 2 when prompted for the week number. Your output should look like that in Figure B-60.

Last Name	First Name	Week #	Hours
Howard	Jane	2	50
Smith	Albert	2	40
Brady	Joseph	2	55

Source: Used with permission from Microsoft Corporation

FIGURE B-60   Number 6 query output

7. Create an Update query that gives certain workers a merit raise. First, you must create an additional table, as shown in Figure B-61.

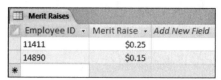

Merit Raises		
Employee ID ▾	Merit Raise ▾	Add New Field
11411	$0.25	
14890	$0.15	
*		

Source: Used with permission from Microsoft Corporation

FIGURE B-61 Merit Raises table

Create a query that adds the Merit Raise to the current Wage Rate for employees who will receive a raise. When you run the query, you should be prompted with *You are about to update two rows.* Check the original Wage Data table to confirm the update. Before creating your query, use the table shown in Figure B-62 to work out your QBE grid on paper.

Field					
Table					
Update to					
Criteria					
Or:					

Source: © Cengage Learning 2014

FIGURE B-62 QBE grid template

CREATING REPORTS

Database packages let you make attractive management reports from a table's records or from a query's output. If you are making a report from a table, the Access report generator looks up the data in the table and puts it into report format. If you are making a report from a query's output, Access runs the query in the background (you do not control it or see it happen) and then puts the output in report format.

There are different ways to make a report. One method is to create one from scratch in Design view, but this tedious process is not explained in this tutorial. A simpler way is to select the query or table on which the report is based and then click Create Report. This streamlined method of creating reports is explained in this tutorial.

Creating a Grouped Report

This tutorial assumes that you already know how to create a basic ungrouped report, so this section teaches you how to make a grouped report. If you do not know how to create an ungrouped report, you can learn by following the first example in the upcoming section.

AT THE KEYBOARD

Suppose you want to create a report from the Hours Worked table. Select the table by clicking it once. Click the Create tab, then click Report in the Reports group. A report appears, as shown in Figure B-63.

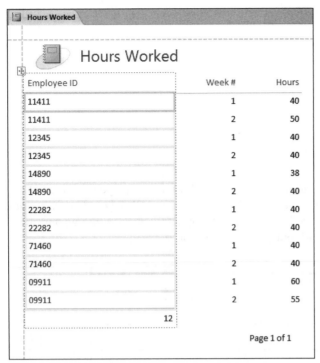

Source: Used with permission from Microsoft Corporation

FIGURE B-63   Initial report based on a table

On the Design tab, select the Group and Sort button in the Grouping and Totals group. Your report will have an additional selection at the bottom, as shown in Figure B-64.

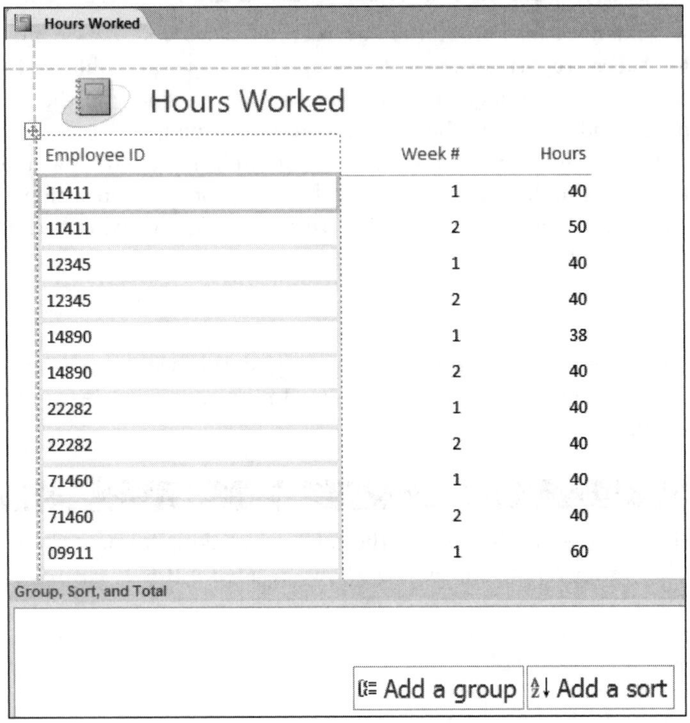

Source: Used with permission from Microsoft Corporation

FIGURE B-64   Report with Grouping and Sorting options

Click the Add a group button at the bottom of the report, and then select Employee ID. Your report will be grouped as shown in Figure B-65.

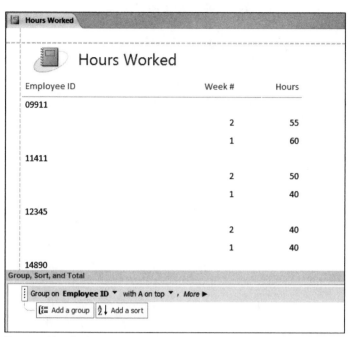

Source: Used with permission from Microsoft Corporation
FIGURE B-65   Grouped report

To complete this report, you need to total the hours for each employee by selecting the Hours column heading. Your report will show that the entire column is selected. On the Design tab, click the Totals button in the Grouping and Totals group, and then choose Sum from the menu, as shown in Figure B-66.

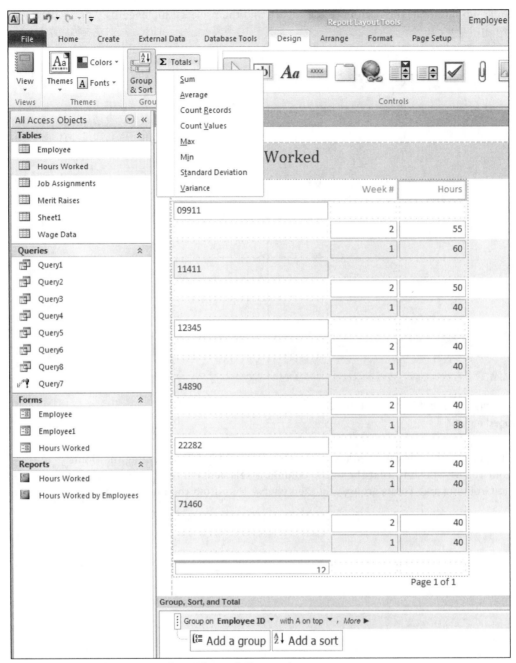

Source: Used with permission from Microsoft Corporation

FIGURE B-66   Hours column selected

Your report will look like the one in Figure B-67.

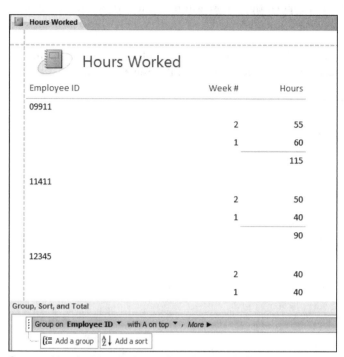

Source: Used with permission from Microsoft Corporation

FIGURE B-67   Completed report

Your report is currently in Layout view. To see how the final report looks when printed, click the Design tab and select Report view from the Views group. Your report looks like the one in Figure B-68, although only a portion is shown in the figure.

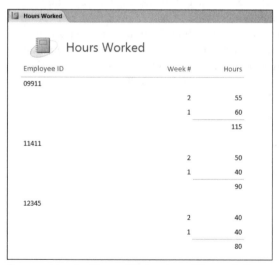

Source: Used with permission from Microsoft Corporation

FIGURE B-68   Report in Report view

NOTE

To change the picture or logo in the upper-left corner of the report, click the notebook symbol and press the Delete key.
You can insert a logo in place of the notebook by clicking the Design tab and then clicking Logo in the Controls group.

Moving Fields in Layout View

If you group records based on more than one field in a report, the report will have an odd "staircase" look or display repeated data, or it will have both problems. Next, you will learn how to overcome these problems in Layout view.

Suppose you make a query that shows an employee's last name, first name, week number, and hours worked, and then you make a report from that query, grouping on last name only. See Figure B-69.

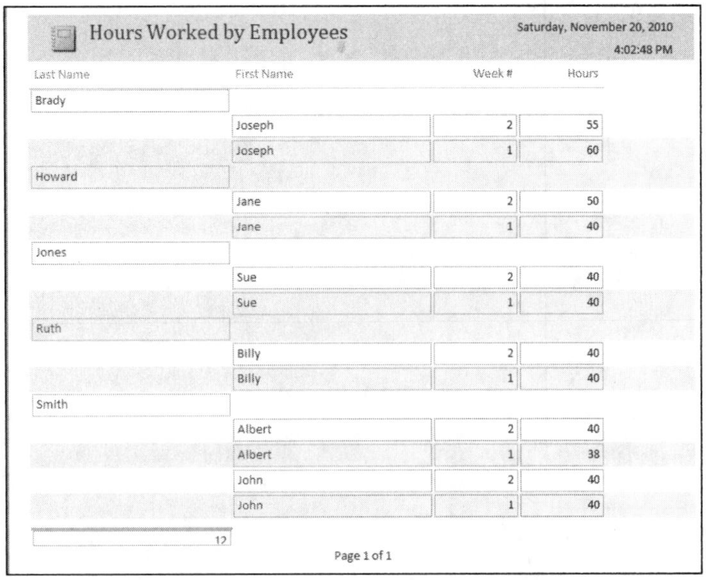

Source: Used with permission from Microsoft Corporation

FIGURE B-69   Query-based report grouped on last name

As you preview the report, notice the repeating data from the First Name field. In the report shown in Figure B-69, notice that the first name repeats for each week worked—hence, the staircase effect. The Week # and Hours fields are shown as subordinate to Last Name, as desired.

Suppose you want the last name and first name to appear on the same line. If so, take the report into Layout view for editing. Click the first record for the First Name (in this case, Joseph), and drag the name up to the same line as the Last Name (in this case, Brady). Your report will now show the First Name on the same line as Last Name, thereby eliminating the staircase look, as shown in Figure B-70.

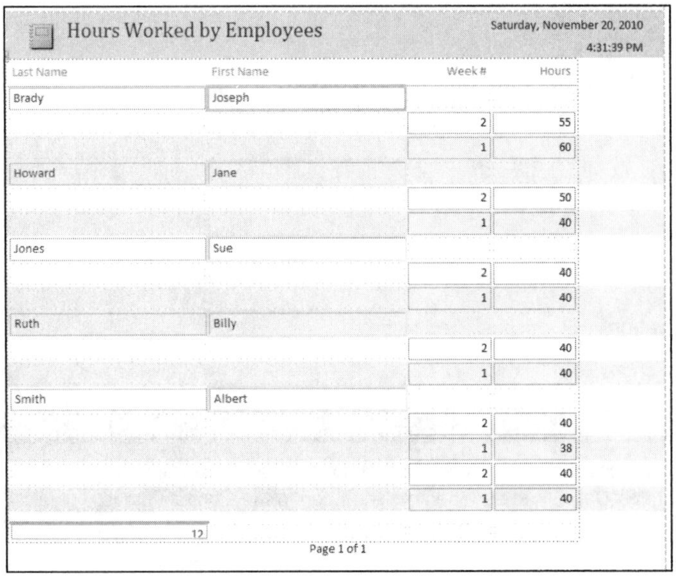

Source: Used with permission from Microsoft Corporation

FIGURE B-70   Report in Layout view with Last Name and First Name on the same line

You can now add the sum of Hours for each group. Also, if you want to add more fields to your report, such as Street Address and Zip, you can repeat the preceding procedure.

IMPORTING DATA

Text or spreadsheet data is easy to import into Access. In business, it is often necessary to import data because companies use disparate systems. For example, assume that your healthcare coverage data is on the human resources manager's computer in a Microsoft Excel spreadsheet. Open the Excel application and then create a spreadsheet using the data shown in Figure B-71.

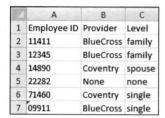

	A	B	C
1	Employee ID	Provider	Level
2	11411	BlueCross	family
3	12345	BlueCross	family
4	14890	Coventry	spouse
5	22282	None	none
6	71460	Coventry	single
7	09911	BlueCross	single

Source: Used with permission from Microsoft Corporation
FIGURE B-71   Excel data

Save the file and then close it. Now you can easily import the spreadsheet data into a new table in Access. With your Employee database open, click the External Data tab, then click Excel in the Import & Link group. Browse to find the Excel file you just created, and make sure the first radio button is selected to import the source data into a new table in the current database (see Figure B-72). Click OK.

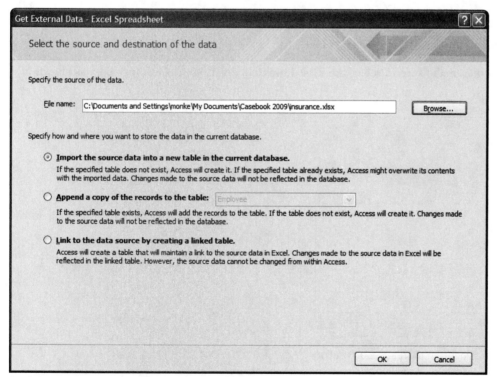

Source: Used with permission from Microsoft Corporation
FIGURE B-72   Importing Excel data into a new table

Choose the correct worksheet. Assuming that you have just one worksheet in your Excel file, your next screen should look like the one in Figure B-73.

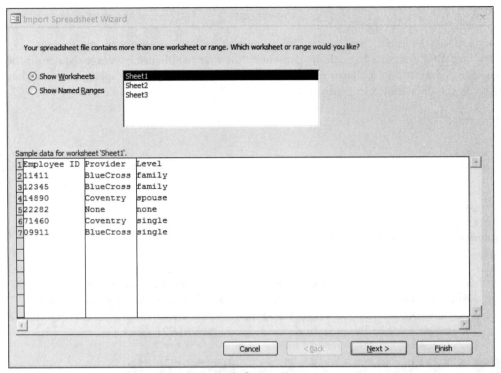

Source: Used with permission from Microsoft Corporation

FIGURE B-73   First screen in the Import Spreadsheet Wizard

Choose Next, and then make sure to select the First Row Contains Column Headings box, as shown in Figure B-74.

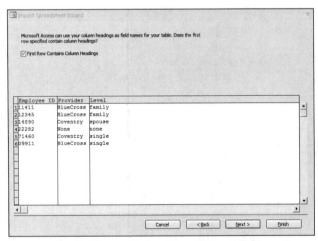

Source: Used with permission from Microsoft Corporation

FIGURE B-74   Choosing column headings in the Import Spreadsheet Wizard

Choose Next. Accept the default setting for each field you are importing on the screen. Each field is assigned a text data type, which is correct for this table. Your screen should look like the one in Figure B-75.

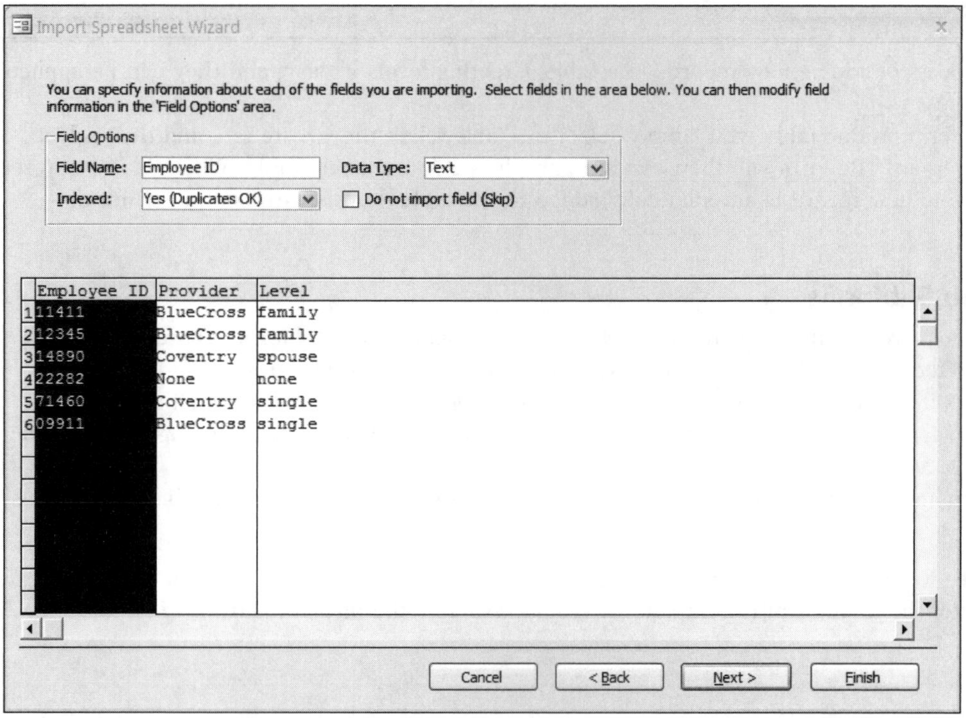

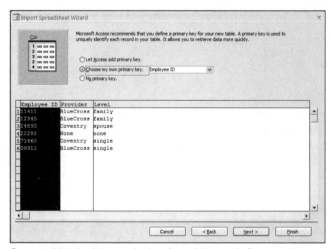

Source: Used with permission from Microsoft Corporation

FIGURE B-75   Choosing the data type for each field in the Import Spreadsheet Wizard

Choose Next. In the next screen of the wizard, you will be prompted to create an index—that is, to define a primary key. Because you will store your data in a new table, choose your own primary key (Employee ID), as shown in Figure B-76.

Source: Used with permission from Microsoft Corporation

FIGURE B-76   Choosing a primary key field in the Import Spreadsheet Wizard

Continue through the wizard, giving your table an appropriate name. After importing the table, take a look at its design by highlighting the Table option and clicking the Design button. Note that each field is very wide. Adjust the field properties as needed.

MAKING FORMS

Forms simplify the process of adding new records to a table. Creating forms is easy, and they can be applied to one or more tables.

When you base a form on one table, you simply select the table, click the Create tab, and then select Form from the Forms group. The form will then contain only the fields from that table. When data is entered into the form, a complete new record is automatically added to the table. Forms with two tables are discussed next.

Making Forms with Subforms

You also can create a form that contains a subform, which can be useful when the form is based on two or more tables. Return to the example Employee database to see how forms and subforms would be useful for viewing all of the hours that each employee worked each week. Suppose you want to show all of the fields from the Employee table; you also want to show the hours each employee worked by including all fields from the Hours Worked table as well.

To create the form and subform, first create a simple one-table form on the Employee table. Follow these steps:

1. Click once to select the Employee table. Click the Create tab, then click Form in the Forms group. After the main form is complete, it should resemble the one in Figure B-77.

Source: Used with permission from Microsoft Corporation

FIGURE B-77   The Employee form

2. To add the subform, take the form into Design view. On the Design tab, make sure that the Use Control Wizards option is selected, scroll to the bottom row of buttons in the Controls group, and click the Subform/Subreport button, as shown in Figure B-78.

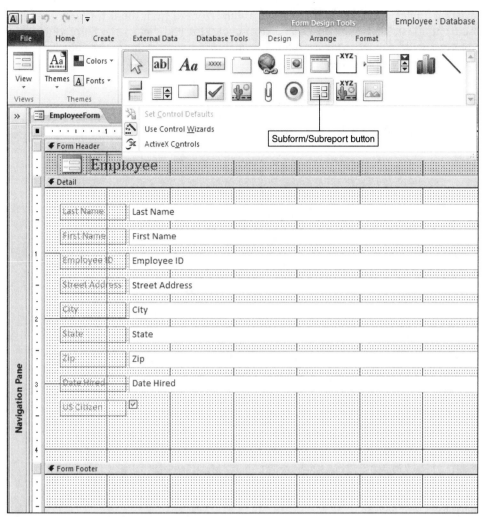

Source: Used with permission from Microsoft Corporation

FIGURE B-78   The Subform/Subreport button

3.   Use your cursor to stretch out the box under your main form. The dialog box shown in Figure B-79 appears.

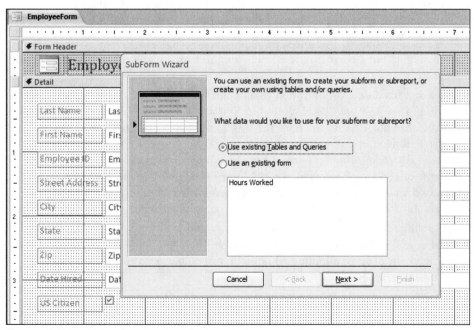

Source: Used with permission from Microsoft Corporation

FIGURE B-79   Adding a subform

4. Select Use existing Tables and Queries, then select Hours Worked from the list. Click Next, select Choose from a List, click Next again, and then click Finished. Select the Form view. Your form and subform should resemble Figure B-80. You may need to stretch out the subform box in Design view if all fields are not visible.

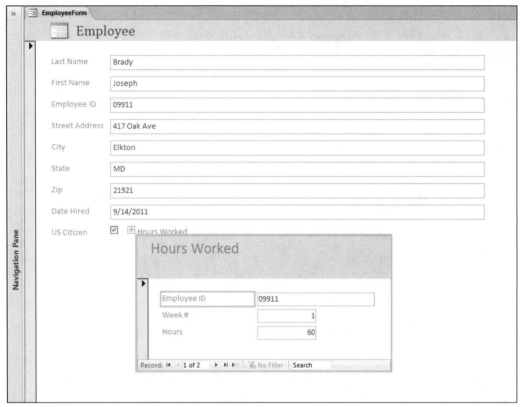

Source: Used with permission from Microsoft Corporation

FIGURE B-80   Form with subform

TROUBLESHOOTING COMMON PROBLEMS

Access is a powerful program, but it is complex and sometimes difficult for new users. People sometimes unintentionally create databases that have problems. Some of these common problems are described below, along with their causes and corrections.

1. *"I saved my database file, but I can't find it on my computer or my external secondary storage medium! Where is it?"*

    You saved your file to a fixed disk or a location other than the Documents folder. Click the Windows Start button, then use the Search option to find all files ending in .accdb (search for *.accdb). If you saved the file, it is on the hard drive (C:\) or a network drive. Your site assistant can tell you the drive designators.

2. *"What is a 'duplicate key field value'? I'm trying to enter records into my Sales table. The first record was for a sale of product X to customer 101, and I was able to enter that one. But when I try to enter a second sale for customer #101, Access tells me I already have a record with that key field value. Am I allowed to enter only one sale per customer?"*

    Your primary key field needs work. You may need a compound primary key—a combination of the customer number and some other field(s). In this case, the customer number, product number, and date of sale might provide a unique combination of values, or you might consider using an invoice number field as a key.

3. *"My query reads 'Enter Parameter Value' when I run it. What is that?"*

This problem almost always indicates that you have misspelled a field name in an expression in a Criteria field or calculated field. Access is very fussy about spelling; for example, it is case-sensitive. Access is also "space-sensitive," meaning that when you insert a space in a field name when defining a table, you must also include a space in the field name when you reference it in a query expression. Fix the typo in the query expression.

4. *"I'm getting an enormous number of rows in my query output—many times more than I need. Most of the rows are duplicates!"*

This problem is usually caused by a failure to link all of the tables you brought into the top half of the query generator. The solution is to use the manual click-and-drag method to link the common fields between tables. The spelling of the field names is irrelevant because the link fields need not have the same spelling.

5. *"For the most part, my query output is what I expected, but I am getting one or two duplicate rows or not enough rows."*

You may have linked too many fields between tables. Usually, only a single link is needed between two tables. It is unnecessary to link each common field in all combinations of tables; it is usually sufficient to link the primary keys. A simplistic explanation for why overlinking causes problems is that it causes Access to "overthink" and repeat itself in its answer.

On the other hand, you might be using too many tables in the query design. For example, you brought in a table, linked it on a common field with some other table, but then did not use the table. In other words, you brought down none of its fields, and/or you used none of its fields in query expressions. In this case, if you got rid of the table, the query would still work. Click the unneeded table's header at the top of the QBE area, and press the Delete key to see if you can make the few duplicate rows disappear.

6. *"I expected six rows in my query output, but I got only five. What happened to the other one?"*

Usually, this problem indicates a data entry error in your tables. When you link the proper tables and fields to make the query, remember that the linking operation joins records from the tables *on common values* (*equal* values in the two tables). For example, if a primary key in one table has the value "123," the primary key or the linking field in the other table should be the same to allow linking. Note that the text string "123" is not the same as the text string "123 "—the space in the second string is considered a character too. Access does not see unequal values as an error. Instead, Access moves on to consider the rest of the records in the table for linking. The solution is to examine the values entered into the linked fields in each table and fix any data entry errors.

7. *"I linked fields correctly in a query, but I'm getting the empty set in the output. All I get are the field name headings!"*

You probably have zero common (equal) values in the linked fields. For example, suppose you are linking on Part Number, which you declared as text. In one field, you have part numbers "001", "002", and "003"; in the other table, you have part numbers "0001," "0002," and "0003." Your tables have no common values, which means that no records are selected for output. You must change the values in one of the tables.

8. *"I'm trying to count the number of today's sales orders. A Totals query is called for. Sales are denoted by an invoice number, and I made that a text field in the table design. However, when I ask the Totals query to 'Sum' the number of invoice numbers, Access tells me I cannot add them up! What is the problem?"*

Text variables are words! You cannot add words, but you can count them. Use the Count Totals operator (not the Sum operator) to count the number of sales, each being denoted by an invoice number.

9. *"I'm doing time arithmetic in a calculated field expression. I subtracted the Time In from the Time Out and got a decimal number! I expected eight hours, and I got the number .33333. Why?"*

[Time Out] – [Time In] yields the decimal percentage of a 24-hour day. In your case, eight hours is one-third of a day. You must complete the expression by multiplying by 24: ([Time Out] – [Time In]) * 24. Don't forget the parentheses.

10. *"I formatted a calculated field for Currency in the query generator, and the values did show as currency in the query output; however, the report based on the query output does not show the dollar sign in its output. What happened?"*

Go to the report Design view. A box in one of the panels represents the calculated field's value. Click the box and drag to widen it. That should give Access enough room to show the dollar sign as well as the number in the output.

11. *"I told the Report Wizard to fit all of my output to one page. It does print to one page, but some of the data is missing. What happened?"*

Access fits all the output on one page by leaving data out. If you can tolerate having the output on more than one page, deselect the Fit to a Page option in the wizard. One way to tighten output is to enter Design view and remove space from each box that represents output values and labels. Access usually provides more space than needed.

12. *"I grouped on three fields in the Report Wizard, and the wizard prints the output in a staircase fashion. I want the grouping fields to be on one line. How can I do that?"*

Make adjustments in Design view and Layout view. See the "Creating Reports" section of this tutorial for instructions on making these adjustments.

13. *"When I create an Update query, Access tells me that zero rows are updating or more rows are updating than I want. What is wrong?"*

If your Update query is not set up correctly (for example, if the tables are not joined properly), Access will either try not to update anything, or it will update all of the records. Check the query, make corrections, and run it again.

14. *"I made a Totals query with a Sum in the Group By row and saved the query. Now when I go back to it, the Sum field reads 'Expression,' and 'Sum' is entered in the field name box. Is that wrong?"*

Access sometimes changes the Sum field when the query is saved. The data remains the same, and you can be assured your query is correct.

15. *"I cannot run my Update query, but I know it is set up correctly. What is wrong?"*

Check the security content of the database by clicking the Security Content button. You may need to enable certain actions.

PRELIMINARY CASE: THE OUTDOOR CLUB

Setting Up a Relational Database to Create Tables, Forms, Queries, and Reports

PREVIEW

In this case, you will create a relational database for a university club that organizes outdoor and adventure events. First, you will create three tables and populate them with data. Next, you will create a form and subform for recording students and their trips. You will create five queries: a select query, a parameter query, an update query, a totals query, and a query used as the basis for a report. Finally, you will create the report from the fifth query.

PREPARATION

- Before attempting this case, you should have some experience using Microsoft Access.
- Complete any part of Tutorial B that your instructor assigns, or refer to the tutorial as necessary.

BACKGROUND

You have always loved the outdoors, so when you started at your university as a freshman, you looked for a club that would fit your interests. The outdoor club on campus was a perfect fit, and you have been an active member for the last year. In fact, you have been elected an officer of the club. As an officer, you have an obligation to improve the club and make it more efficient. Up to this point, the club has kept all its records on paper. Because you are proficient in Microsoft Access from a recent university class, you decide to computerize the record keeping and create a database for the club.

Your first tasks are to design the database, create the tables, and populate them with current data. You have decided to keep the database simple at first, so your design includes only three tables, as shown in Figures 1-1, 1-2, and 1-3:

- Student Members, which keeps track of each student ID number, the student's full name, telephone number, e-mail address, birth date, and whether the student has signed a release form
- Trips, which keeps track of trips planned for the year by displaying the trip ID number, the dates for the trip, its destination and location, and its cost
- Sign-ups, which records the trip ID number, student ID number, and the amount that each student has paid so far for the trip

After the database tables are complete and populated with data, you want to computerize several common tasks. First, you need a streamlined way to enroll new students in the club and to sign up new and current members for trips. You can accomplish both tasks by creating a form and subform that includes the Student Members table and the Sign-ups table.

No students are allowed to travel on trips unless they have a release form on file with the university outdoor club. This release form is a legal requirement so that if an accident occurs on a trip, the university is not liable for injuries. You decide that a select query is needed so that a club member can run it before each trip to see if any students have not filled out the release form.

A common challenge for students is trying to find the best possible trip at the lowest price. You realize that you can use a parameter query that prompts for the upper limit of the trip's cost. For example, if a club member came into the office and asked which trips were available for less than $150, the parameter query would help answer the question quickly.

Another common issue involves insurance for trips. The club is negotiating with a local agency to insure all members for each trip. The cost of insurance will add $5 or $10 to the cost of the trip, but the actual cost is still not known. You decide to set up an update query that prompts the user to enter the additional insurance amount for each trip. Once you know the insurance rate, you will run the query and enter the insurance amount when prompted; the database will then add that amount to the cost of each trip.

An officer of the club is always curious to know which trips are the most popular. You decide to create a query so that the officer can monitor the most popular trips whenever he wants.

Finally, an important piece of information is a weekly listing of the remaining money each student owes for each trip. You think a query that feeds into a report is the best solution. The report will list each student who owes additional money and display the trips for which the student has signed up.

ASSIGNMENT 1: CREATING TABLES

Use Microsoft Access to create tables that contain the fields shown in Figures 1-1 through 1-3; you learned about these tables in the Background section. Populate the database tables as shown. Add your name to the Student Members table with a fictitious ID number; complete the entry by adding your phone number, e-mail address, and birth date, and assume that you do not have a release form on file. Add more records to the Sign-ups table as your instructor requests.

This database contains the following three tables:

Student Members

Student ID	Last Name	First Name	Telephone	Email	Birthdate	Release Form?	Click to Add
2101	Stanzione	Luke	(610)292-1779	luke@uni.edu	12/1/1995	☑	
2102	Black	Leon	(610)366-1929	leon@uni.edu	3/16/1993	☑	
2103	Delgado	Luis	(610)366-0159	luis@uni.edu	4/2/1996	☐	
2104	Spurkowski	Lauren	(610)369-9550	lauren@uni.edu	11/28/1991	☑	
2105	Wyatt	Sally	(302)738-5785	sally@uni.edu	7/11/1994	☑	
2106	Winoski	Harry	(302)369-0643	harry@uni.edu	8/1/1994	☑	
2107	Patel	Patti	(610)544-0923	patti@uni.edu	9/27/1995	☐	
2108	Poplawski	Meredith	(410)433-0927	meredith@uni.edu	9/27/1993	☑	
2109	McMillan	Maria	(302)655-1829	maria@uni.edu	11/1/1993	☑	
2110	Laurie	Gregory	(610)655-0936	gregory@uni.edu	2/12/1992	☑	
2111	Teaster	Emily	(610)685-9541	emily@uni.edu	1/2/1995	☑	
2112	Trego	Ted	(302)874-9632	ted@uni.edu	2/12/1996	☑	
2113	Hillyard	Eleanor	(610)541-1845	eleanor@uni.edu	3/14/1992	☐	
2114	Dahlstrom	Emma	(302)655-5541	emma@uni.edu	5/8/1993	☑	
2115	Almazan	Dorie	(610)964-8512	dorie@uni.edu	4/11/1992	☑	
2116	Balbuena	Sandy	(610)852-1547	sandy@uni.edu	6/19/1996	☑	
2117	Valenzula	Darren	(610)369-8521	darren@uni.edu	7/4/1995	☑	
2118	Desoto	Tameka	(302)685-4711	tameka@uni.edu	9/10/1994	☑	
2119	Hanning	Tyrone	(610)221-8564	tyrone@uni.edu	11/30/1994	☐	
2120	Beam	Bobbie	(610)365-8854	bobbie@uni.edu	9/7/1991	☑	
*						☐	

Source: Used with permission from Microsoft Corporation

FIGURE 1-1   The Student Members table

Trips

Trip ID	Start Date	End Date	Destination	Location	Cost	Click to Add
101	2/26/2014	2/26/2014	Stairway to Heaven Day Hike	Hewitt, NJ	$25.00	
102	3/3/2014	3/6/2014	Skiing/Snowboarding	Killington, VT	$245.00	
103	3/10/2014	3/14/2014	Beginner Backpacking	Appalachian Trail, VA	$210.00	
104	3/14/2014	3/14/2014	North Bay Adventure Course	Northeast, MD	$55.00	
105	3/20/2014	3/22/2014	Skiing/Snowboarding	Poconos, PA	$195.00	
106	4/8/2014	4/10/2014	Old Rag Backpacking	Shenandoah, VA	$165.00	
107	4/15/2014	4/17/2014	Canoeing/Camping	Delaware Water Gap, PA	$190.00	
108	4/15/2014	4/17/2014	Climbing	New River Gorge, WV	$245.00	
109	4/22/2014	4/24/2014	Backpacking	West Rim Trail, PA	$135.00	
110	5/5/2014	5/5/2014	Canoe Day Trip	Brandywine State Park, DE	$30.00	
*						

Source: Used with permission from Microsoft Corporation

FIGURE 1-2   The Trips table

Trip ID	Student ID	Amount Paid	Click to Add
103	2117	$125.00	
103	2120	$100.00	
104	2101	$55.00	
104	2102	$55.00	
104	2105	$55.00	
104	2107	$55.00	
104	2109	$55.00	
104	2113	$55.00	
105	2110	$100.00	
105	2111	$100.00	
105	2112	$195.00	
105	2113	$195.00	
105	2114	$195.00	
105	2115	$195.00	
106	2116	$165.00	
106	2117	$165.00	
106	2118	$50.00	
106	2119	$25.00	
106	2120	$75.00	
107	2105	$190.00	
107	2106	$190.00	
107	2107	$100.00	
107	2108	$90.00	
107	2109	$100.00	
108	2101	$50.00	
108	2102	$75.00	
109	2101	$135.00	
109	2103	$135.00	
109	2104	$100.00	
109	2105	$25.00	
109	2107	$25.00	
109	2109	$25.00	
110	2110	$30.00	
110	2111	$30.00	
110	2112	$30.00	
110	2113	$30.00	
110	2114	$30.00	
110	2115	$30.00	
110	2116	$30.00	
110	2117	$30.00	
110	2118	$30.00	

Source: Used with permission from Microsoft Corporation

FIGURE 1-3   The Sign-ups table

ASSIGNMENT 2: CREATING A FORM, QUERIES, AND A REPORT

Assignment 2A: Creating a Form

Create a form for easy recording of new students and any trips for which members have signed up. The main form should be based on the Student Members table, and the subform should be inserted with the fields from the Sign-ups table. Save the form as Student Members. View one record; if required by your instructor, print the record. Your output should resemble that in Figure 1-4.

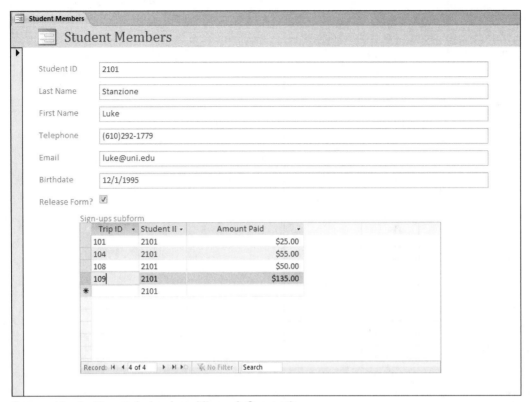

Source: Used with permission from Microsoft Corporation

FIGURE 1-4   The Student Members form with subform

Assignment 2B: Creating a Select Query

Create a query that shows all students who do not have a release form on file. Display columns for the student's Last Name, First Name, Telephone, and Email. Save the query as Students Without Release Forms. Your output should resemble that shown in Figure 1-5. Print the output if desired.

Last Name	First Name	Telephone	Email
Delgado	Luis	(610)366-0159	luis@uni.edu
Patel	Patti	(610)544-0923	patti@uni.edu
Hillyard	Eleanor	(610)541-1845	eleanor@uni.edu
Hanning	Tyrone	(610)221-8564	tyrone@uni.edu

Source: Used with permission from Microsoft Corporation

FIGURE 1-5   Students Without Release Forms query

Assignment 2C: Creating a Parameter Query

Create a parameter query that prompts for trips below a certain cost and then displays columns for the Trip ID, Start Date, End Date, Destination, Location, and Cost. Save the query as Trips by Price. Your output should resemble Figure 1-6 when you enter $150 at the prompt.

Trips by Price					
Trip ID	Start Date	End Date	Destination	Location	Cost
101	2/26/2014	2/26/2014	Stairway to Heaven Day Hike	Hewitt, NJ	$25.00
104	3/14/2014	3/14/2014	North Bay Adventure Course	Northeast, MD	$55.00
109	4/22/2014	4/24/2014	Backpacking	West Rim Trail, PA	$135.00
110	5/5/2014	5/5/2014	Canoe Day Trip	Brandywine State Park, DE	$30.00
*					

Source: Used with permission from Microsoft Corporation

FIGURE 1-6   Trips by Price query

Assignment 2D: Creating an Update Query

Create a query that updates the cost of each trip by prompting the user to enter an increased insurance rate and then incorporating the rate into the cost. Click the Run button to test the query. When prompted to change 10 records, answer "No." Save the query as Update Cost.

Assignment 2E: Creating a Totals Query

Create a totals query that counts the number of students on each trip. In the output, display columns for Trip ID, Start Date, Destination, and Number of Students. Sort your output so that the most popular trip is listed at the top and the least popular trip is shown at the bottom. Note that the Number of Students heading is a column heading change from the default setting provided by the query generator. Save the query as Popular Trips. Your output should resemble that in Figure 1-7. Print the output if desired.

Popular Trips			
Trip ID	Start Date	Destination	Number of Students
110	5/5/2014	Canoe Day Trip	11
102	3/3/2014	Skiing/Snowboarding	7
109	4/22/2014	Backpacking	6
105	3/20/2014	Skiing/Snowboarding	6
104	3/14/2014	North Bay Adventure Course	6
101	2/26/2014	Stairway to Heaven Day Hike	6
107	4/15/2014	Canoeing/Camping	5
106	4/8/2014	Old Rag Backpacking	5
103	3/10/2014	Beginner Backpacking	5
108	4/15/2014	Climbing	2

Source: Used with permission from Microsoft Corporation

FIGURE 1-7   Popular Trips query

Assignment 2F: Generating a Report

Generate a report named Money Due based on a query. The query should display columns for Last Name, First Name, Email, Trip ID, Destination, and Money Due, which is a calculated field. Save the query as Money Due. The query output should resemble that in Figure 1-8.

Money Due					
Last Name	First Name	Email	Trip ID	Destination	Money Due
Black	Leon	leon@uni.edu	102	Skiing/Snowboarding	$95.00
Wyatt	Sally	sally@uni.edu	102	Skiing/Snowboarding	$145.00
Poplawski	Meredith	meredith@uni.edu	102	Skiing/Snowboarding	$145.00
Laurie	Gregory	gregory@uni.edu	102	Skiing/Snowboarding	$120.00
Almazan	Dorie	dorie@uni.edu	102	Skiing/Snowboarding	$120.00
Beam	Bobbie	bobbie@uni.edu	103	Beginner Backpacking	$110.00
Valenzula	Darren	darren@uni.edu	103	Beginner Backpacking	$85.00
Teaster	Emily	emily@uni.edu	103	Beginner Backpacking	$110.00
Laurie	Gregory	gregory@uni.edu	105	Skiing/Snowboarding	$95.00
Teaster	Emily	emily@uni.edu	105	Skiing/Snowboarding	$95.00
Desoto	Tameka	tameka@uni.edu	106	Old Rag Backpacking	$115.00
Hanning	Tyrone	tyrone@uni.edu	106	Old Rag Backpacking	$140.00
Beam	Bobbie	bobbie@uni.edu	106	Old Rag Backpacking	$90.00
Patel	Patti	patti@uni.edu	107	Canoeing/Camping	$90.00
Poplawski	Meredith	meredith@uni.edu	107	Canoeing/Camping	$100.00
McMillan	Maria	maria@uni.edu	107	Canoeing/Camping	$90.00
Stanzione	Luke	luke@uni.edu	108	Climbing	$195.00
Black	Leon	leon@uni.edu	108	Climbing	$170.00
Spurkowski	Lauren	lauren@uni.edu	109	Backpacking	$35.00
Wyatt	Sally	sally@uni.edu	109	Backpacking	$110.00
Patel	Patti	patti@uni.edu	109	Backpacking	$110.00
McMillan	Maria	maria@uni.edu	109	Backpacking	$110.00

Source: Used with permission from Microsoft Corporation

FIGURE 1-8 Money Due query

From that query, create a report that displays the total amount of money due for each student and a grand total at the bottom. Make sure that all fields and data are visible, and enter "Money Due" as the title at the top of the report. Your report output should resemble that in Figure 1-9.

Source: Used with permission from Microsoft Corporation

FIGURE 1-9   Money Due report

If you are working with a portable storage disk or USB thumb drive, make sure that you remove it *after* closing the database file.

DELIVERABLES

Assemble the following deliverables for your instructor, either electronically or in printed form:

1. Three tables
2. Form and subform: Student Members
3. Query 1: Students Without Release Forms
4. Query 2: Trips by Price
5. Query 3: Update Cost
6. Query 4: Popular Trips
7. Query 5: Money Due
8. Report: Money Due
9. Any other required printouts or electronic media

Staple all the pages together. Write your name and class number at the top of the page. If required, make sure that your electronic media are labeled.

TIES TO GO DATABASE

Designing a Relational Database to Create Tables, Forms, Queries, and Reports

PREVIEW

In this case, you will design a relational database for a business that rents designer neckties. After your design is completed and correct, you will create database tables and populate them with data. Then you will produce one form with a subform, three queries, and two reports. The queries will address the following questions: What ties are available from a specified manufacturer? How many ties are in inventory? How many ties has each customer rented this month? Your reports will display the ties available for rent by manufacturer and the amount of money each customer owes for his monthly plan.

PREPARATION

- Before attempting this case, you should have some experience in database design and in using Microsoft Access.
- Complete any part of Tutorial A that your instructor assigns.
- Complete any part of Tutorial B that your instructor assigns, or refer to the tutorial as necessary.
- Refer to Tutorial F as necessary.

BACKGROUND

Many men need good neckties for job interviews or to work in certain industries. Because a designer tie can cost about $90, a man starting in the working world might have to make a considerable investment in ties. You and a college friend realize that a small business can fill this niche. You were watching a Netflix movie a few nights ago and had the idea: Why not rent high-quality ties through a Netflix-type business model? Customers can sign up for different plans and rent ties to suit their fashion needs.

Ties to Go is the company name you have chosen. The company model is similar to that of Netflix. Customers can sign up for four different plans: one tie per month at $10.99, two ties per month at $18.99, three ties per month at $24.99, or five ties per month at $29.99. The ties are sent with return bags and prepaid postage to help ensure that the company gets its ties back. The ties are cleaned every time they are returned. If ties are too stained to clean, then the customer is charged for the tie. If the stain is stubborn and requires extra cleaning, the customer is charged as well.

Because you have learned about database design and have worked with Microsoft Access, you decide to create a prototype database to help you understand the business. You will use this prototype during a presentation to your banker to request funds for the business start-up. You decide to focus on a few key parts of the business for the presentation.

First, you need to keep track of all the ties that are available for renting. Your business partner's father has a large collection of ties that he arranges in his closet by manufacturer, design, and color. You will use a similar system to begin to organize your inventory. Your database must record customer information as well. You want customers to be able to order ties on a secure Web site where they can browse for ties and place the ones they like in an order queue.

When customers register for the service, they provide their e-mail address, regular mailing address, and credit card number. (All transactions are conducted with credit cards.) Your database must record each tie rental along with the date rented and date returned.

As customers request ties, you want to be able to enter the order information directly into the database. You need a form with a subform to record these orders. Again, you want this form to be available on the Web so customers can do their leasing online.

You can imagine that customers will request ties from certain manufacturers. Your business partner wants to be able to enter a manufacturer's name and then see a list of available ties. You will create a query to generate this list.

You need a way to count how many ties customers have rented at a given time so they do not exceed their monthly limit. For marketing purposes, you would also like to know how many ties you have in inventory from each manufacturer. This information will be used for marketing purposes. Again, you can create queries to produce this information.

Finally, you need to create two reports. The first report lists the ties available by manufacturer and will be used for advertising purposes. The second report will calculate and list the amount of money each customer is spending on a monthly basis.

ASSIGNMENT 1: CREATING THE DATABASE DESIGN

In this assignment, you design your database tables using a word-processing program. Pay close attention to the logic and structure of the tables. Do not start developing your Access database in Assignment 2 before getting feedback from your instructor on Assignment 1. Keep in mind that you need to examine the requirements in Assignment 2 to design your fields and tables properly. It is good programming practice to look at the required outputs before beginning your design. When designing the database, observe the following guidelines:

- First, determine the tables you will need by listing the name of each table and the fields it should contain. Avoid data redundancy. Do not create a field if it can be created by a calculated field in a query.
- You will need transaction tables. Think about the business events that occur with each customer's actions. Avoid duplicating data.
- Document your tables using the table feature of your word processor. Your tables should resemble the format shown in Figure 2-1.
- You must mark the appropriate key field(s) by entering an asterisk (*) next to the field name. Keep in mind that some tables might need a compound primary key to uniquely identify a record within a table.
- Print the database design.

Table Name	
Field Name	Data Type (text, numeric, currency, etc.)
...	...
...	...

Source: © Cengage Learning 2014

FIGURE 2-1 Table design

NOTE

Have your design approved before beginning Assignment 2; otherwise, you may need to redo Assignment 2.

ASSIGNMENT 2: CREATING THE DATABASE, QUERIES, AND REPORTS

In this assignment, you first create database tables in Access and populate them with data. Next, you create a form, three queries, and two reports.

Assignment 2A: Creating Tables in Access

In this part of the assignment, you create your tables in Access. Use the following guidelines:

- Enter at least 20 records for ties from four different manufacturers. Use the Internet to find tie manufacturers and ideas for design and color.
- Enter records for at least eight customers, including their names, addresses, telephone numbers, e-mail addresses, and fictional credit card numbers. Enter your own name and information as a customer. Assign a rental plan to each customer.
- Each tie should be rented at least once. Each customer should rent two ties, and at least two customers should rent more than two.
- Make sure that some ties have not been returned yet. Leave the return date blank for these instances.
- Appropriately limit the size of the text fields; for example, a telephone number does not need the default length of 255 characters.
- Print all tables if your instructor requires it.

Assignment 2B: Creating Forms, Queries, and Reports

You must generate one form with a subform, three queries, and two reports, as outlined in the Background section of this case.

Form

Create a form and subform based on your Ties table and Rentals table (or whatever you named these tables). Save the form as Ties. Your form should resemble that in Figure 2-2.

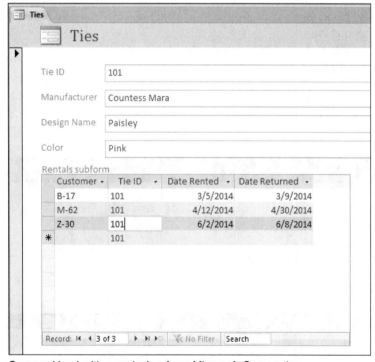

Source: Used with permission from Microsoft Corporation

FIGURE 2-2 Ties form and Rentals subform

Query 1

Create a parameter query called Ties by Manufacturer that prompts for the manufacturer's name and then lists the design name of the tie and its color. In the example shown in Figure 2-3, the manufacturer Countess Mara was entered at the prompt. Your output should resemble that in Figure 2-3, although your data will be different.

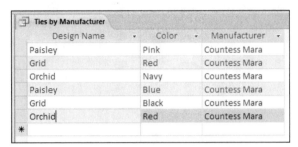

Design Name	Color	Manufacturer
Paisley	Pink	Countess Mara
Grid	Red	Countess Mara
Orchid	Navy	Countess Mara
Paisley	Blue	Countess Mara
Grid	Black	Countess Mara
Orchid	Red	Countess Mara

Source: Used with permission from Microsoft Corporation

FIGURE 2-3   Ties by Manufacturer query

Query 2

Create a query called Number of Ties Outstanding that calculates the number of ties currently rented by each customer. The query should include columns for Last Name, First Name, Email Address, and Number of Ties Outstanding. Sort the output from most ties rented to the least. Note that the Number of Ties Outstanding heading is a column heading change from the default setting provided by the query generator. Your output should look like that in Figure 2-4, although your data will be different.

Last Name	First Name	Email Address	Number of Ties Outstanding
Berry	Andrew	aberry@hotmail.com	4
Murray	Alistair	belle@comcast.net	3
Zern	Joseph	zern@comcast.net	2
Pao	John	pao@comcast.net	2
Smith	Patrick	patti1@gmail.com	1
Quinn	Samuel	quinn45@gmail.com	1
Lopato	Martin	mrl@hotmail.com	1
Franco	George	gf59@gmail.com	1

Source: Used with permission from Microsoft Corporation

FIGURE 2-4   Number of Ties Outstanding query

Query 3

Create a query called Inventory Totals that counts the number of ties in inventory from each manufacturer. The query should include columns for Manufacturer and Number of Ties. Note the column heading change from the default setting provided by the query generator. Your output should resemble the format shown in Figure 2-5, but the data will be different.

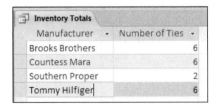

Manufacturer	Number of Ties
Brooks Brothers	6
Countess Mara	6
Southern Proper	2
Tommy Hilfiger	6

Source: Used with permission from Microsoft Corporation

FIGURE 2-5   Inventory Totals query

Report 1

Create a report named Ties. The report's output should include headings for Manufacturer, Tie ID, Design Name, and Color. Group the report by Manufacturer. Depending on your data, the output should resemble that in Figure 2-6. Note that only a portion of the output appears in the figure.

Ties			

Ties

Tuesday, August 21, 2012
11:30:32 AM

Manufacturer	Tie ID	Design Name	Color
Brooks Brothers			
	117	Pine Needles	Green
	116	Fern	Teal
	115	Dot	Orange
	104	Dot	Camel
	105	Fern	Green
	106	Pine Needles	Blue
Countess Mara			
	102	Grid	Red
	103	Orchid	Navy
	114	Orchid	Red
	113	Grid	Black
	112	Paisley	Blue
	101	Paisley	Pink

Source: Used with permission from Microsoft Corporation

FIGURE 2-6   Ties report

Report 2

Create a report called Rental Costs per Month that includes columns for Last Name, First Name, and Price Per Month. You need to create a query for this report first. Depending on your data, the report should resemble that in Figure 2-7.

Rental Costs per Month		

Rental Costs per Month

Tuesday, August 21, 2012
11:31:02 AM

Last Name	First Name	Price Per Month
Berry	Andrew	$29.90
Franco	George	$10.99
Lopato	Martin	$10.99
Murray	Alistair	$24.99
Pao	John	$18.99
Quinn	Samuel	$10.99
Smith	Patrick	$10.99
Zern	Joseph	$18.99
		$136.83

Page 1 of 1

Source: Used with permission from Microsoft Corporation

FIGURE 2-7   Rental Costs per Month report

ASSIGNMENT 3: MAKING A PRESENTATION

Create a presentation that explains the database to the banker who will approve your company's finances and loans. Include the design of your database tables and instructions for using the database. Discuss future improvements to the database, such as the ability to track customers' favorite tie manufacturers and colors. Your presentation should take 10 minutes or less, including a brief question-and-answer period.

DELIVERABLES

Assemble the following deliverables for your instructor, either electronically or in printed form:

1. Word-processed design of tables
2. Tables created in Access
3. Form and subform: Ties
4. Query 1: Ties by Manufacturer
5. Query 2: Number of Ties Outstanding
6. Query 3: Inventory Totals
7. Report 1: Ties
8. Query for Report 2
9. Report 2: Rental Costs per Month
10. Presentation materials
11. Any other required printouts or electronic media

Staple all the pages together. Write your name and class number at the top of each page. Make sure that your electronic media are labeled, if required.

SWEET SNACKS DATABASE

Designing a Relational Database to Create Tables, Forms, Queries, and Reports

PREVIEW

In this case, you will design a relational database for a company that sells and delivers cookies and other sweets to students at the local university. After your design is completed and correct, you will create database tables and populate them with data. You will produce one form with a subform that allows you to record incoming orders, and you will produce seven queries and a report. Five queries will address the following questions: Which products that you sell contain chocolate? Which orders are arriving late at night? Which products cost less than a specified price limit? Which products are the most popular? What is the average delivery time? An additional query will update the prices of your products. Your report, based on another query, will display all the orders and the money they bring in on a specific date.

PREPARATION

- Before attempting this case, you should have some experience in database design and in using Microsoft Access.
- Complete any part of Tutorial A that your instructor assigns.
- Complete any part of Tutorial B that your instructor assigns, or refer to the tutorial as necessary.
- Refer to Tutorial F as necessary.

BACKGROUND

You and your college roommate have always loved to bake cookies and make other sweet desserts. During the evenings, especially around final exams, you yearn for good, sweet cookies and hot drinks. After talking with your friends, you realize that you are not alone: Most college students love to eat sweets, especially late at night. Your roommate, who is always scheming and coming up with new ideas, wants to start a company that creates delicious cookies and candies and then sells and delivers them to fellow students at your university during the evenings. You agree that the idea has great business potential, and the two of you begin to draw up your business plan.

You both want to start small and then expand if the business is successful. To begin, you will do all the baking and other preparations in your university apartment and enlist friends to take orders and make deliveries. The first step is to test the recipes! Your favorite cookie recipe is for dark chocolate oatmeal cookies, but your roommate prefers your macaroons. One of your friends shares her recipes for hot drink sticks—a hot chocolate stick and a butterscotch stick. These blocks of candy on a stick dissolve in hot milk and create a delicious sweet drink—perfect for a study break during final exams! All in all, you anticipate offering about 10 products in your business, which you will call Sweet Snacks.

You are taking a database class and have learned about database design and implementation with Microsoft Access. You think it would be a good idea to begin creating your small-business plan using Access. After a trial run, you can replace the database with a larger one that can be accessed by multiple users and ported to the Internet.

When considering your database design, you know you need to record customer information such as names, addresses, telephone numbers, and e-mail addresses. You also need to track products, which will be sold by the dozen and have varying prices depending on the cost of the ingredients. Finally, you need to record order information, including the date of the order and the time it was received and delivered.

You and your roommate want to make sure that all customers are satisfied and that orders are delivered promptly. Keep in mind that students who crave sweets at night often order more than one product. At first, you will run your company as a cash-only business. All deliveries are free, but payment is due when the goods are delivered to the student.

Next, you need to consider how to process customer, product, and order information in the most effective way for the company. First, you need to create a form that records incoming orders. After the company grows, you plan to move this form to the company's Web site so hungry students can place orders quickly online.

The database can answer many questions, so the next step is to create queries. First, many of your friends love chocolate, so you want to be able to quickly list all the products that contain chocolate. Also, many college students are on a tight budget, so you want to be able to quickly tell customers which products cost less than $10. To help friends who are delivering your products, you want to list all the orders placed after 11 p.m.

To monitor how the business is doing, you want to create a query that shows the most popular products on a given day. This query can help confirm your impression that students order more of certain products during the stress of final exams. Also, you want to calculate the average delivery time of orders to make sure customers are not waiting too long. Because you think your pricing can be flexible, you want to create a query that increases or decreases all product prices by 2 percent. You can use flexible pricing to encourage sales during slumps or to increase your revenue as the business takes off.

Finally, you need to create a summary report that lists all sales for a specified day. The report should specify each customer, a list of what they ordered, and the quantities of each item in the order. This information is important for tracking revenue and for tax purposes.

ASSIGNMENT 1: CREATING THE DATABASE DESIGN

In this assignment, you design your database tables using a word-processing program. Pay close attention to the logic and structure of the tables. Do not start developing your Access database in Assignment 2 before getting feedback from your instructor on Assignment 1. Keep in mind that you need to examine the requirements in Assignment 2 to design your fields and tables properly. It is good programming practice to look at the required outputs before beginning your design. When designing the database, observe the following guidelines:

- First, determine the tables you will need by listing the name of each table and the fields it should contain. Avoid data redundancy. Do not create a field if it can be created by a calculated field in a query.
- You will need transaction tables. Think about the business events that occur with each student's order. Avoid duplicating data.
- Keep in mind that students may request more than one product within an order.
- Document your tables using the table feature of your word processor. Your tables should resemble the format shown in Figure 3-1.
- You must mark the appropriate key field(s) by entering an asterisk (*) next to the field name.
- Print the database design if your instructor requires it.

Table Name	
Field Name	Data Type (text, numeric, currency, etc.)
...	...
...	...

Source: © Cengage Learning 2014

FIGURE 3-1 Table design

NOTE

Have your design approved before beginning Assignment 2; otherwise, you may need to redo Assignment 2.

ASSIGNMENT 2: CREATING THE DATABASE, QUERIES, AND REPORT

In this assignment, you first create database tables in Access and populate them with data. Next, you create a form, queries, and a report.

Assignment 2A: Creating Tables in Access

In this part of the assignment, you create your tables in Access. Use the following guidelines:

- Create at least nine customer records and add yourself as the tenth customer.
- Create 10 different types of products to sell.
- Create at least 10 orders for a single day. Make sure that most customers order more than one product, and make sure that some products have orders for multiple dozens.
- Appropriately limit the size of the text fields; for example, a telephone number does not need the default length of 255 characters.
- Print all tables if your instructor requires it.

Assignment 2B: Creating Forms, Queries, and a Report

You will generate one form with a subform, seven queries, and one report, as outlined in the Background section of this case.

Form

Create a form and subform based on the tables you developed. Save the form as Orders. Your form should resemble that in Figure 3-2.

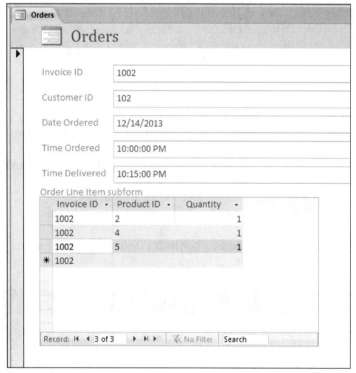

Source: Used with permission from Microsoft Corporation

FIGURE 3-2   Orders form and Order Line Item subform

Query 1

Create a query called Chocolate Products that displays the product name and price per dozen of all products that contain chocolate. Consider using a wildcard to filter data for this query. Your output should resemble that in Figure 3-3, although your data will be different.

Source: Used with permission from Microsoft Corporation

FIGURE 3-3 Chocolate Products query

Query 2

Create a query called Orders At or After 11pm. Your query should filter orders for a specific date and list the last name of each customer, their addresses and telephone numbers, and the name and quantity of each product in the order. Your output should look like that in Figure 3-4, although your data will be different.

Last Name	Address	Telephone	Product Name	Quantity
Smith	34 Redback Road	(512)998-0675	Dark Chocolate Oatmeal	3
Smith	34 Redback Road	(512)998-0675	Lemon Daisies	2
O'Hara	20 Cheswold Blvd	(307)887-0176	Sticky Toffee Crunchies	2
Isaacs	10010 N. Barrett Dr	(201)584-1028	Hot Chocolate Sticks	2
Isaacs	10010 N. Barrett Dr	(201)584-1028	Butterscotch Cremes	2
Isaacs	10010 N. Barrett Dr	(201)584-1028	Butterscotch Sticks	2
Downing	200 Mac Duff Rd	(801)912-6564	Pecan Clusters	3
Downing	200 Mac Duff Rd	(801)912-6564	Macaroons	2

Source: Used with permission from Microsoft Corporation

FIGURE 3-4 Orders At or After 11pm query

Query 3

Create a query called Products by Price that prompts for an upper monetary limit so you can tell budget-minded customers which products cost less than $10. The query output should include columns for Product Name and Price Per Dozen. If you run this query and enter $10 as the upper limit, your output should resemble the format shown in Figure 3-5, but the data will be different.

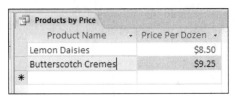

Source: Used with permission from Microsoft Corporation

FIGURE 3-5 Products by Price query

Query 4

Create a query to update the prices of all products. For example, if you want to decrease each price by 2 percent, test the query by running it and checking the prices in your product table. Save the query as Updated Prices.

Query 5

Create a query named Popular Products that lists the number of products sold on a specific date. The output should include columns for Product Name and Number Ordered. Sort your output so the most popular product is listed first and the least popular product is shown last. Note the column heading change from the default setting provided by the query generator. Your output should resemble the format shown in Figure 3-6, but the data will be different.

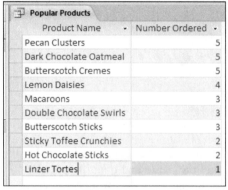

Source: Used with permission from Microsoft Corporation

FIGURE 3-6   Popular Products query

Query 6

Create a query that calculates the average delivery time in minutes for all orders on a specific date. Display columns for Date Ordered and Length of Delivery Time in Minutes. Note the column heading change from the default setting provided by the query generator. Save the query as Delivery Time. Your output should resemble the format shown in Figure 3-7, but the data will be different.

Source: Used with permission from Microsoft Corporation

FIGURE 3-7   Delivery Time query

Report

Create a report based on a query that lists all sales on a specific date. The report should display columns for Last Name, First Name, Address, Product Name, Quantity, and Total Price, which is a calculated field. Give the report a title of "Sales for *specific date*," as shown at the top of Figure 3-8. Make sure that all data and column headings are visible, and format your output appropriately, such as including currency signs. Depending on your data, the output should resemble that in Figure 3-8; only a portion of the report is shown. Save the report file as Today's Sales.

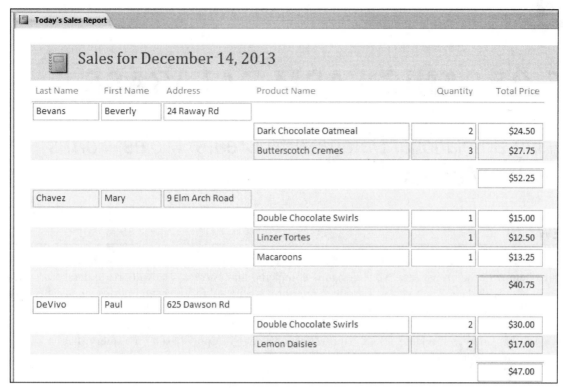

Source: Used with permission from Microsoft Corporation

FIGURE 3-8   Today's Sales report

ASSIGNMENT 3: MAKING A PRESENTATION

Create a presentation that explains the database to your business partner. Demonstrate how she can use the database by running the queries and generating a report. Discuss future improvements and additions to the database and how you might use it to expand your company. Your presentation should take 10 minutes or less.

DELIVERABLES

Assemble the following deliverables for your instructor, either electronically or in printed form:

1. Word-processed design of tables
2. Tables created in Access
3. Form and subform: Orders
4. Query 1: Chocolate Products
5. Query 2: Orders At or After 11pm
6. Query 3: Products by Price
7. Query 4: Updated Prices
8. Query 5: Popular Products
9. Query 6: Delivery Time
10. Query 7: For Report
11. Report: Today's Sales
12. Presentation materials
13. Any other required printouts or electronic media

Staple all the pages together. Include your name and class number at the top of each page. Make sure that your electronic media are labeled, if required.

THE ORGANIC FARM DATABASE

Designing a Relational Database to Create Tables, Forms, Queries, and Reports

PREVIEW

In this case, you will design a relational database for an organic farm that sells to local grocers and enlists volunteers to help with the crops. After your design is completed and correct, you will create database tables and populate them with data. Then you will produce one form with a subform, seven queries, and one report. The queries will answer the following questions: Are the volunteers over 18 years old? What products are sold? What is the lot number of specific products? What hours do the volunteers work? Who are the best customers? Another query will update prices. You will also produce a report based on a query that summarizes the farm's sales.

PREPARATION

- Before attempting this case, you should have some experience in database design and in using Microsoft Access.
- Complete any part of Tutorial A that your instructor assigns.
- Complete any part of Tutorial B that your instructor assigns, or refer to the tutorial as necessary.
- Refer to Tutorial F as necessary.

BACKGROUND

Your older sister, Joan, and her husband, Joe, inherited a small farm from Joe's grandmother. The couple is environmentally conscientious and would like to farm the land organically. Because Joan and Joe are not experienced farmers, they need help planting crops, tilling the soil, and starting the business in general. You are proficient with databases and their design, so you agree to help Joan and Joe by designing a database that will help them run the farm. In return, you will receive free room and board and a small salary. You look forward to the challenge of developing the system and to meeting new people. Joan and Joe are joining a consortium of farmers who host volunteers to work in exchange for free room and board and a chance to learn organic farming. You will share your living accommodations with volunteers from all over the world.

When you arrive at the farm, your first task is to learn how the volunteers work. Farm volunteers must work at least 35 hours per week and must keep a log of their hours to avoid being dropped from the program. Because the volunteers tend to be free spirits and itinerant, many do not have home addresses. You assume that you will have to keep track of the volunteers' names, cell phone numbers, and e-mail addresses. You realize that some of the volunteers are siblings and might have the same last name. Also, each organic farm that uses volunteers as workers requires that they be at least 18 years old, so you must record a birth date in the database as well. The organic farm consortium requires an annual report for each volunteer who has worked on the farm in the past year, along with the volunteer's age.

You talk with a volunteer, Serena, who has worked at several organic farms. Serena says that most farms have old-fashioned time clocks that require workers to "punch in" and stamp a card with the beginning date and time of their work shift. When workers finish their shift, they repeat the procedure by inserting their card into the time clock for the ending date and time. Serena mentions that by law, the volunteers are allowed to work only one shift a day. Once a week, Joan will post a list of hours worked by volunteers next to the time clock. Volunteers can check to see if they are working enough hours to continue the program.

Once you understand what volunteer information you need, you move on to study the farming operation. You talk with Joan and Joe about the products they want to sell. Joan shows you a proposed list of crops and a bushel price for each crop. (A bushel is about 35 liters, or eight gallons of dry crops.) Joe explains the detailed record keeping required to allow the farm to label its products *organic*. For each harvested bushel, a farmer must identify the field where it was picked, the row number, the date of harvest, and who picked the bushel. This information is combined to form a specific "lot" ID, which is important for tracking product safety and tracing any problems back to the source.

Many consumers buy organic foods, so several local grocery stores are ready to buy from Joan and Joe's farm. Joan gives you her contact address book, which lists the name of each store, along with its address, contact phone number, e-mail address, and contact person.

Even though the farm is not yet running at full capacity, you decide to set up a prototype of the ordering system. Here's how you envision it working: Joan will take telephone calls from customers and fill out a computer form that records the date of the order, the customer name, the products ordered, and the quantity of each product. At this point, you are not concerned with recording payment amounts and lot IDs.

Joan and Joe have questions that they want the database to be able to answer. For example, they must submit a form each year to the organic farmers' consortium to verify that their volunteer workers are at least 18 years old. Joan and Joe need to be able to create this list easily. They also want a quick way to find the price of a specific product when clients call for information. Likewise, if agricultural inspectors call with questions about a product's safety and need information about a particular lot of produce, Joan and Joe need to be able to find the information quickly. The farm also needs to be able to update prices quickly because farmers use surpluses and shortages in the market to raise and lower prices.

As previously mentioned, workers must put in a certain amount of hours per week to maintain their volunteer status. Instead of adding all the volunteer hours manually, Joan would like a quick way to determine how many hours each volunteer worked in a given week.

Finally, Joe would like to track the farm's best customers and be able to create a report that summarizes sales.

ASSIGNMENT 1: CREATING THE DATABASE DESIGN

In this assignment, you design your database tables using a word-processing program. Pay close attention to the logic and structure of the tables. Do not start developing your Access code in Assignment 2 before getting feedback from your instructor on Assignment 1. Keep in mind that you need to examine the requirements in Assignment 2 to design your fields and tables properly. It is good programming practice to look at the required outputs before beginning your design. When designing the database, observe the following guidelines:

- First, determine the tables you will need by listing the name of each table and the fields it should contain. Avoid data redundancy. Do not create a field if it can be created by a calculated field in a query.
- You will need transaction tables. Think about the business events that occur with each hour worked and with each order. Avoid duplicating data.
- Document your tables using the table feature of your word processor. Your tables should resemble the format shown in Figure 4-1.
- You must mark the appropriate key field(s) by entering an asterisk (*) next to the field name. Keep in mind that some tables might need a compound primary key to uniquely identify a record.
- Print the database design.

Table Name	
Field Name	Data Type (text, numeric, currency, etc.)
…	…
…	…

Source: © Cengage Learning 2014

FIGURE 4-1   Table design

N O T E

Have your design approved before beginning Assignment 2; otherwise, you may need to redo Assignment 2.

ASSIGNMENT 2: CREATING THE DATABASE, QUERIES, AND REPORT

In this assignment, you first create database tables in Access and populate them with data. Next, you create a form, seven queries, and a report.

Assignment 2A: Creating Tables in Access

In this part of the assignment, you create your tables in Access. Use the following guidelines:

- Enter data for at least 10 volunteers, including hours worked for some volunteers. Use your name as one of the volunteers.
- Create records for sales of at least nine different products to eight different grocers.
- Create many orders that include multiple items. Consider using the random number generator in Excel and importing the data into Access to populate your tables quickly. If necessary, ask your instructor for help.
- Add at least three different lots of produce picked at various times by different volunteers.
- Appropriately limit the size of the text fields; for example, a mobile phone number does not need the default length of 255 characters.
- Print all tables if your instructor requires it.

Assignment 2B: Creating Forms, Queries, and Reports

You will generate one form with a subform, seven queries, and one report, as outlined in the Background section of this case.

Form

Create a form and subform based on your Orders table and Order Line Item table (or whatever you named the tables). Save the form as Orders. Your form should resemble that in Figure 4-2.

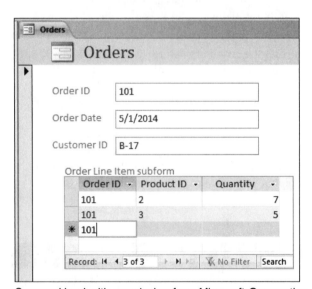

Source: Used with permission from Microsoft Corporation

FIGURE 4-2   Orders form and subform

Query 1

Create a select query called Volunteers Over 18 Years that includes headings for Last Name, First Name, Telephone, and Email. Your output should resemble that in Figure 4-3, although your data will be different.

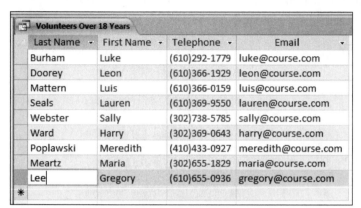

Source: Used with permission from Microsoft Corporation

FIGURE 4-3   Volunteers Over 18 Years query

Query 2

Create a parameter query called Product Listing that prompts for a product name and then displays headings for Product Name and Price Per Bushel. If you enter "Pears" when prompted, the output should resemble that in Figure 4-4, although the price data will be different.

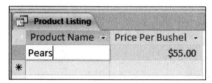

Source: Used with permission from Microsoft Corporation

FIGURE 4-4   Product Listing query

Query 3

Create a parameter query called Lot Identification that displays headings for Product Name, Row, Date Picked, and Last Name of the volunteer who picked the crop. The query should prompt the user to enter the Lot ID number. Your output should resemble the format shown in Figure 4-5, but the data will be different.

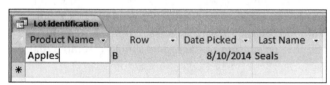

Source: Used with permission from Microsoft Corporation

FIGURE 4-5   Lot Identification query

Query 4

Recent good weather has helped create a bumper crop of pears that must be sold before they spoil. Create a query that updates the price of pears by making them cheaper by $5 per bushel. Save the query as Update Pear Price. Test the query by clicking the Run button and inspecting the Product table.

Query 5

Create a query called Volunteer Hours Worked that includes headings for Last Name, First Name, and Total Time, which is a calculation. Restrict the output to a specific week that the user defines. Note the column heading change from the default setting provided by the query generator. Your output should resemble the format shown in Figure 4-6, but the data will be different.

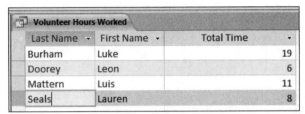

Source: Used with permission from Microsoft Corporation

FIGURE 4-6   Volunteer Hours Worked query

Query 6

Create a query called Best Customers that includes headings for Store Name and Number of Orders. The query should list the store with the most orders at the top and the store with the fewest orders at the bottom. Note the column heading change from the default setting provided by the query generator. Your output should resemble that in Figure 4-7.

Store Name	Number of Orders
Green Grocer	15
Crisp and Fresh	14
R U Healthy	13
Shop Healthy	12
Patty's Produce	12
Heart-wise	10
Apples and More	10
Anna's Produce	10

Source: Used with permission from Microsoft Corporation

FIGURE 4-7   Best Customers query

Report

Create a report called Sales that includes headings for Store Name, Order Date, Product Name, and Total Price. First, create a query to calculate the total price of each product sold. Bring the query into a report and group the sales by store name. Create a Total Price subtotal for each store. Make sure that all column headings and data are visible. Depending on your data, the top portion of the report should resemble that in Figure 4-8.

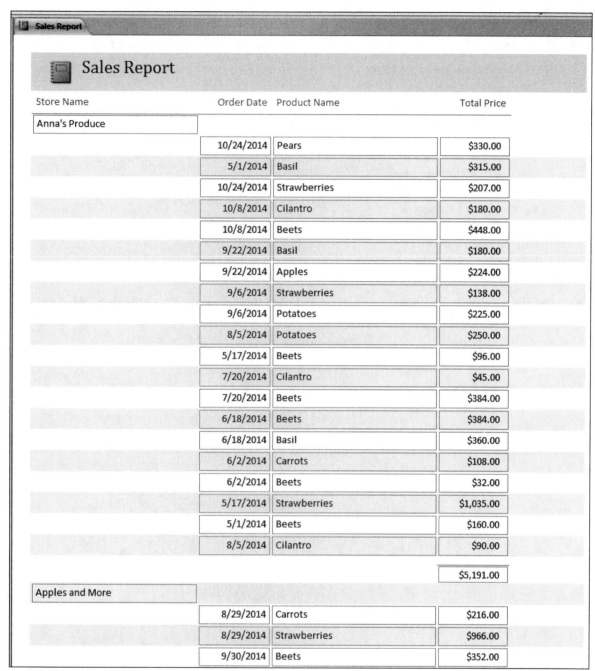

Source: Used with permission from Microsoft Corporation

FIGURE 4-8   Sales report

ASSIGNMENT 3: MAKING A PRESENTATION

Create a presentation that explains the database to Joan and Joe. Describe the design of the tables and include instructions for using the database. Discuss future improvements to the database, such as moving various parts of the system online. Your presentation should take less than 10 minutes, including a question-and-answer period.

DELIVERABLES

Assemble the following deliverables for your instructor, either electronically or in printed form:

1. Word-processed design of tables
2. Tables created in Access
3. Form and subform: Orders
4. Query 1: Volunteers Over 18 Years
5. Query 2: Product Listing
6. Query 3: Lot Identification
7. Query 4: Update Pear Price
8. Query 5: Volunteer Hours Worked
9. Query 6: Best Customers
10. Query 7: Query for Report
11. Report: Sales
12. Presentation materials
13. Any other required printouts or electronic media

Staple all the pages together. Include your name and class number at the top of the page. Make sure that your electronic media are labeled, if required.

CASE **5**

CREATIVE CAKES DATABASE

Designing a Relational Database to Create Tables, Forms, Queries, and Reports

PREVIEW

In this case, you will design a relational database for a bakery featured on a reality TV show. The bakery sells cakes and pies and conducts cake decorating classes. After you have designed and created the database tables, you will populate the database and create one form, 10 queries, and two reports. The form will help the bakery take orders from local customers. The queries will address the following questions: Who are the out-of-town customers? Who is attending the cake decorating classes? Which cakes and pies are the most popular products? The queries will address the following questions for orders of custom wedding cakes: What kind of cake was ordered, who ordered it, and how much will it cost? What is the additional shipping charge for these wedding cakes? A query will also help the bakery change the prices of in-store products. Two reports based on queries will summarize the custom orders and in-store orders.

PREPARATION

- Before attempting this case, you should have some experience in database design and in using Microsoft Access.
- Complete any part of Tutorial A that your instructor assigns.
- Complete any part of Tutorial B that your instructor assigns, or refer to the tutorial as necessary.
- Refer to Tutorial F as necessary.

BACKGROUND

You enjoy watching television in your spare time and are particularly fond of reality programs and cooking shows. You were excited to find a show on a YouTube channel that combines the two features. The new show, *Creative Cakes*, features a family-owned bakery in Connecticut, on the outskirts of New York City. The show has a small following, but you realize that it has the potential to be a big hit. One reason you are drawn to the show is that you live in the same town as the bakery and remember going there as a child for delicious cakes and pies. Because you are searching for a summer internship, you decide to e-mail the owner of the bakery, Celia O'Mara, and ask if she will hire you in some capacity. You explain that you have been learning about databases at the university. Celia is delighted by your e-mail and says she has been thinking about hiring someone to help with the overwhelming number of orders they have received since the show began to air. She is so busy with the show that she hires you to begin work immediately when you finish classes for the year.

You arrive home in May and rush over to Creative Cakes to interview Celia's cousin, Carrie, who is dealing with the mounting orders without a computer. Celia gives you a tour of the bakery and explains the different offerings. Some items, such as cakes, pies, and cupcakes, are sold in the store; they are ordered by local customers who come from the tri-state area of New Jersey, New York, and Connecticut. The bakery has also begun to offer cake decorating classes. Since the launch of the reality show online, the demand for classes is increasing. For example, students can take classes to create fondant layer cakes and "sculptured toppers," which are hand-crafted figures for the tops of wedding cakes. The third and most profitable product line is custom cakes, which are ordered mostly for weddings. Many of these custom cakes are ordered by customers outside the tri-state area. They have seen the reality show and feel that a custom cake would be perfect for their big event.

Pricing for the different product lines varies. For the in-store products, cakes and pies are priced individually and cupcakes are sold by the half dozen. A few other items are priced in different quantities; you note all of this information in preparation for creating your database tables. The classes run for different periods of time and are priced accordingly. For example, a two-hour fondant cake decorating class costs $125. The custom cakes are priced per slice, with a minimum of 50 slices per order. The custom cakes include a basic cake with buttercream frosting, cakes with basic or elaborate fondant icing, and cakes with special fillings and sculptured fondant toppers.

Carrie takes orders in a large notebook. The top corner of each page contains an order number. To take an order, she first records the day's date and then writes the customer's name, address, phone number, e-mail address, and credit card number. She is frustrated with having to rewrite information for repeat customers, whose numbers are increasing as the show gains popularity. She would like to be able to assign each customer a number and use it to record an order.

In the notebook, Carrie writes down all the items that the customer orders, being careful to segregate each type of product on the page. For example, she records all orders for in-store items near the top of the page. Carrie lists each item ordered and the quantity requested. In the middle of the page, Carrie lists any cake decorating classes that the customer wants to take. At the bottom of the page, she records the details of custom cake orders, including extra options that a customer requests and the number of slices required. All custom orders are shipped 10 days after they are placed, and customers who want a wedding cake are made aware of the timing.

Carrie is open to hearing your suggestions for making her job easier. You begin by suggesting a database form that she can use to record incoming orders. You will create a form for in-store orders that Carrie will test for effectiveness.

Carrie then explains several aspects of her job that you realize can be handled with queries. First, Carrie is compiling a list of all customers who live outside the tri-state area. The list will help the show producers determine where the market is strongest for the program. Carrie also keeps a list of all students in the cake decorating classes, and she spends a lot of time tallying up sales totals for popular in-store products. When you tell her that you can develop queries to handle all these tasks, she gratefully gives you a red velvet cupcake to encourage more ideas!

Carrie's brother, Carl, who does much of the baking and ordering of supplies, talks to you about the raw materials needed for baking. He complains about the rising costs of flour and sugar and realizes that he needs to increase the prices of all in-store products to make a decent profit. You tell him that you can easily automate the process using an update query. Carl also requests a listing of custom orders so he can make sure he has enough raw materials on hand to prepare all orders within the 10-day lead time. Again, you explain that you can easily accommodate his request with a query.

Carl's younger brother, Cecil, is in charge of shipping the cakes and handling payments. Cecil has several requests for information that he needs from the database. First, he needs to be able to determine the total cost of a custom cake. The basic cost is based on the number of slices, with charges added for extra options requested, such as fondant and cream filling. Then, depending on the size of the cake, shipping charges are added. Cecil would like to be able to determine the total cake cost, including shipping charges. The bakery is becoming busier, so any automation of these calculations would be welcome. You will design three queries to generate the information that Cecil needs.

Finally, you suggest creating summary reports for custom orders and in-store orders so that Celia can monitor the success of the bakery without spending extra time looking through the notebooks.

ASSIGNMENT 1: CREATING THE DATABASE DESIGN

In this assignment, you design your database tables using a word-processing program. Pay close attention to the logic and structure of the tables. Do not start developing your Access code in Assignment 2 before getting feedback from your instructor on Assignment 1. Keep in mind that you need to examine the requirements in Assignment 2 to design your fields and tables properly. It is good programming practice to look at the required outputs before beginning your design. When designing the database, observe the following guidelines:

- First, determine the tables you will need by listing the name of each table and the fields it should contain. Avoid data redundancy. Do not create a field if it can be created by a calculated field in a query.
- You will need a few transaction tables for ordering each type of product. Avoid duplicating data.

- Document your tables using the table feature of your word processor. Your tables should resemble the format shown in Figure 5-1.
- You must mark the appropriate key field(s) by entering an asterisk (*) next to the field name. Keep in mind that some tables might need a compound primary key to uniquely identify a record.
- Print the database design.

Table Name	
Field Name	Data Type (text, numeric, currency, etc.)
...	...
...	...

Source: © Cengage Learning 2014

FIGURE 5-1   Table design

NOTE

Have your design approved before beginning Assignment 2; otherwise, you may need to redo Assignment 2.

ASSIGNMENT 2: CREATING THE DATABASE, FORM, QUERIES, AND REPORTS

In this assignment, you first create database tables in Access and populate them with data. Next, you create a form with a subform, 10 queries, and two reports.

Assignment 2A: Creating Tables in Access

In this part of the assignment, you create your tables in Access. Use the following guidelines:

- Create entries for 10 in-store products after looking on the Internet or visiting your favorite bakery to find different items for sale. Use current prices for the products to make the project realistic.
- Create records for five different cake decorating or baking classes. Assign a cost to each class, along with a date, a start time, and an end time.
- Create five different options for the custom cakes: basic buttercream, basic fondant, elaborate fondant, cream filling, and sculpted topper, all at varying prices per slice.
- Create at least 10 customer records, including one for yourself. At least half the customers should be from out of town.
- Each customer should order several items. The local customers should order more than one item from the in-store product line and sign up for at least one cake decorating class. Customers from outside the tri-state area should order a wedding cake with special options.
- Appropriately limit the size of the text fields; for example, a customer ID number does not need the default length of 255 characters.
- Print all tables if your instructor requires it.

Assignment 2B: Creating a Form, Queries, and Reports

You will create a form with a subform, 10 queries, and two reports, as outlined in the Background section of this case.

Form

Create a form and subform based on your tables so bakery employees can take orders for the in-store products. Use all the fields in your tables to create the form and subform. Name the form In-store Orders. Your data will vary, but the output should resemble that in Figure 5-2.

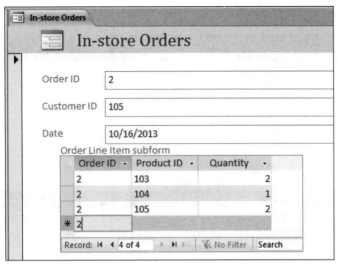

Source: Used with permission from Microsoft Corporation

FIGURE 5-2 In-store Orders form

Query 1

Create a query called Out-of-town Customers for customers who do not live in the tri-state area. The query should display headings for Last Name, First Name, Address, City, and State. Your data will differ, but the output should resemble that in Figure 5-3.

Last Name	First Name	Address	City	State
Dickerson	Allen	138 Woodlawn Ave	Seattle	WA
Hearn	Arthur	26 Julie Court	Media	PA
Sunzar	Sam	103 Chadd Rd	Owings Mills	MD
Turner	Cynthia	1502 Valley Stream Lane	Salt Lake City	UT
Wills	Billy	25 Brown Lane	Centerville	OH

Source: Used with permission from Microsoft Corporation

FIGURE 5-3 Out-of-town Customers query

Query 2

Create a query called Class Attendees that displays headings for Class Name, Class Date, Length of Class, and Last Name and First Name of students. The Length of Class column is a series of calculated fields. Your data will differ, but the output should resemble that in Figure 5-4.

Class Attendees

Class Name	Class Date	Length of Class	Last Name	First Name
Fondant Layer Cake	11/1/2013	2	Trapp	John
Elaborate Fondant	12/5/2013	2	Trapp	John
Pies and Candy	12/17/2013	1.5	Trapp	John
Sculptured Toppers	12/1/2013	3	Trapp	John
Sculptured Toppers	12/1/2013	3	Nelson	Janice
Pies and Candy	12/17/2013	1.5	Nelson	Janice
Sculptured Toppers	12/1/2013	3	Faber	Dale
Elaborate Fondant	12/5/2013	2	Faber	Dale
Pies and Candy	12/17/2013	1.5	Faber	Dale

Source: Used with permission from Microsoft Corporation

FIGURE 5-4   Class Attendees query

Query 3

Create a query called Popular In-store Products that displays headings for Product Name, Product Quantity, Product Price, and Total Ordered. Sort your output to display the most popular product at the top of the list. Your data will differ, but your output should resemble that in Figure 5-5. Note the column heading change from the default setting provided by the query generator.

Popular In-store Products

Product Name	Product Quantity	Product Price	Total Ordered
Cheesecake	6 inches	$16.95	4
White Chocolate Mousse	7 inches	$24.95	2
Strawberry Shortcake	7 inches	$24.95	2
Individual Tiramisu	2 pieces, 3 inches	$13.00	2
Chocolate Fudge Cake	7 inches	$24.95	2
Vanilla Buttercream Cupcakes	1/2 dozen	$13.50	1
Snickers Cake	7 inches	$29.95	1
Red Velvet Cupcakes	1/2 dozen	$18.00	1
Red and White Fondant Cake	7 inches	$46.95	1
Cannoli	1/2 dozen	$18.00	1

Source: Used with permission from Microsoft Corporation

FIGURE 5-5   Popular In-store Products query

Query 4

Create an update query called Update In-store Product Price that increases the price of all in-store products by 10 percent. Run the query to test it and check the table to make sure the query works correctly.

Query 5

Create a query called Custom Order Round-up. Display headings for Custom Product Description, Price Per Slice, Total Number of Slices, and Total Money In. The last field is a calculated field that determines how much money is generated from each type of custom product. Sort the output to show the largest amount of money at the top of the list. Your data will differ, but your output should resemble that in Figure 5-6.

Custom Order Round-up

Custom Product Description	Price Per Slice	Total Number of Slices	Total Money In
Basic buttercream	$7.00	1,485	$10,395.00
Cream filling	$1.00	1,185	$1,185.00
Basic fondant	$1.00	1,035	$1,035.00
Elaborate fondant	$2.00	515	$1,030.00

Source: Used with permission from Microsoft Corporation

FIGURE 5-6   Custom Order Round-up query

Query 6

Create a query called Total Custom Cake Cost that calculates the total cost of each custom cake, not including the delivery charge. Display headings for Customer ID, Last Name, and Total Cost. Your data will differ, but your output should resemble that in Figure 5-7.

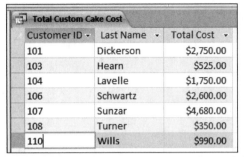

Total Custom Cake Cost		
Customer ID	Last Name	Total Cost
101	Dickerson	$2,750.00
103	Hearn	$525.00
104	Lavelle	$1,750.00
106	Schwartz	$2,600.00
107	Sunzar	$4,680.00
108	Turner	$350.00
110	Wills	$990.00

Source: Used with permission from Microsoft Corporation

FIGURE 5-7 Total Custom Cake Cost query

Query 7

Create a query called Additional Shipping Charge. To calculate the charge, use an IF statement and the following rule: For any order of 150 slices or less, charge $50; otherwise, charge $85. Display headings for Customer ID, Last Name, First Name, Address, City, State, Zip, and Additional Shipping Charge. Your data will differ, but your output should resemble that in Figure 5-8.

Additional Shipping Charge							
Customer ID	Last Name	First Name	Address	City	State	Zip	Additional Shipping Charge
101	Dickerson	Allen	138 Woodlawn Ave	Seattle	WA	98119	$85.00
103	Hearn	Arthur	26 Julie Court	Media	PA	19063	$50.00
104	Lavelle	Shirley	4001 Birch Street	Peekskill	NY	10566	$85.00
106	Schwartz	Byron	2 Waverly Rd	Deep River	CT	09776	$85.00
107	Sunzar	Sam	103 Chadd Rd	Owings Mills	MD	21117	$85.00
108	Turner	Cynthia	1502 Valley Stream Lane	Salt Lake City	UT	84109	$50.00
110	Wills	Billy	25 Brown Lane	Centerville	OH	45459	$50.00

Source: Used with permission from Microsoft Corporation

FIGURE 5-8 Additional Shipping Charge query

Query 8

Create a query called Grand Total Custom Cost. Use the previous two queries to calculate the grand total, which is the sum of the custom cake's cost and the additional shipping charge. Display headings for Last Name, First Name, Address, City, State, Zip, and Grand Total. Your data will differ, but your output should resemble that in Figure 5-9.

Grand Total Custom Cost						
Last Name	First Name	Address	City	State	Zip	Grand Total
Dickerson	Allen	138 Woodlawn Ave	Seattle	WA	98119	$2,835.00
Hearn	Arthur	26 Julie Court	Media	PA	19063	$575.00
Lavelle	Shirley	4001 Birch Street	Peekskill	NY	10566	$1,835.00
Schwartz	Byron	2 Waverly Rd	Deep River	CT	09776	$2,685.00
Sunzar	Sam	103 Chadd Rd	Owings Mills	MD	21117	$4,765.00
Turner	Cynthia	1502 Valley Stream Lane	Salt Lake City	UT	84109	$400.00
Wills	Billy	25 Brown Lane	Centerville	OH	45459	$1,040.00

Source: Used with permission from Microsoft Corporation

FIGURE 5-9 Grand Total Custom Cost query

Report 1

Create a report called Custom Orders. First create a query that displays headings for Last Name, Custom Product Description, Price Per Slice, and Number of Slices. Also create a column that calculates and displays the total cost of the order, not including the shipping charge. Save the query as Custom Orders, and then bring the query into a report that has the same title. Sum each customer's order, as shown. Make sure that all field headings are visible and all data is formatted correctly. Your data will differ, but the output should resemble that in Figure 5-10. Only a portion of the output is shown in the figure.

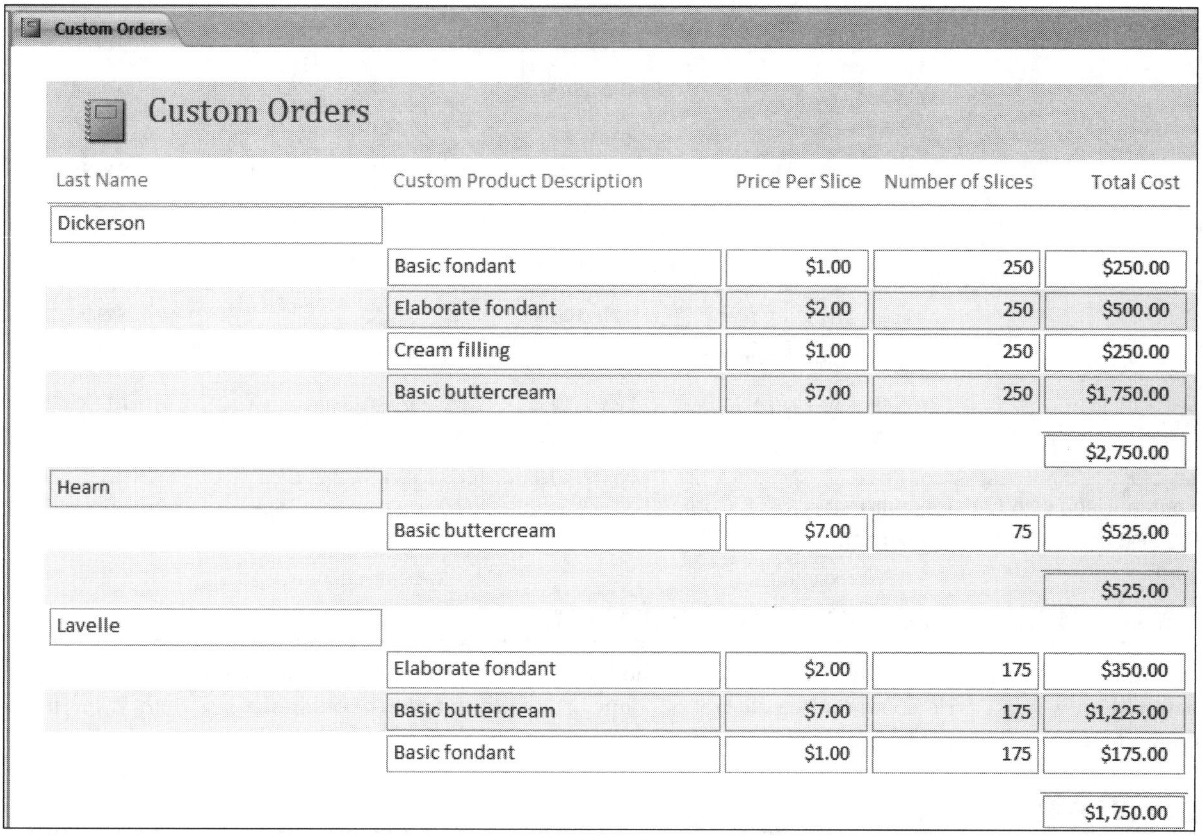

Custom Orders

Last Name	Custom Product Description	Price Per Slice	Number of Slices	Total Cost
Dickerson				
	Basic fondant	$1.00	250	$250.00
	Elaborate fondant	$2.00	250	$500.00
	Cream filling	$1.00	250	$250.00
	Basic buttercream	$7.00	250	$1,750.00
				$2,750.00
Hearn				
	Basic buttercream	$7.00	75	$525.00
				$525.00
Lavelle				
	Elaborate fondant	$2.00	175	$350.00
	Basic buttercream	$7.00	175	$1,225.00
	Basic fondant	$1.00	175	$175.00
				$1,750.00

Source: Used with permission from Microsoft Corporation
FIGURE 5-10   Custom Orders report

Report 2

Create an In-store Orders report based on a query that displays headings for Last Name, Date (of order), Product Name, Product Quantity, Price, Quantity (ordered), and Total Amount Due. The Total Amount Due field is a calculated field. Save the query as In-store Orders. Group the report on Last Name, and then adjust the output so that the date is on the same line as the last name. Sum the total amount due for each customer. Your data will differ, but the output should resemble that in Figure 5-11. Give the report a title of In-store Orders and save the report with the same name. All data and headings should be visible and all data should be formatted correctly, as shown in Figure 5-11.

In-store Orders

In-store Orders

Last Name	Date	Product Name	Product Quantity	Price	Quantity	Total Amount Due
Faber	10/17/2013					
		Cheesecake	6 inches	$16.95	4	$67.80
		Individual Tiramisu	2 pieces, 3 inches	$13.00	2	$26.00
		Cannoli	1/2 dozen	$18.00	1	$18.00
		Snickers Cake	7 inches	$29.95	1	$29.95
		Red and White Fondant Cake	7 inches	$46.95	1	$46.95
						$188.70
Nelson	10/16/2013					
		Strawberry Shortcake	7 inches	$24.95	2	$49.90
		Vanilla Buttercream Cupcakes	1/2 dozen	$13.50	1	$13.50
		White Chocolate Mousse	7 inches	$24.95	2	$49.90
						$113.30

Source: Used with permission from Microsoft Corporation

FIGURE 5-11 In-store Orders report

ASSIGNMENT 3: MAKING A PRESENTATION

Create a presentation for Creative Cakes. Discuss how your system could be a prototype for a Web-based database in which Carrie could handle all orders online. Your presentation should take less than 15 minutes, including a brief question-and-answer period.

DELIVERABLES

Assemble the following deliverables for your instructor, either electronically or in printed form:

1. Word-processed design of tables
2. Tables created in Access
3. Form and subform: In-store Orders
4. Query 1: Out-of-town Customers
5. Query 2: Class Attendees
6. Query 3: Popular In-store Products
7. Query 4: Update In-store Product Price
8. Query 5: Custom Order Round-up
9. Query 6: Total Custom Cake Cost
10. Query 7: Additional Shipping Charge
11. Query 8: Grand Total Custom Cost
12. Query 9: Custom Orders
13. Query 10: In-store Orders
14. Report 1: Custom Orders
15. Report 2: In-store Orders

Staple all the pages together. Include your name and class number at the top of the page. Make sure that your electronic media are labeled, if required.

PART 2

DECISION SUPPORT CASES USING EXCEL SCENARIO MANAGER

TUTORIAL C

BUILDING A DECISION SUPPORT SYSTEM IN EXCEL

Decision Support Systems (DSS) are computer programs used to help managers solve complex business problems. DSS programs are commonly found in large, integrated packages called enterprise resource planning software that provide information services to an organization. Software packages such as SAP™, Microsoft Dynamics™, and PeopleSoft™ offer sophisticated DSS capabilities. However, many business problems can be modeled for solutions using less complex tools, such as Visual Basic, Microsoft Access, and Microsoft Excel.

A DSS program is actually a model representing a quantitative business problem. The problem can range from finding a desired product mix to sales forecasts to risk analysis, but almost all of the problems examine *financial outcomes*. The model itself contains the data and the algorithms (mathematical processes) needed to solve the problem.

In a DSS program, the user manually inputs data or the program accesses data from a file in the system. The program runs the data through its algorithms and displays output formatted as information; the manager uses this data to decide what action to take to solve the problem. Some sophisticated DSS programs display multiple solutions and recommend one based on predefined parameters.

Managers often find the Excel spreadsheet program particularly useful for their DSS needs. Excel contains hundreds of built-in arithmetic, statistical, logical, and financial functions. It can import data in numerous formats from large database programs, and it can be set up to display well-organized, visually appealing tables and graphs from the output.

This tutorial is organized into four sections:

1. **Spreadsheet and DSS Basics**—This section lets you "get your feet wet" by creating a DSS program in Excel. The program is a cash flow model for a small business looking to expand. You will get an introduction to spreadsheet design, building a DSS, and using financial functions.
2. **Scenario Manager**—Here you will learn how to use the Excel Scenario Manager. A DSS typically gives you one set of answers based on one set of inputs—the real value of the tool lies in its ability to play "what if" and take a comparative look at all the solutions based on all combinations of the inputs. Rather than inputting and running the DSS several times manually, you can use Scenario Manager to run and display the outputs from all possible combinations of the inputs. The output is summarized on a separate worksheet in the Excel workbook.
3. **Practice Using Scenario Manager**—Next, you will be given a new problem to model as a DSS, using Scenario Manager to display your solutions.
4. **Review of Excel Basics**—This section reviews additional information that will help you complete the spreadsheet cases that follow this tutorial. You will learn some basic operations, logical functions, and cash flow calculations.

SPREADSHEET AND DSS BASICS

You are the owner of a thrift shop that resells clothing and housewares in a university town. Many of your customers are college students. Your business is unusual in that sales actually increase during an economic recession. Your cost of obtaining used items basically follows the consumer price index. It is the end of 2013, and business has been very good due to the continuing recession. You are thinking of expanding your business to an adjacent storefront that is for sale, but you will have to apply for a business loan to finance the purchase. The bank requires a projection of your profit and cash flows for the next two years before it will loan you the money to expand, so you have to determine your net income (profit) and cash flows for 2014 and 2015. You decide that your forecast should be based on four factors: your 2013 sales dollars, your cost of goods sold per sales dollar, your estimates of the underlying economy, and the business loan payment amount and interest rate.

Because you will present this model to your prospective lenders, you decide to use an Income and Cash Flow Statements framework. You will input values for two possible states of the economy for 2014 and 2015: R for a continuing recession and B for a "boom" (recovery). Your sales in the recession were growing at 20% per year. If the recession continues and you expand the business, you expect sales to continue growing at 30% per year. However, if the economy recovers, some of your customers will switch to buying "new," so you expect sales growth for your thrift shop to be 15% above the previous year (only 5% growth plus 10% for the business expansion). If you do not expand, your recession or boom growth percentages will only be 20% and 5%, respectively. To determine the cost of goods sold for purchasing your merchandise, which is currently 70% of your sales, you will input values for two possible consumer price outlooks: H for high inflation (1.06 multiplied by the average cost of goods sold) and L for low inflation (1.02 multiplied by the cost of goods sold).

You currently own half the storefront and will need to borrow $100,000 to buy and renovate the other half. The bank has indicated that, depending on your forecast, it may be willing to loan you the money for your expansion at 5% interest during the current recession with a 10-year repayment compounded annually ("R"). However, if the prime rate drops at the start of 2014 because of an economic turnaround ("B"), the bank can drop your interest rate to 4% with the same repayment terms.

As an entrepreneur, an item of immediate interest is your cash flow position with the additional burden of a loan payment. After all, one of your main objectives is to make a profit (Net Income After Taxes). You can use the DSS model to determine if it is more profitable *not* to expand the business.

Organization of the DSS Model

A well-organized spreadsheet will make the design of your DSS model easier. Your spreadsheet should have the following sections:

- Constants
- Inputs
- Summary of Key Results
- Calculations (with separate calculations for Expansion vs. No Expansion)
- Income and Cash Flow Statements (with separate statements for Expansion vs. No Expansion)

Figures C-1 and C-2 illustrate the spreadsheet setup for the DSS model you want to build.

	A	B	C	D
1	**Tutorial Exercise--Collegetown Thrift Shop**			
2				
3	**Constants**	**2013**	**2014**	**2015**
4	Tax Rate	NA	33%	35%
5	Loan Amount for Store Expansion	NA	$100,000	NA
6				
7	**Inputs**	**2013**	**2014**	**2015**
8	Economic Outlook (R=Recession, B=Boom)	NA		NA
9	Inflation Outlook (H=High, L=Low)	NA		NA
10				
11	**Summary of Key Results**	**2013**	**2014**	**2015**
12	Net Income after Taxes (Expansion)	NA		
13	End-of-year Cash on Hand (Expansion)	NA		
14	Net Income after Taxes (No Expansion)	NA		
15	End-of-year Cash on Hand (No Expansion)	NA		
16				
17	**Calculations (Expansion)**	**2013**	**2014**	**2015**
18	Total Sales Dollars	$350,000		
19	Cost of Goods Sold	$245,000		
20	Cost of Goods Sold (as a percent of Sales)	70%		
21	Interest Rate for Business Loan		NA	NA
22				
23	**Calculations (No Expansion)**	**2013**	**2014**	**2015**
24	Total Sales Dollars	$350,000		
25	Cost of Goods Sold	$245,000		
26	Cost of Goods Sold (as a percent of Sales)	70%		

Source: Used with permission from Microsoft Corporation

FIGURE C-1 Tutorial skeleton 1

	A	B	C	D
28	**Income and Cash Flow Statements (Expansion)**	**2013**	**2014**	**2015**
29	Beginning-of-year Cash on Hand	NA		
30	Sales (Revenue)	NA		
31	Cost of Goods Sold	NA		
32	*Business Loan Payment*	NA		
33	Income before Taxes	NA		
34	Income Tax Expense	NA		
35	Net Income after Taxes	NA		
36	End-of-year Cash on Hand	$15,000		
37				
38	**Income and Cash Flow Statements (No Expansion)**	**2013**	**2014**	**2015**
39	Beginning-of-year Cash on Hand	NA		
40	Sales (Revenue)	NA		
41	Cost of Goods Sold	NA		
42	Income before Taxes	NA		
43	Income Tax Expense	NA		
44	Net Income after Taxes	NA		
45	End-of-year Cash on Hand	$15,000		

Source: Used with permission from Microsoft Corporation

FIGURE C-2   Tutorial skeleton 2

Each spreadsheet section is discussed in detail next.

The Constants Section

This section holds values that are needed for the spreadsheet calculations. These values are usually given to you, and generally do not change for the exercise. However, you can change these values later if necessary; for example, you might need to borrow more or less money for your business expansion (cell C5). For this tutorial, the constants are the Tax Rate and the Loan Amount.

The Inputs Section

The Inputs section in Figure C-1 provides a place to designate the two possible economic outlooks and the two possible inflation outlooks. If you wanted to make these outlooks change by business year, you could leave blanks under both business years. However, as you will see later when you use Scenario Manager, this approach would greatly increase the complexity of interpreting the results. For simplicity's sake, assume that the same outlooks will apply to both 2014 and 2015.

The Summary of Key Results Section

This section summarizes the Year 2 and 3 Net Income after Taxes (profit) and the End-of-year Cash on Hand both for expanding the business and for not expanding. These cells are copied from the Income and Cash Flow Statements section at the bottom of the sheet. Summary sections are frequently placed near the top of a spreadsheet to allow managers to see a quick "bottom line" summary without having to scroll down the spreadsheet to see the final result. Summary sections can also make it easier to select cells for charting.

The Calculations Sections (Expansion and No Expansion)

The following areas are used to compute the following necessary results:

- The Total Sales Dollars, which is a function of the Year 2013 value and the Economic Outlook input
- The Cost of Goods Sold, which is the Total Sales Dollars multiplied by the Cost of Goods Sold (as a percent of Sales)
- The Cost of Goods Sold (as a percent of Sales), which is a function of the Year 2013 value and the Inflation Outlook input
- In addition, the Calculations section for the expansion includes the interest rate, which is also a function of the Economic Outlook input. This interest rate will be used to determine the Business Loan Payment in the Income and Cash Flow Statements section.

You could make these formulas part of the Income and Cash Flow Statements section. However, it makes more sense to use the approach shown here because it makes the formulas in the Income and Cash Flow

Statements less complicated. In addition, when you create other DSS models that include unit costing and pricing calculations, you can enter the formulas in this section to facilitate managerial accounting cost analysis.

The Income and Cash Flow Statements Sections (Expansion and No Expansion)

These sections are the financial or accounting "body" of the spreadsheet. They contain the following values:

- Beginning-of-year Cash on Hand, which equals the *prior* year's End-of-year Cash on Hand.
- Sales (Revenue), which in this tutorial is simply the results of the Total Sales Dollars copied from the Calculations section.
- Cost of Goods Sold, which also is copied from the Calculations section.
- Business Loan Payment, which is calculated using the PMT (Payment) function and the inputs for loan amount and interest rate from the Constants and Calculations sections. Note that only the Income and Cash Flow Statement for Expansion includes a value for Business Loan Payment. If you do not expand, you do not need to borrow the money.
- Income before Taxes, which is Sales minus the Cost of Goods Sold; for the expansion scenarios, you also subtract the Business Loan Payment.
- Income Tax Expense, which is zero when there is no income or the income is negative; otherwise, this value is the Income before Taxes multiplied by the Tax Rate from the Constants section.
- Net Income after Taxes, which is Income before Taxes minus Income Tax Expense.
- End-of-year Cash on Hand, which is Beginning-of-year Cash on Hand plus Net Income after Taxes.

Note that this Income and Cash Flow Statement is greatly simplified. It does not address the issues of changes in Inventories, Accounts Payable, and Accounts Receivable, nor any period expenses such as Selling and General Administrative expenses, utilities, salaries, real estate taxes, insurance, or depreciation.

Construction of the Spreadsheet Model

Next, you will work through three steps to build the spreadsheet model:

1. Make a skeleton or "shell" of the spreadsheet. Save it with a name you can easily recognize, such as TUTC.xlsx or Tutorial C *YourName*.xlsx. When submitting electronic work to an instructor or supervisor, include your last name and first initial in the filename.
2. Fill in the "easy" cell formulas.
3. Then enter the "hard" spreadsheet formulas.

Making a Skeleton or "Shell"

The first step is to set up the skeleton worksheet. The skeleton should have headings, text labels, and constants. Do not enter any formulas yet.

Before you start entering data, you should first try to visualize a sensible structure for your worksheet. In Figures C-1 and C-2, the seven sections are arranged vertically down the page; the item descriptions are in the first column (A), and the time periods (years) are in the next three columns (B, C, and D). This is a widely accepted business practice, and is commonly called a "horizontal analysis." It is used to visually compare financial data side by side through successive time periods.

Because your key results depend on the Income and Cash Flow Statements, you usually set up that section first, and then work upward to the top of the sheet. In other words, you set up the Income and Cash Flow Statements section, then the Calculations section, and then the Summary of Key Results, Inputs, and Constants sections. Some might argue that the Income and Cash Flow Statements should be at the top of the sheet, but when you want to change values in the Constants or Inputs section or examine the Summary of Key Results, it does not make sense to have to scroll to the bottom of the worksheet. When you run the model, you do not enter anything in the Income and Cash Flow Statements—they are all calculations. So, it makes sense to put them last.

Here are some other general guidelines for designing effective DSS spreadsheets:

- Decide which items belong in the Calculations section. A good rule of thumb is that if your items have formulas but do not belong in the Income and Cash Flow Statements, put them in the

Calculations section. Good examples are intermediate calculations such as unit volumes, costs and prices, markups, or changing interest rates.

- The Summary of Key Results section should be just that—*key* results. These outputs help you make good business decisions. Key results frequently include net income before taxes (profit) and end-of-year cash on hand (how much cash your business has). However, if you are creating a DSS model on alternative capital projects, your key results can also include cost savings, net present value of a project, or rate of return for an investment.
- The Constants section holds known values needed to perform other calculations. You use a Constants section rather than just including the values in formulas so that you can input new values if they change. This approach makes your DSS model more flexible.

AT THE KEYBOARD

Enter the Excel skeleton shown in Figures C-1 and C-2.

NOTE

When you see NA (Not Applicable) in a cell, do not enter any values or formulas in the cell. The cells that contain values in the 2013 column are used by other cells for calculations. In this example, you are mainly interested in what happens in 2014 and 2015. The rest of the cells are "Not Applicable."

Filling in the "Easy" Formulas

The next step in building a spreadsheet is to fill in the "easy" formulas. To begin, format all the cells that will contain monetary values as Currency with zero decimal places:

- Constants—C5
- Summary of Key Results—C12 to C15, D12 to D15
- Calculations (Expansion)—C18, C19, D18, D19
- Calculations (No Expansion)—C24, C25, D24, D25
- Income and Cash Flow Statements (Expansion)—B36, C29 to C36, D29 to D36
- Income and Cash Flow Statements (No Expansion)—B45, C39 to C45, D39 to D45

NOTE

With the insertion point in cell C12 (where the $0 appears), note the editing window—the white space at the top of the spreadsheet to the right of the f_x symbol. The cell's contents, whether it is a formula or value, should appear in the editing window. In this case, the window shows =C35.

The Summary of Key Results section (see Figure C-3) will contain the values you calculate in the Income and Cash Flow Statements sections. To copy the cell contents for this section, move your cursor to cell C12, click the cell, type =C35, and press Enter. If you formatted your money cells properly, a $0 should appear in cell C12.

C12		f_x =C35		
	A	B	C	D
11	**Summary of Key Results**	**2013**	**2014**	**2015**
12	Net Income after Taxes (Expansion)	NA	$0	
13	End-of-year Cash on Hand (Expansion)	NA		
14	Net Income after Taxes (No Expansion)	NA		
15	End-of-year Cash on Hand (No Expansion)	NA		

Source: Used with permission from Microsoft Corporation

FIGURE C-3   Value from cell C35 (Net Income after Taxes) copied to cell C12

Because cell C35 does not contain a value yet, Excel assumes that the empty cell has a numerical value of 0. When you put a formula in cell C35 later, cell C12 will echo the resulting answer. Because Net Income after Taxes (Expansion) for 2015 (cell D35) and its corresponding cell in Summary of Key Results (cell D12) are both directly to the right of the values for 2014, you can either type =D35 into cell D12 or copy cell C12 to D12. To perform the copy operation:

1. Click in a cell or click and drag to select the range of cells you want to copy.
2. Hold down the Control key and press C (Ctrl+C).
3. A moving dashed box called a *marquee* should now be animated over the cell(s) selected for copying.
4. Select the cell(s) where you want to copy the data.
5. Hold down the Control key and press V (Ctrl+V). Cell D12 should now contain $0, but actually it has a reference to cell D35. Click cell D12 and look again at the editing window; it should display =D35.

Cells C14, C15, D14, and D15 represent Net Income after Taxes and End-of-year Cash on Hand for both years of No Expansion; these cells are mirrors of cells C44, C45, D44, and D45 in the last section. Select cell C14, type =C44, and press Enter. Select cell C14 again, use the Copy command, and paste the contents into cell D14 (see Figure C-4).

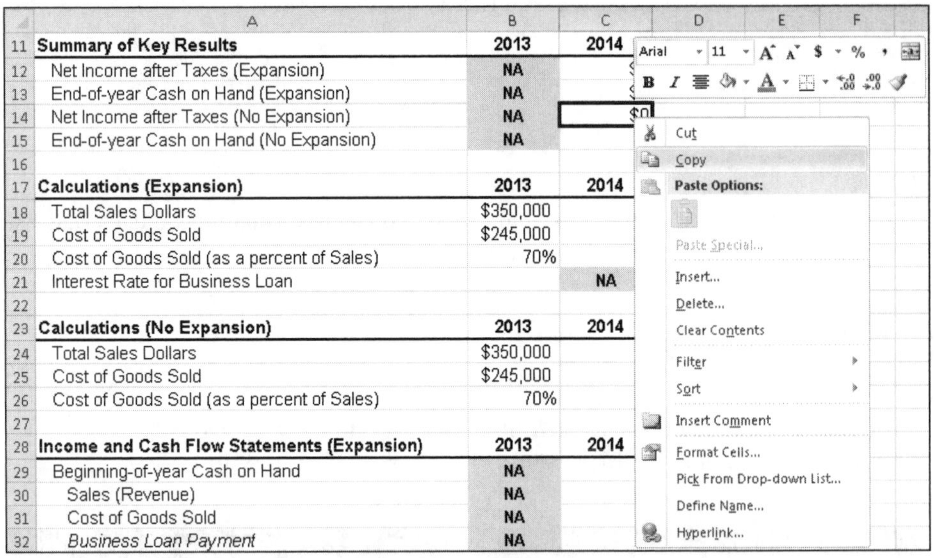

Source: Used with permission from Microsoft Corporation

FIGURE C-4   Copying the formula from cell C14 to cell D14

Because Excel uses *relative* cell references by default, copying cell C14 into cell D14 will copy and paste the contents of cell D44 (the cell adjacent to C44) into cell D14. See Figure C-5.

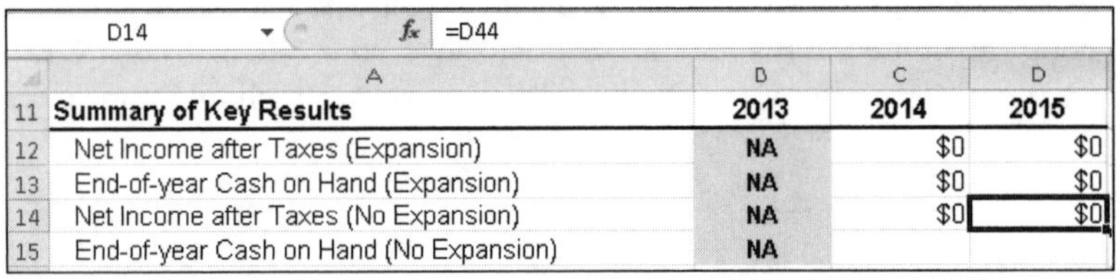

Source: Used with permission from Microsoft Corporation

FIGURE C-5   Formula from cell D44 pasted into cell D14

Use the Copy command again, this time downward from cells C14 and D14, to complete cells C15 and D15. If you are successful, the formula in the editing window for cell C15 will be "=C45" and for cell D15 will display "=D45."

You will create the formulas for the two Calculations sections last because they are the hardest formulas. Next, you will create the formulas for the two Income and Cash Flow Statements sections; all the cells in these two sections should be formatted as Currency with zero decimal places.

As shown in Figure C-6, the Beginning-of-year Cash on Hand for 2014 is the End-of-year Cash on Hand for 2013. In cell C29, type =B36. A handy shortcut is to type the "=" sign, immediately move your mouse pointer to the cell you want to designate, and then click the left mouse button. Excel will enter the cell location into the formula for you. This shortcut is especially useful if you want to avoid making a typing error.

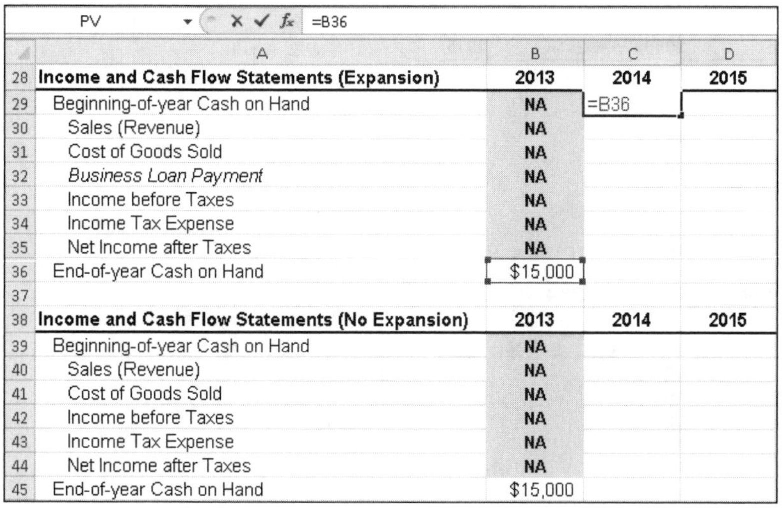

PV		X ✓ ƒₓ	=B36		
	A		B	C	D
28	**Income and Cash Flow Statements (Expansion)**		**2013**	**2014**	**2015**
29	Beginning-of-year Cash on Hand		NA	=B36	
30	Sales (Revenue)		NA		
31	Cost of Goods Sold		NA		
32	*Business Loan Payment*		NA		
33	Income before Taxes		NA		
34	Income Tax Expense		NA		
35	Net Income after Taxes		NA		
36	End-of-year Cash on Hand		$15,000		
37					
38	**Income and Cash Flow Statements (No Expansion)**		**2013**	**2014**	**2015**
39	Beginning-of-year Cash on Hand		NA		
40	Sales (Revenue)		NA		
41	Cost of Goods Sold		NA		
42	Income before Taxes		NA		
43	Income Tax Expense		NA		
44	Net Income after Taxes		NA		
45	End-of-year Cash on Hand		$15,000		

Source: Used with permission from Microsoft Corporation

FIGURE C-6   End-of-year Cash on Hand for 2013 copied to Beginning-of-year Cash on Hand for 2014

Likewise, copy the other three End-of-year Cash on Hand cells to the Beginning-of-year Cash on Hand cells for both Income and Cash Flow Statements (cells D29, C39, and D39).

The Sales (Revenue) cells C30, D30, C40, and D40 are simply copies of cells C18, D18, C24, and D24, respectively, from the Calculations sections (both Expansion and No Expansion). Use the shortcut method to copy these cells. Note that all four cells will display $0 until you enter the formulas in the Calculations sections (see Figure C-7).

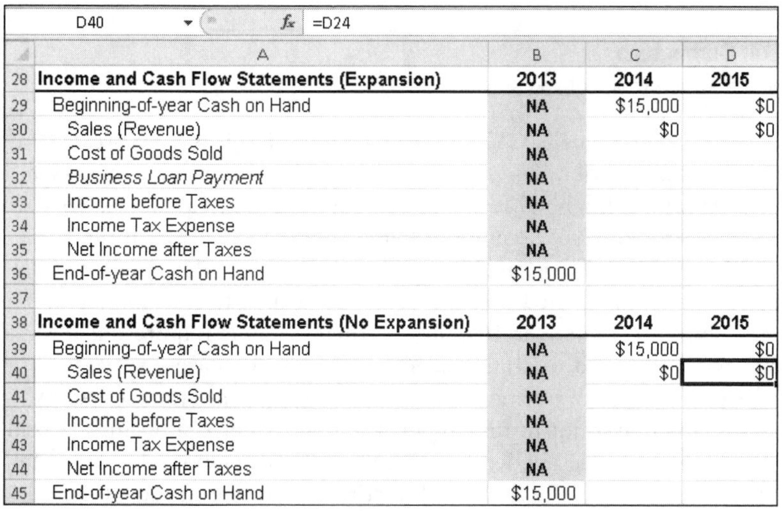

D40		ƒₓ	=D24		
	A		B	C	D
28	**Income and Cash Flow Statements (Expansion)**		**2013**	**2014**	**2015**
29	Beginning-of-year Cash on Hand		NA	$15,000	$0
30	Sales (Revenue)		NA	$0	$0
31	Cost of Goods Sold		NA		
32	*Business Loan Payment*		NA		
33	Income before Taxes		NA		
34	Income Tax Expense		NA		
35	Net Income after Taxes		NA		
36	End-of-year Cash on Hand		$15,000		
37					
38	**Income and Cash Flow Statements (No Expansion)**		**2013**	**2014**	**2015**
39	Beginning-of-year Cash on Hand		NA	$15,000	$0
40	Sales (Revenue)		NA	$0	$0
41	Cost of Goods Sold		NA		
42	Income before Taxes		NA		
43	Income Tax Expense		NA		
44	Net Income after Taxes		NA		
45	End-of-year Cash on Hand		$15,000		

Source: Used with permission from Microsoft Corporation

FIGURE C-7   Sales Revenue cells copied from the Calculations sections

The Cost of Goods Sold cells C31, D31, C41, and D41 are simply copies of the contents of cells C19, D19, C25, and D25, respectively, from the Calculations sections. Because the cells in both locations are directly below the Sales cells in the four locations, you can use the Copy command to fill those cells easily. As you can see in Figure C-8, you can drag your mouse pointer over both cells C40 and D40, right-click to see the floating toolbar, and select Copy. Move your mouse pointer to select cells C41 and D41, right-click the mouse, and select Paste. If you are uncomfortable copying and pasting with the mouse, you can type =C19, =D19, =C25, and =D25 in cells C31, D31, C41, and D41.

	A	B	C	D	E	F	G	H
28	**Income and Cash Flow Statements (Expansion)**	**2013**	**2014**	**2015**				
29	Beginning-of-year Cash on Hand	NA	$15,000	$0				
30	Sales (Revenue)	NA	$0	$0				
31	Cost of Goods Sold	NA	$0	$0				
32	*Business Loan Payment*	NA						
33	Income before Taxes	NA						
34	Income Tax Expense	NA						
35	Net Income after Taxes	NA						
36	End-of-year Cash on Hand	$15,000						
37								
38	**Income and Cash Flow Statements (No Expansion)**	**2013**	**2014**	**2015**				
39	Beginning-of-year Cash on Hand	NA	$15,000					
40	Sales (Revenue)	NA	$0	$0				
41	Cost of Goods Sold	NA						
42	Income before Taxes	NA						
43	Income Tax Expense	NA						
44	Net Income after Taxes	NA						
45	End-of-year Cash on Hand	$15,000						
46								
47								
48								
49								
50								
51								
52								
53								
54								
55								
56								
57								
58								

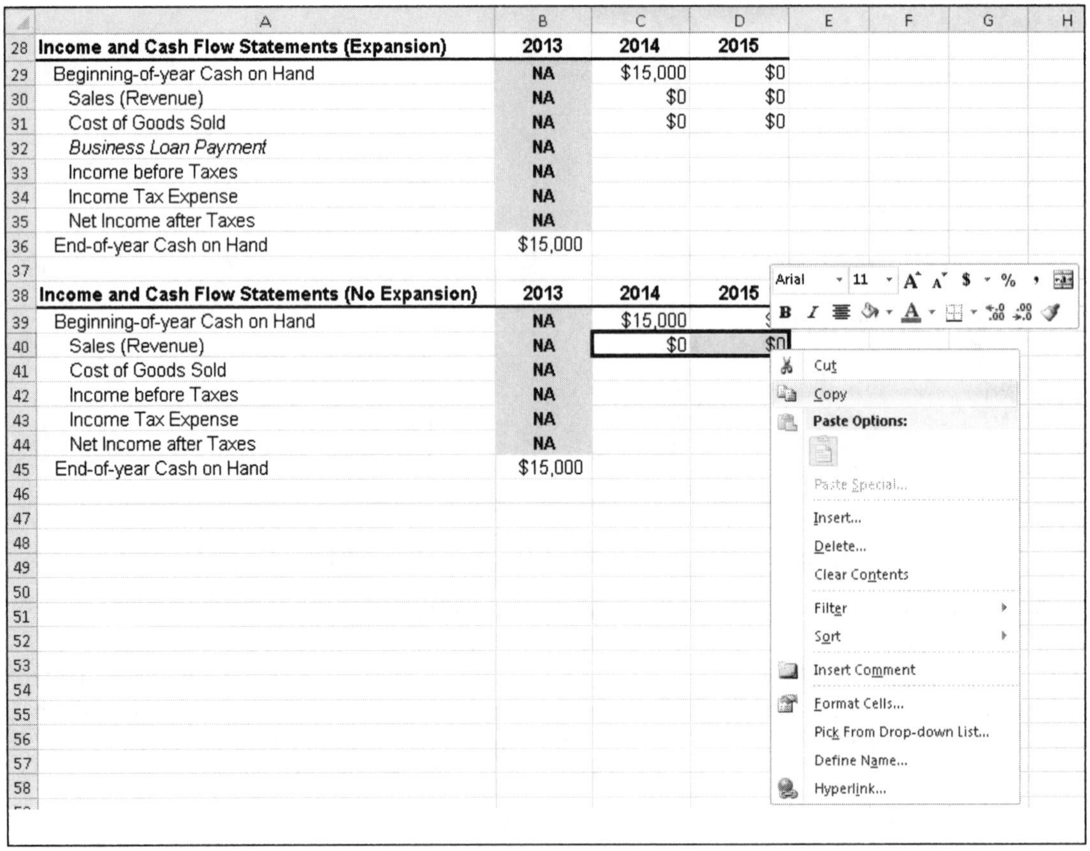

Source: Used with permission from Microsoft Corporation

FIGURE C-8   Cost of Goods Sold cells copied from the Calculations sections

Next you determine the Business Loan Payment for cells C32 and D32—notice that it is only present in the Income and Cash Flow Statements (Expansion) section, because if you do not expand the business, you do not need the business loan of $100,000. Excel has financial formulas to figure out loan payments. To determine a loan payment, you need to know three things: the amount being borrowed (cell C5 in the Constants section), the interest rate (cell B21 in the Calculations-Expansion section), and the number of payment periods. At the beginning of the tutorial, you learned that the bank was willing to loan money at either 5% or 4% interest compounded annually, to be paid over 10 years. Normally, banks require businesses to make monthly payments on their loans and compound the interest monthly, in which case you would enter 120 (12 months/year × 10 years) for the number of payments and divide the annual interest rate by 12 to get the period interest rate. This formula is important to remember when you enter the business world, but for now you will simplify the calculation by specifying one loan payment per year compounded annually. To put in the payment formula, click cell C32, then click the f_x symbol next to the editing window (circled in

Figure C-9). The Payment function is called PMT, so type PMT in the Insert Function window—you will immediately see a short description of the function with its arguments, as shown in Figure C-9.

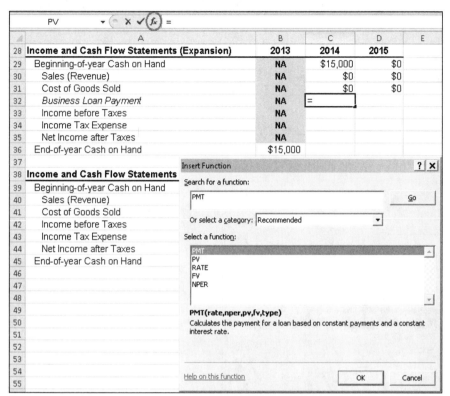

Source: Used with permission from Microsoft Corporation

FIGURE C-9 Accessing the PMT function in Excel for cell C32

NOTE

Rate is the interest rate per period of the loan, Nper is an abbreviation for the number of loan periods, and Pv is an abbreviation for Present Value, the amount of money you are borrowing "today." The PMT function can determine a series of equal loan payments necessary to pay back the amount borrowed, plus the accumulated compound interest over the life of the loan.

When you click OK, the resulting window allows you to enter the cells or values needed in the function arguments (see Figure C-10). In the Rate text box, enter B21, which is the cell that will contain the calculated interest rate. In the Nper text box, enter 10 (for 10 years). In the Pv text box, enter C5, which is the cell that contains the loan amount.

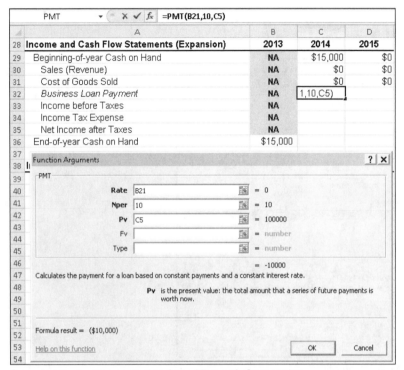

Source: Used with permission from Microsoft Corporation

FIGURE C-10   The Function Arguments dialog box for the PMT function with the values filled in

> **NOTE**
>
> Be careful if you decide to copy the PMT formula from cell C32 into cell D32, because the Copy command will change the cells in the formula arguments to the next adjacent cells. To make the Copy command work correctly, you have two options. First, you can change the Rate and Pv cells in the cell C32 formula from *relative reference* (B21, C5) to *absolute reference* ($B$21, $C$5). Your other option is to re-insert the PMT function into cell D32 and type the same arguments as before in the boxes. Absolute referencing of a cell (using $ signs in front of the Column and Row designators) "anchors" the cell so that when the Copy command is used, the destination cell will refer back to the same cells that the source cell used. If necessary, consult the Excel online Help for an explanation of relative and absolute cell references.

When you click OK ($10,000) should appear in cell C32. Payments in Excel always appear as negative numbers, which is why the number has parentheses around it. (Depending on your cell formatting, the number may also appear in red.) Next, you need to have the same payment amount in cell D32 (for 2015). Because the PMT function creates equal payments over the life of the loan, you can simply type =C32 into cell D32.

The next line in the Income and Cash Flow Statements is Income before Taxes, which is an easy calculation. It is the Sales minus the Cost of Goods Sold, minus the Business Loan Payment. However, because the PMT function shows the loan payment as a negative number, you will instead add the Business Loan Payment. In cell C33, enter =C30-C31+C32. Again, a negative $10,000 should be displayed, as the cells other than the loan payment currently have zero in them. Copy cell C33 to cell D33. In cell C42 of the next section below (No Expansion), enter =C40-C41. (There is no loan payment in this section to put in the calculation.) Next, copy cell C42 to cell D42. At this point, your Income and Cash Flow Statements should look like Figure C-11.

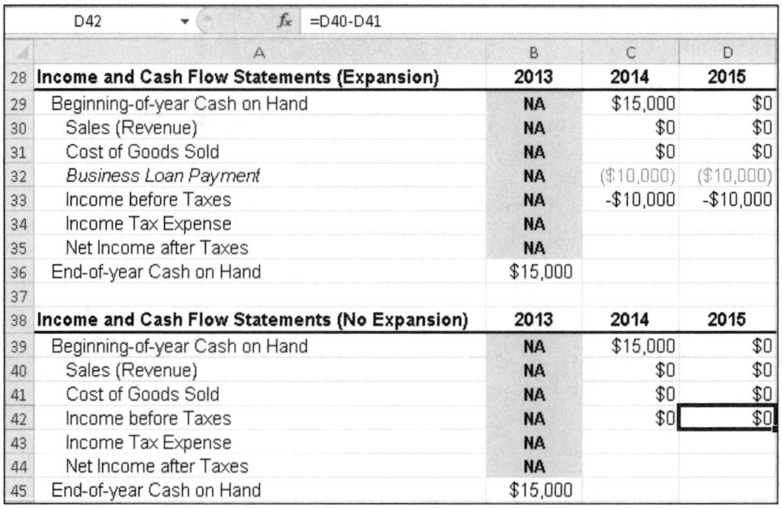

D42		fx	=D40-D41		

	A	B	C	D
28	**Income and Cash Flow Statements (Expansion)**	**2013**	**2014**	**2015**
29	Beginning-of-year Cash on Hand	NA	$15,000	$0
30	Sales (Revenue)	NA	$0	$0
31	Cost of Goods Sold	NA	$0	$0
32	*Business Loan Payment*	NA	($10,000)	($10,000)
33	Income before Taxes	NA	-$10,000	-$10,000
34	Income Tax Expense	NA		
35	Net Income after Taxes	NA		
36	End-of-year Cash on Hand	$15,000		
37				
38	**Income and Cash Flow Statements (No Expansion)**	**2013**	**2014**	**2015**
39	Beginning-of-year Cash on Hand	NA	$15,000	$0
40	Sales (Revenue)	NA	$0	$0
41	Cost of Goods Sold	NA	$0	$0
42	Income before Taxes	NA	$0	$0
43	Income Tax Expense	NA		
44	Net Income after Taxes	NA		
45	End-of-year Cash on Hand	$15,000		

Source: Used with permission from Microsoft Corporation

FIGURE C-11   The Income and Cash Flow Statements completed up to Income before Taxes

Income Tax Expense is the most complex formula for these sections. Because you do not pay income tax when you have no income or a loss, you must use a formula that allows you to enter 0 if there is no income or a loss, or to calculate the tax rate on a positive income. You can use the IF function in Excel to enter one of two results in a cell, depending on whether a defined logical statement is true or false. To create an IF function, select cell C34, then click the *fx* symbol next to the cell editing window (circled in Figure C-12). When the Insert Function dialog box appears, type IF in the "Search for a function" text box, and click the Go button if necessary. The IF function should appear. When you click OK, the Function Arguments dialog box appears (see Figure C-13).

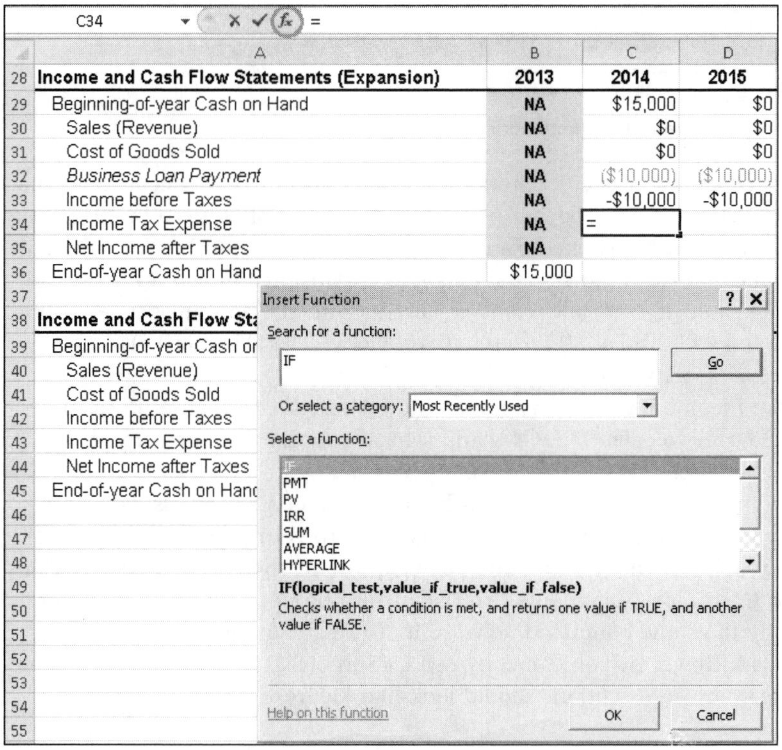

Source: Used with permission from Microsoft Corporation

FIGURE C-12   The IF function

Type the following in the Function Arguments dialog box:

- Next to Logical_test, type C33<=0.
- Next to Value_if_true, type 0.
- Next to Value_if_false, type C33*C4 (the Income before Taxes multiplied by the Tax Rate for 2014).

As you fill in the arguments, Excel writes the formula for you in the formula editing window (circled in Figure C-13).

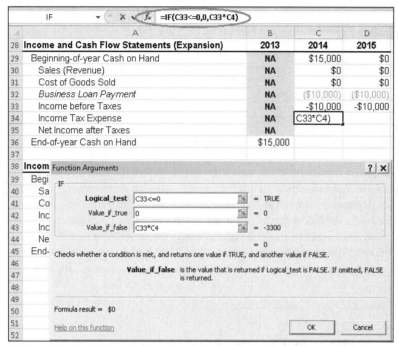

Source: Used with permission from Microsoft Corporation

FIGURE C-13 The Function Arguments dialog box with the arguments filled in

Once you have entered the arguments, click OK; Excel enters the formula into the cell. Because you had negative income, the cell should display a zero for now. Because the same formula will be used in 2015 (but with the 2015 tax rate), you can simply copy and paste the formula from cell C34 to cell D34. You also have to calculate the income tax for the Income and Cash Flow Statements (No Expansion). In cell D43, use the same IF function, but in the Logical_test, Value_if_true, and Value_if_false arguments, you must type C42<=0, 0, and C42*C4, respectively. Again, the cell will display $0 for an answer. Copy cell C43 to cell D43 to complete the Income Tax Expense line for No Expansion.

Net Income after Taxes is simply the Income before Taxes minus the Income Tax Expense. Enter the formula into cell C35, then copy cell C35 over to cells D35, C44, and D44. If you did this correctly, cells C35 and D35 will display a negative $10,000, and cells C44 and D44 will display $0.

End-of-year Cash on Hand, the last line in both Income and Cash Flow Statements sections, is not difficult either. Conceptually, the cash you have at the end of the year is equal to your Beginning-of-year Cash on Hand plus your Net Income after Taxes. Enter the formula into cell C36, then copy cell C36 over to cell D36. Note that because the Income and Cash Flow Statements (No Expansion) do not have a line item for Business Loan Payment, you cannot copy the same command down to it. You have to enter the formula manually for cell C45, which is =C39+C44. However, you can copy cell C45 to cell D45 to finish the Income and Cash Flow Statements sections. The completed sections should look like Figure C-14.

		f_x	=D39+D44			

	A	B	C	D
28	**Income and Cash Flow Statements (Expansion)**	**2013**	**2014**	**2015**
29	Beginning-of-year Cash on Hand	NA	$15,000	$5,000
30	Sales (Revenue)	NA	$0	$0
31	Cost of Goods Sold	NA	$0	$0
32	*Business Loan Payment*	NA	($10,000)	($10,000)
33	Income before Taxes	NA	-$10,000	-$10,000
34	Income Tax Expense	NA	$0	$0
35	Net Income after Taxes	NA	-$10,000	-$10,000
36	End-of-year Cash on Hand	$15,000	$5,000	-$5,000
37				
38	**Income and Cash Flow Statements (No Expansion)**	**2013**	**2014**	**2015**
39	Beginning-of-year Cash on Hand	NA	$15,000	$15,000
40	Sales (Revenue)	NA	$0	$0
41	Cost of Goods Sold	NA	$0	$0
42	Income before Taxes	NA	$0	$0
43	Income Tax Expense	NA	$0	$0
44	Net Income after Taxes	NA	$0	$0
45	End-of-year Cash on Hand	$15,000	$15,000	$15,000

Source: Used with permission from Microsoft Corporation

FIGURE C-14   The completed Income and Cash Flow Statements sections

Filling in the "Hard" Formulas

To finish the spreadsheet, you will enter values in the Inputs section and write the formulas in both Calculations sections.

AT THE KEYBOARD

In cell C8, enter an R for Recession, and in cell C9, enter H for High Inflation. You could enter any values here, but these two values will work with the IF functions you will write later. Recall that you did not use separate inputs for 2014 and 2015. You are assuming that the economic outlook or inflation rate that exists for 2014 will extend into 2015. However, because you are using the same inputs from these two locations, you must remember to use *absolute* cell references to both cells C8 and C9 in the various IF statements if you want to use a Copy command for adjacent cells. Your Inputs section should look like the one in Figure C-15.

	A	B	C	D
7	**Inputs**	**2013**	**2014**	**2015**
8	Economic Outlook (R=Recession, B=Boom)	NA	R	NA
9	Inflation Outlook (H=High, L=Low)	NA	H	NA

Source: Used with permission from Microsoft Corporation

FIGURE C-15   The Inputs section with values entered in cells C8 and C9

Remember that you referred to cell addresses in both Calculations sections in your formulas in the Income and Cash Flow Statements sections. Now you will enter formulas for these calculations. If necessary, format the four Total Sales Dollars cells and the four Cost of Goods Sold cells in the Calculations sections as Currency with no decimal places.

As described at the beginning of the tutorial, the forecast for Total Sales Dollars is a function of both the Economic Outlook and whether you expand the business. The following table lists the predicted sales growth percentages:

Sales Growth Forecast—Collegetown Thrift Shop

	Business Expansion	**No Business Expansion**
Recession-R	30%	20%
Boom-B	15%	5%

You will use IF formulas to forecast Total Sales Dollars. Click cell C18, then bring up the IF function and type the following in the text boxes:

Logical_test: $C$8="R" (Note that you must use absolute cell referencing for cell B8 and quotation marks for Excel to recognize a text string.)

Value_if_true: B18*1.3 (the 2013 sales multiplied by 1.3 for 30% sales growth)

Value_if_false: B18*1.15 (the 2013 sales multiplied by 1.15 for 15% sales growth)

Compare your entries to Figure C-16.

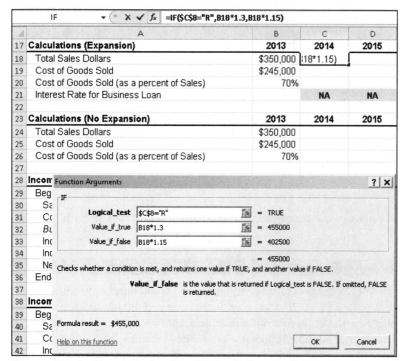

Source: Used with permission from Microsoft Corporation

FIGURE C-16   Using the IF function to enter the Total Sales Dollars forecast for 2014

When you click OK, cell C18 should display $455,000, because 30% of $350,000 is $105,000, and $350,000 plus $105,000 equals $455,000. So, it appears that the formula returned a "true" value with an R inserted in cell C8. Because you "anchored" cell C8 by entering $C$8, copy this formula over to cell D18 for the year 2015.

Once you complete the Total Sales Dollars cells for the Expansion scenario, go down to the Calculations (No Expansion) section and use IF statements to enter formulas for the Total Sales Dollars. Use a 20% sales growth factor for Recession and 5% for Boom. You can copy the formula from cell C18 into cell C24, but you then will have to use the editing window to change the values in the true and false arguments from 1.3 and 1.15 to 1.2 and 1.05, respectively, to reflect the fact that you did not expand the business. See Figure C-17.

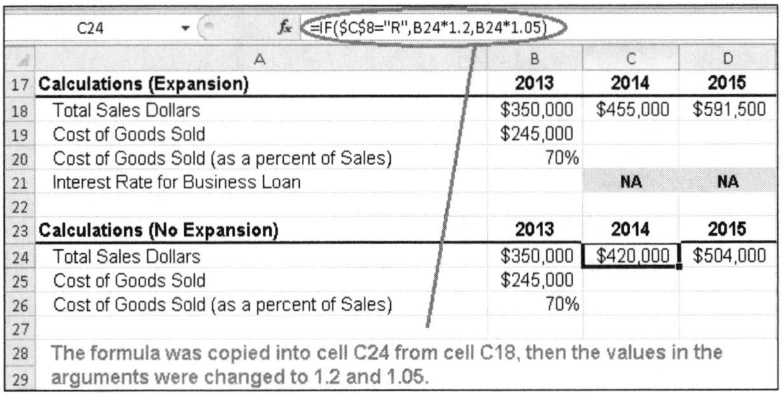

	C24	▾	f_x =IF($C$8="R",B24*1.2,B24*1.05)		
	A		B	C	D
17	**Calculations (Expansion)**		**2013**	**2014**	**2015**
18	Total Sales Dollars		$350,000	$455,000	$591,500
19	Cost of Goods Sold		$245,000		
20	Cost of Goods Sold (as a percent of Sales)		70%		
21	Interest Rate for Business Loan			NA	NA
22					
23	**Calculations (No Expansion)**		**2013**	**2014**	**2015**
24	Total Sales Dollars		$350,000	$420,000	$504,000
25	Cost of Goods Sold		$245,000		
26	Cost of Goods Sold (as a percent of Sales)		70%		
27					
28	The formula was copied into cell C24 from cell C18, then the values in the				
29	arguments were changed to 1.2 and 1.05.				

Source: Used with permission from Microsoft Corporation

FIGURE C-17   Copying cell C18 into cell C24 and then editing the IF function arguments to change the sales growth percentages

As before, you can now copy cell C24 to cell D24. You have completed the Total Sales Dollars calculations.

The Cost of Goods Sold (cells C19, D19, C25, and D25) is the Total Sales Dollars multiplied by the Cost of Goods Sold as a percent of Sales. In cell C19, type =C18*C20 and press Enter. Copy cell C19 and paste the contents into cells D19, C25, and D25. Your answers will be $0 until you enter the formulas for the Cost of Goods Sold as a percent of Sales.

The Cost of Goods Sold as a percent of Sales (cells C20, D20, C26, and D26) was 70% in 2013. In variety merchandising for resold items, it is easier to use an aggregate measure such as Cost of Goods Sold as a percent of Sales rather than trying to capture an individual Cost of Goods Sold for each item. From the 2013 data, you determined that for every dollar of sales you collected in 2013, you spent 70 cents purchasing the item and preparing it for resale. You will use that percentage as a basis for forecasting Cost of Goods Sold as a percent of Sales, applying an appropriate inflation factor for the cost of acquiring the stock for sale. The following table lists the predicted inflation percentages for Cost of Goods Sold.

Cost of Goods Sold Forecast—Collegetown Thrift Shop

	Business Expansion	No Business Expansion
High Inflation	6%	6%
Low Inflation	2%	2%

As with Total Sales Dollars previously, you will again use the IF function to calculate the Cost of Goods Sold as a percent of Sales. Now that you are familiar with the IF function, you can probably enter the function without using the dialog boxes. In cell C20, type the following:

=IF($C$9="H",B20*1.06,B20*1.02)

This expression means that if the text string in cell C9 is the letter H, you multiply the value in cell B20 by 1.06 (6% inflation). If the value in cell C9 is not an H, multiply the value in cell B20 by 1.02 (2% inflation). The value in cell B20 was the baseline Cost of Goods Sold as a percent of Sales in 2013, which was 70%. You can now copy cell C20 and paste the contents into cell D20.

Because the inflation percentages were exactly the same for both the Expansion and No Expansion calculations, you can also copy cell C20 and paste the contents into cells C26 and D26. Your Calculations sections should now look like Figure C-18.

	D26	▼	*fx* =IF($C$9="H",C26*1.06,C26*1.02)		

	A	B	C	D
17	**Calculations (Expansion)**	**2013**	**2014**	**2015**
18	Total Sales Dollars	$350,000	$455,000	$591,500
19	Cost of Goods Sold	$245,000	$337,610	$465,227
20	Cost of Goods Sold (as a percent of Sales)	70%	74%	79%
21	Interest Rate for Business Loan		**NA**	**NA**
22				
23	**Calculations (No Expansion)**	**2013**	**2014**	**2015**
24	Total Sales Dollars	$350,000	$420,000	$504,000
25	Cost of Goods Sold	$245,000	$311,640	$396,406
26	Cost of Goods Sold (as a percent of Sales)	70%	74%	79%

Source: Used with permission from Microsoft Corporation

FIGURE C-18   Calculations sections nearly complete

The last item in the Calculations section is the Interest Rate for Business Loan (cell B21). Remember the bank's statement that if the economy recovers, it could lower the interest rate from 5% to 4%. So, you will need one more IF function to insert into cell B21 based on the economic outlook. If the economic outlook is for a Recession (R), then the interest rate will be 5% annually; if the outlook is for a Boom (B), then the interest rate will be 4% annually. Now that you are familiar with the IF function, you can simply type the expression into the cell yourself. Click cell B21, type =IF(C8="R",5%,4%), and press Enter.

You will immediately notice that 5% appears in the cell because you have R in the input cell for Economic Outlook. You may also notice that you now have a negative $12,950 in the Business Loan Payment cells (C32 and D32). See Figure C-19 to compare your results.

	B21	▼	*fx* =IF(C8="R",5%,4%)		

	A	B	C	D
7	**Inputs**	**2013**	**2014**	**2015**
8	Economic Outlook (R=Recession, B=Boom)	**NA**	R	**NA**
9	Inflation Outlook (H=High, L=Low)	**NA**	H	**NA**
10				
11	**Summary of Key Results**	**2013**	**2014**	**2015**
12	Net Income after Taxes (Expansion)	**NA**	$69,974	$73,660
13	End-of-year Cash on Hand (Expansion)	**NA**	$84,974	$158,634
14	Net Income after Taxes (No Expansion)	**NA**	$72,601	$69,936
15	End-of-year Cash on Hand (No Expansion)	**NA**	$87,601	$157,537
16				
17	**Calculations (Expansion)**	**2013**	**2014**	**2015**
18	Total Sales Dollars	$350,000	$455,000	$591,500
19	Cost of Goods Sold	$245,000	$337,610	$465,227
20	Cost of Goods Sold (as a percent of Sales)	70%	74%	79%
21	Interest Rate for Business Loan	5%	**NA**	**NA**
22				
23	**Calculations (No Expansion)**	**2013**	**2014**	**2015**
24	Total Sales Dollars	$350,000	$420,000	$504,000
25	Cost of Goods Sold	$245,000	$311,640	$396,406
26	Cost of Goods Sold (as a percent of Sales)	70%	74%	79%
27				
28	**Income and Cash Flow Statements (Expansion)**	**2013**	**2014**	**2015**
29	Beginning-of-year Cash on Hand	**NA**	$15,000	$84,974
30	Sales (Revenue)	**NA**	$455,000	$591,500
31	Cost of Goods Sold	**NA**	$337,610	$465,227
32	Business Loan Payment	**NA**	($12,950)	($12,950)
33	Income before Taxes	**NA**	$104,440	$113,323
34	Income Tax Expense	**NA**	$34,465	$39,663
35	Net Income after Taxes	**NA**	$69,974	$73,660
36	End-of-year Cash on Hand	$15,000	$84,974	$158,634
37				
38	**Income and Cash Flow Statements (No Expansion)**	**2013**	**2014**	**2015**
39	Beginning-of-year Cash on Hand	**NA**	$15,000	$87,601
40	Sales (Revenue)	**NA**	$420,000	$504,000
41	Cost of Goods Sold	**NA**	$311,640	$396,406
42	Income before Taxes	**NA**	$108,360	$107,594
43	Income Tax Expense	**NA**	$35,759	$37,658
44	Net Income after Taxes	**NA**	$72,601	$69,936
45	End-of-year Cash on Hand	$15,000	$87,601	$157,537

Source: Used with permission from Microsoft Corporation

FIGURE C-19   The finished spreadsheet

You can change the economic inputs in four different combinations: R-H, R-L, B-H, B-L. This allows you to see the impact on your net income and cash on hand both for expanding and not expanding. However, you have another more powerful way to do this. In the next section, you will learn how to tabulate the financial results of the four possible combinations using an Excel tool called Scenario Manager.

SCENARIO MANAGER

You are now ready to evaluate the four possible outcomes for your DSS model. Because this is a simple, four-outcome model, you could have created four different spreadsheets, one for each set of outcomes, and then transferred the financial information from each spreadsheet to a Summary Report.

In essence, Scenario Manager performs the same task. It runs the model for all the requested outcomes and presents a tabular summary of the results. This summary is especially useful for reports and presentations needed by upper managers, financial investors, or in this case, the bank.

To review, the four possible combinations of input values are: R-H (Recession and High Inflation), R-L (Recession and Low Inflation), B-H (Boom and High Inflation), and B-L (Boom and Low Inflation). You could consider each combination of inputs a separate scenario. For each of these scenarios, you are interested in four outputs: Net Income after Taxes for Expansion and No Expansion, and End-of-year Cash on Hand for Expansion and No Expansion.

Scenario Manager runs each set of combinations and then records the specified outputs as a summary into a separate worksheet. You can use these summary values as a table of numbers and print it, or you can copy them into a Microsoft Word document or a PowerPoint presentation. You can also use the data table to build a chart or graph, which you can put into a report or presentation.

When you define a scenario in Scenario Manager, you name it and identify the input cells and input values. Then you identify the output cells so Scenario Manager can capture the outputs in a summary sheet.

AT THE KEYBOARD

To start, click the Data tab on the Ribbon. In the Data Tools group, click the What-If Analysis button, then click Scenario Manager from the menu that appears (see Figure C-20).

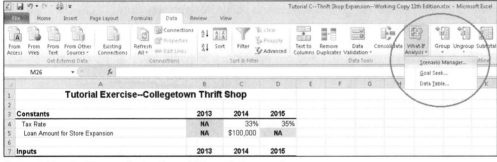

Source: Used with permission from Microsoft Corporation

FIGURE C-20   Scenario Manager option in the What-If Analysis menu

Scenario Manager appears (see Figure C-21), but no scenarios are defined. Use the dialog box to add, delete, or edit scenarios.

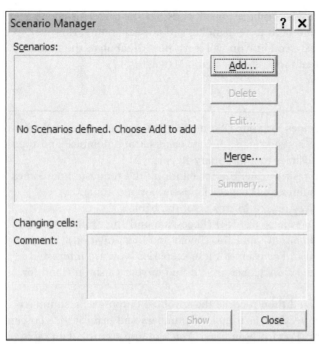

Source: Used with permission from Microsoft Corporation

FIGURE C-21   Initial Scenario Manager dialog box

N O T E

When working with the Scenario Manager dialog box and any following dialog boxes, do not use the Enter key to navigate. Use mouse clicks to move from one step to the next.

To define a scenario, click the Add button. The Edit Scenario dialog box appears. Enter Recession-High Inflation in the field under Scenario name. Then type the input cells in the Changing cells field (in this case, C8: C9). Better yet, you can use the button next to the field to select the cells in your spreadsheet. If you do, Scenario Manager changes the cell references to absolute cell references, which is acceptable (see Figure C-22).

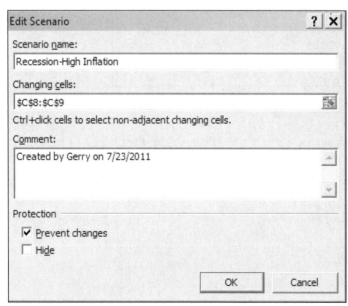

Source: Used with permission from Microsoft Corporation

FIGURE C-22   Defining a scenario name and input cells

Click OK to open the Scenario Values dialog box. Enter the input values for the scenario. In the case of Recession and High Inflation, the values will be R and H for cells C8 and C9, respectively (see Figure C-23). Note that if you already have entered values in the spreadsheet, the dialog box will display the current values. Make sure to enter the correct values.

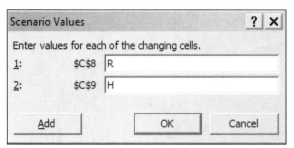

Source: Used with permission from Microsoft Corporation

FIGURE C-23   Entering values for the input cells

Click OK to return to the Scenario Manager dialog box. Enter the other three scenarios: Recession-Low Inflation, Boom-High Inflation, and Boom-Low Inflation (R-L, B-H, and B-L), and their related input values. When you finish, you should see the names and changing cells for the four scenarios (see Figure C-24).

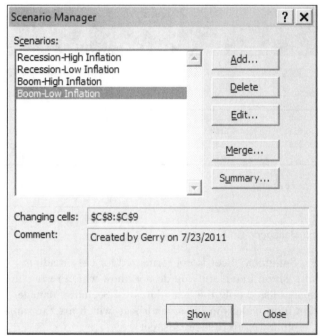

Source: Used with permission from Microsoft Corporation

FIGURE C-24   Scenario Manager dialog box with all four scenarios entered

You can now create a summary sheet that displays the results of running the four scenarios. Click the Summary button to open the Scenario Summary dialog box, as shown in Figure C-25. You must now enter the output cell addresses in Excel—they will be the same for all four scenarios. Recall that you created a section in your spreadsheet called Summary of Key Results. You are primarily interested in the results at the end of 2015, so you will choose the four cells that represent the Net Income after Taxes and End-of-year Cash on Hand, and then use them for both the expansion scenario and the non-expansion scenario. These cells are D12 to D15 in your spreadsheet. Either type D12:D15 or use the button next to the Result cells field and select those cells in the spreadsheet.

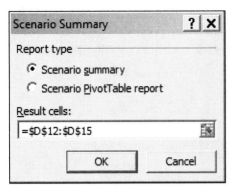

FIGURE C-25   Scenario Summary dialog box with Result cells entered

Another good reason for having a Summary of Key Results section is that it provides a contiguous range of cells to define for summary output. However, if you want to add output from other cells in the spreadsheet, simply separate each cell or range of cells in the dialog box with a comma. Next, click OK. Excel runs each set of inputs in the background, collects the results from the result cells, and then creates a new sheet called Scenario Summary (the name on the sheet's lower tab), as shown in Figure C-26.

	A	B	C	D	E	F	G	H
1								
2		Scenario Summary						
3				Current Values:	Recession-High Inflation	Recession-Low Inflation	Boom-High Inflation	Boom-Low Inflation
5		Changing Cells:						
6		$C$8	R		R	R	B	B
7		$C$9	H		H	L	H	L
8		Result Cells:						
9		$D$12		$73,660	$73,660	$96,052	$56,216	$73,738
10		$D$13		$158,634	$158,634	$189,562	$132,531	$157,605
11		$D$14		$69,936	$69,936	$89,015	$53,545	$68,152
12		$D$15		$157,537	$157,537	$184,496	$132,071	$153,573
13		Notes: Current Values column represents values of changing cells at						
14		time Scenario Summary Report was created. Changing cells for each						
15		scenario are highlighted in gray.						

FIGURE C-26   Scenario Summary sheet created by Scenario Manager

As you can see, the output created by the Scenario Summary sheet is not formatted for easy reading. You do not know which results are the net income and cash on hand, and you do not know which results are for Expansion vs. No Expansion, because Scenario Manager listed only the cell addresses. Scenario Manager also listed a separate column (column D) for the current input values in the spreadsheet, which are the same as the values in column E. It also left a blank column (column A) in the spreadsheet.

Fortunately, it is fairly easy to format the output. Delete columns D and A, put in the labels for cell addresses in the new column A, and then retitle the Scenario Summary as Collegetown Thrift Shop Financial Forecast, End of Year 2015 (because you are looking only at Year 2015 results). You can also make the results columns narrower by breaking the column headings into two lines; place your cursor in the editing window where you want to break the words, and then press Alt+Enter. Add a heading for column B (Cell Address). Finally, merge and center the title, and center the column headings and the input cell values (R, B, H, and L). Leave your financial data right-justified to keep the numbers lined up correctly. Finally, put some border boxes around each column of results. Figure C-27 shows a formatted Scenario Summary worksheet.

	A	B	C	D	E	F
1						
2	Scenario Summary--Collegetown Thrift Shop Financial Forecast--End of Year 2015					
3		Cell Address	Recession- High Inflation	Recession- Low Inflation	Boom- High Inflation	Boom- Low Inflation
5	**Changing Cells:**					
6	**Economic Outlook: R-Recession, B-Boom**	$C$8	R	R	B	B
7	**Inflation: H-High, L-Low**	$C$9	H	L	H	L
8	**Result Cells:**					
9	Net Income after Taxes (Expansion)	$D$12	$73,660	$96,052	$56,216	$73,738
10	End-of-year Cash on Hand (Expansion)	$D$13	$158,634	$189,562	$132,531	$157,605
11	Net Income after Taxes (No Expansion)	$D$14	$69,936	$89,015	$53,545	$68,152
12	End-of-year Cash on Hand (No Expansion)	$D$15	$157,537	$184,496	$132,071	$153,573
13	Notes: Current Values column represents values of changing cells at					
14	time Scenario Summary Report was created. Changing cells for each					
15	scenario are highlighted in gray.					

Source: Used with permission from Microsoft Corporation

FIGURE C-27   Scenario Summary worksheet after formatting and adding labels

Interpreting the Results

Now that you have good data, what do you do with it? Remember, you wanted to see if taking a $100,000 business loan to expand the thrift shop was a good financial decision. This is a relatively simple business case, and the shop's success so far ($350,000 of sales in 2013) would seem to make expansion a good risk. But how good a risk is the expansion?

After building the spreadsheet and doing the analysis, you can make comparisons and interpret the results. Regardless of the economic outlook or inflation, all four scenarios indicate that expanding the business should provide greater Net Income After Taxes and End-of-year Cash on Hand (after two years) than not expanding. So, the DSS model not only provides a quantitative basis for expanding, it provides an analysis that you can present to prospective lenders.

What decision would you make about expansion if you looked only at the 2014 forecast? You could go back to the original spreadsheet and look at the figures for 2014, or you can go to Scenario Manager and create a new summary, specifying the 2014 cells C12 through C15. See Figure C-28.

	A	B	C	D	E	F	G	H
7	**Inputs**	2013	2014	2015				
8	Economic Outlook (R=Recession, B=Boom)	NA	R	NA				
9	Inflation Outlook (H=High, L=Low)	NA	H	NA				
10								
11	**Summary of Key Results**	2013	2014	2015				
12	Net Income after Taxes (Expansion)	NA	$69,974	$73,660				
13	End-of-year Cash on Hand (Expansion)	NA	$84,974	$158,634				
14	Net Income after Taxes (No Expansion)	NA	$72,601	$69,936				
15	End-of-year Cash on Hand (No Expansion)	NA	$87,601	$157,537				
16								
17	**Calculations (Expansion)**	2013	2014	2015				
18	Total Sales Dollars	$350,000	$455,000	$591,500				
19	Cost of Goods Sold	$245,000	$337,610	$465,227				
20	Cost of Goods Sold (as a percent of Sales)	70%	74%	79%				

Scenario Summary dialog box:
Report type
- Scenario summary
- Scenario PivotTable report

Result cells:
=$C$12:$C$15

OK   Cancel

Source: Used with permission from Microsoft Corporation

FIGURE C-28   Creating a new Scenario Summary for 2014 instead of 2015

When you click OK, Excel creates a second Scenario Summary (appropriately named Scenario Summary 2), but this time the output values come from 2014, not 2015. After editing and formatting, the 2014 Scenario Summary should look like Figure C-29.

	A	B	C	D	E	F
1						
2	Scenario Summary 2--Collegetown Thrift Shop Financial Forecast--End of Year 2014					
3		Cell Address	Recession-High Inflation	Recession-Low Inflation	Boom-High Inflation	Boom-Low Inflation
5	Changing Cells:					
6	Economic Outlook: R-Recession, B-Boom	$C$8	R	R	B	B
7	Inflation: H-High, L-Low	$C$9	H	L	H	L
8	Result Cells:					
9	Net Income after Taxes (Expansion)	$C$12	$69,974	$78,510	$61,316	$68,867
10	End-of-year Cash on Hand (Expansion)	$C$13	$84,974	$93,510	$76,316	$83,867
11	Net Income after Taxes (No Expansion)	$C$14	$72,601	$80,480	$63,526	$70,420
12	End-of-year Cash on Hand (No Expansion)	$C$15	$87,601	$95,480	$78,526	$85,420
13	Notes: Current Values column represents values of changing cells at					
14	time Scenario Summary Report was created. Changing cells for each					
15	scenario are highlighted in gray.					

Source: Used with permission from Microsoft Corporation

FIGURE C-29   Scenario Summary for End of Year 2014

As you can see, *not* expanding the business yields slightly better financial results at the end of 2014. As the original Scenario Summary points out, it will take two years for the business expansion to start making more money when compared with not expanding. You can also revise the original spreadsheet to copy the columns out to 2016, 2017, and beyond to forecast future income and cash flows. However, note that the accuracy of a forecast gets worse as you extend it in time.

Managers must also maintain a healthy skepticism about the validity of their assumptions when formulating a DSS model. Most assumptions about economic outlooks, inflation, and interest rates are really educated guesses. For example, who could have predicted the economic meltdown in 2007? Business DSS models for investments, new product launches, business expansion, or major capital projects commonly look at three possible outcomes: best case, most likely, and worst case. The most likely outcome is based on previous years' data already collected by the firm. The best-case and worst-case outcomes are formulated based on some percentage of performance that falls above or below the most likely scenario. At least these are data-driven forecasts, or what people in the business world call "guessing—with data."

So, how do you reduce risk when making financial decisions based on DSS model results? It helps to formulate the model based on valid data and to use conservative estimates for success. More importantly, collecting pertinent data and tracking the business results *after* deciding to invest or expand can help reduce the risk of failure for the enterprise.

Summary Sheets

When you start working on the Scenario Manager spreadsheet cases later in this book, you will need to know how to manipulate summary sheets and their data. Some of these operations are explained in the following sections.

Rerunning Scenario Manager

The Scenario Summary sheet does not update itself when you change formulas or inputs in the spreadsheet. To get an updated Scenario Summary, you must rerun Scenario Manager, as you did when changing the outputs from 2015 to 2014. Click the Summary button in the Scenario Manager dialog box, and then click OK. Another summary sheet is created; Excel numbers them sequentially (Scenario Summary, Scenario Summary2, etc.), so you do not have to worry about Excel overwriting any of your older summaries. That is why you should rename each summary with a description of the changes.

Deleting Unwanted Scenario Manager Summary Sheets

When working with Scenario Manager, you might produce summary sheets you do not want. To delete an unwanted sheet, move your mouse pointer to the group of sheet tabs at the bottom of the screen and *right*-click the tab of the sheet you want to delete. Click Delete from the menu that appears (see Figure C-30).

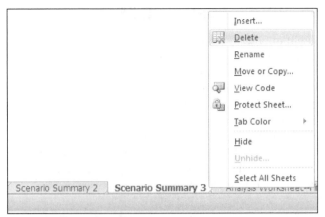

FIGURE C-30   Deleting unwanted worksheets

Charting Summary Sheet Data

You can easily chart Summary Sheet results using the Charts group in the Insert tab, as discussed in Tutorial F. Figure C-31 shows a clustered column chart prepared from the data in the Scenario Summary for 2015. Charts are useful because they provide a visual comparison of results. As the chart shows, the best economic climate for the thrift shop is a Recession with Low Inflation.

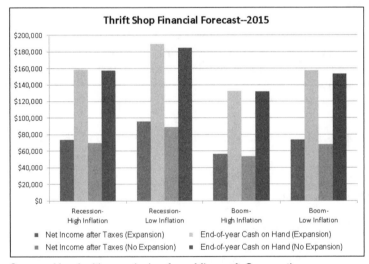

FIGURE C-31   Clustered column chart displaying data from the Summary Sheet

Copying Summary Sheet Data to the Clipboard

As you can with almost everything else in Microsoft Office, you can copy summary sheet data to other Office applications (a Word document or PowerPoint slide, for example) by using the Clipboard. Follow these steps:

1. Select the data range you want to copy.
2. Right-click the mouse and select Copy from the resulting menu.
3. Open the Word document or PowerPoint presentation into which you want to copy.
4. Click your cursor where you want the upper-left corner of the copied data to be displayed.
5. Right-click the mouse and select Paste from the resulting menu. The data should now appear on your document.

PRACTICE EXERCISE—TED AND ALICE'S HOUSE PURCHASE DECISION

Ted and Alice are a young couple who have been living in an apartment for the first two years of their marriage. They would like to buy their first house, but do not know whether they can afford it. Ted works as a carpenter's apprentice, and Alice is a customer service specialist at a local bank. In 2013, Ted's "take home" wages were $24,000 after taxes and deductions, and Alice's take-home salary was $30,000. Ted gets a 2% raise every year, and Alice gets a 3% raise. Their apartment rent is $1,200 per month ($14,400 per year), but the lease is up for renewal and the landlord said he needs to increase the rent for the next lease.

Ted and Alice have been looking at houses and have found one they can buy, but they will need to borrow $200,000 for a mortgage. Their parents are helping them with the down payment and closing costs. After talking to several lenders, Ted and Alice have learned that the state legislature is voting on a first-time home buyers' mortgage bond. If the bill passes, they will be able to get a 30-year fixed mortgage at 3% interest. Otherwise, they will have to pay 6% interest on the mortgage.

Because of the depressed housing market, Ted and Alice are not figuring equity value into their calculations. In addition, although the mortgage interest and real estate taxes will be deductible on their income taxes, these deductions will not be higher than the standard allowable tax deduction, so they are not figuring on any savings there either. Ted and Alice's other living expenses (such as car payments, food, and medical bills), the utilities expenses for either renting or buying, and estimated house maintenance expenses are listed in the Constants section (see Figure C-32).

Ted and Alice's primary concern is their cash on hand at the end of years 2014 and 2015. They are thinking of starting a family, but they know it will be difficult without adequate savings.

Getting Started on the Practice Exercise

If you closed Excel after the first tutorial exercise, start Excel again—it will automatically open a new workbook for you. If your Excel workbook from the first tutorial is still open, you may find it useful to start a new worksheet in the same workbook. Then you can refer back to the first tutorial when you need to structure or format the spreadsheet; the formatting of both exercises in this tutorial is similar. Set up your new worksheet as explained in the following sections.

Constants Section

Your spreadsheet should have the constants shown in Figure C-32. An explanation of the line items follows the figure.

	A	B	C	D
1	**Tutorial Exercise Skeleton--Ted and Alice's House Decision**			
2				
3	**Constants**	**2013**	**2014**	**2015**
4	Non-Housing Living Expenses (Cars, Food, Medical, etc)	NA	$36,000	$39,000
5	Mortgage Amount for Home Purchase	NA	$200,000	NA
6	Real Estate Taxes and Insurance on Home	NA	$3,000	$3,150
7	Utilities Expense (Heat & Electric)--Apartment	NA	$2,000	$2,200
8	Utilities Expense(Heat, Electric, Water, Trash)--House	NA	$2,500	$2,600
9	House Repair and Maintenance Expenses	NA	$1,200	$1,400

Source: Used with permission from Microsoft Corporation

FIGURE C-32   Constants section

- Non-Housing Living Expenses—This value represents Ted and Alice's estimate of all their other living expenses for 2014 and 2015.
- Mortgage Amount for Home Purchase
- Real Estate Taxes and Insurance on Home—A lender has given Ted and Alice estimates for these values; they are usually paid monthly with the house mortgage payment. The money is placed in an escrow account and then paid by the mortgage company to the state or county and insurance company.
- Utilities Expense—Apartment—This value is Ted and Alice's estimate for 2014 and 2015 based on their 2013 bills.

- Utilities Expense—House—Currently the apartment rent includes fees for water, sewer, and trash disposal. If they get a house, Ted and Alice expect the utilities to be higher.
- House Repair and Maintenance Expenses—In an apartment, the landlord is responsible for repair and maintenance. Ted and Alice will have to budget for repair and maintenance on the house.

Inputs Section

Your spreadsheet should have the inputs shown in Figure C-33. An explanation of line items follows the figure.

	A	B	C	D
11	**Inputs**	**2013**	**2014**	**2015**
12	Rental Occupancy (H=High, L=Low)	NA		NA
13	First Time Buyer Bond Loans Available (Y=Yes, N=No)	NA		NA

Source: Used with permission from Microsoft Corporation

FIGURE C-33   Inputs section

- Rental Occupancy (H=High, L=Low)—When the housing market is depressed (in other words, people are not buying homes), rental housing occupancy percentages are high, which allows landlords to charge higher rents when leases are renewed. Ted and Alice think their rent will increase in 2014. The amount of the increase depends on the Rental Occupancy. If the occupancy is high, Ted and Alice expect to see a 10% increase in rent in both 2014 and 2015. If occupancy is low, they only expect a 3% increase for each year.
- First Time Buyer Bond Loans Available (Y=Yes, N=No)—As described earlier, when housing markets are depressed, local governments will frequently pass a bond bill to provide low-interest mortgage money to first-time home buyers. If the bond loans are available, Ted and Alice can obtain a 30-year fixed mortgage at only 3%, which is half the interest rate they would otherwise pay for a conventional mortgage.

Summary of Key Results Section

Figure C-34 shows what key results Ted and Alice are looking for. They want to know their End-of-year Cash on Hand for both 2014 and 2015 if they decide to stay in the apartment and if they decide to purchase the house.

	A	B	C	D
15	**Summary of Key Results**	**2013**	**2014**	**2015**
16	End-of-year Cash on Hand (Rent)	NA		
17	End-of-year Cash on Hand (Buy)	NA		

Source: Used with permission from Microsoft Corporation

FIGURE C-34   Summary of Key Results section

These results are copied from the End-of-year Cash on Hand sections of the Income and Cash Flow Statements sections (for both renting and buying).

Calculations Section

Your spreadsheet will need formulas to calculate the apartment rent, house payments, and interest rate for the mortgage (see Figure C-35). You will use the rent and house payments later in the Income and Cash Flow Statements for both renting and buying.

	A	B	C	D
19	**Calculations**	**2013**	**2014**	**2015**
20	Apartment Rent	$14,400		
21	House Payments	NA		
22	Interest Rate for House Mortgage		NA	NA

Source: Used with permission from Microsoft Corporation

FIGURE C-35   Calculations section

- Apartment Rent—The 2013 amount is given. Use IF formulas to increase the rent by 10% if occupancy rates are high, or by 3% if occupancy rates are low.
- House Payments—This value is the total of the 12 monthly payments made on the mortgage. An important point to note is that house mortgage interest is always compounded *monthly*, not annually, as in the thrift shop tutorial. To properly calculate the house payments for the year, you divide the annual interest rate by 12 to determine the monthly interest. You also have to multiply a 30-year mortgage by 12 to get 360 payments, and then multiply the PMT formula by 12 to get the total amount for your annual house payments. Also, you will precede the PMT function with a negative sign to make the payment amount a positive number. Your formula should look like the following:
      =–PMT(B22/12,360,C5)*12
- Interest Rate for House Mortgage—Use the IF formula to enter a 3% interest rate if the bond money is available, and a 6% interest rate if no bond money is available.

Income and Cash Flow Statements Sections

As with the thrift shop tutorial, you want to see the Income and Cash Flow Statements for two scenarios—in this case, for continuing to rent and for purchasing a house. Each section begins with cash on hand at the end of 2013. As you can see in Figure C-36, Ted and Alice have only $4,000 in their savings.

	A	B	C	D
24	**Income and Cash Flow Statement (Continue to Rent)**	**2013**	**2014**	**2015**
25	Beginning-of-year Cash on Hand	NA		
26	Ted's Take Home Wages	$24,000		
27	Alice's Take Home Salary	$30,000		
28	Total Take Home Income	$54,000		
29	*Apartment Rent*	NA		
30	Utilities (Apartment)	NA		
31	Non-Housing Living Expenses	NA		
32	Total Expenses	NA		
33	End-of-year Cash on Hand	$4,000		
34				
35	**Income and Cash Flow Statement (Purchase House)**	**2013**	**2014**	**2015**
36	Beginning-of-year Cash on Hand	NA		
37	Ted's Take Home Wages	$24,000		
38	Alice's Take Home Salary	$30,000		
39	Total Take Home Income	$54,000		
40	*House Payments*	NA		
41	*Real Estate Taxes and Insurance*	NA		
42	Utilities (House)	NA		
43	*House Repair and Maintenance Expense*	NA		
44	Non-Housing Living Expenses	NA		
45	Total Expenses	NA		
46	End-of-year Cash on Hand	$4,000		

Source: Used with permission from Microsoft Corporation

FIGURE C-36   Income and Cash Flow Statements sections (for both rent and purchase)

- Beginning-of-year Cash on Hand—This value is the End-of-year Cash on Hand from the previous year.
- Ted's Take Home Wages—This value is given for 2013. To get values for 2014 and 2015, increase Ted's wages by 2% each year.
- Alice's Take Home Salary—This value is given for 2013. To get values for 2014 and 2015, increase Alice's salary by 3% each year.
- Total Take Home Income—The sum of Ted and Alice's pay.
- Apartment Rent—The rent is copied from the Calculations section.
- House Payments—The house payments are also copied from the Calculations section.

- Real Estate Taxes and Insurance, Utilities (Apartment or House), House Repair and Maintenance Expense, and Non-Housing Living Expenses—These values all are copied from the Constants section.
- Total Expenses—This value is the sum of all the expenses listed above. Note that the house payment is now a positive number, so you can sum it normally with the other expenses.
- End-of-year Cash on Hand—This value is the Beginning-of-year Cash on Hand plus the Total Take Home Income minus the Total Expenses.

Scenario Manager Analysis

When you have completed the spreadsheet, set up Scenario Manager and create a Scenario Summary sheet. Ted and Alice want to look at their End-of-year Cash on Hand in 2015 for renting or buying under the following four scenarios:

- High occupancy and bond money available
- High occupancy and no bond money available
- Low occupancy and bond money available
- Low occupancy and no bond money available

If you have done your spreadsheet and Scenario Manager correctly, you should get the results shown in Figure C-37.

	A	B	C	D	E	F
1	Scenario Summary--Ted & Alice's House Purchase Decision--2015					
2			Hi Occ-Bond $	Hi Occ-No Bond $	Lo Occ-Bond $	Lo Occ-No Bond $
4	Changing Cells:					
5	Rental Occupancy (H-High, L-Low)	$C$12	H	H	L	L
6	Bond Mortgage Available (Y or N)	$C$13	Y	N	Y	N
7	Result Cells:					
8	End of Year Cash on Hand (Rent)	$D$16	$3,713	$3,713	$6,868	$6,868
9	End of Year Cash on Hand (Buy)	$D$17	$7,090	($1,452)	$7,090	($1,452)
10	Notes: Current Values column represents values of changing cells at					
11	time Scenario Summary Report was created. Changing cells for each					
12	scenario are highlighted in gray.					

Source: Used with permission from Microsoft Corporation

FIGURE C-37 Scenario Summary results

Interpreting the Results

Based on the Scenario Summary results, what should Ted and Alice do? At first glance, it looks like the safe decision is to stay in the apartment. Actually, their decision hinges on whether they can get the lower-interest mortgage from the first-time buyers' bond issue. If they can, and if occupancy levels in apartments stay high, purchasing a house will give them about $3,300 more in savings at the end of 2015 than if they continued renting. Some other intangible factors are that home owners do not need permission to have pets, detached houses are quieter than apartments, and homes usually have a yard for pets and children to play in. Also, for the purposes of this exercise, you did not consider the tax benefits of home ownership. Depending on the amount of mortgage interest and real estate taxes Ted and Alice have to pay, they may be able to itemize their deductions and pay less income tax. If the income tax savings are more than $1,500, they can purchase the house even at the higher interest rate. In any case, because you did the DSS model for them, Ted and Alice now have a quantitative basis to help them make a good decision.

Visual Impact: Charting the Results

Charts and graphs often add visual impact to a Scenario Summary. Using the data from the Scenario Summary output table, try to create a chart similar to the one in Figure C-38 to illustrate the financial impact of each outcome.

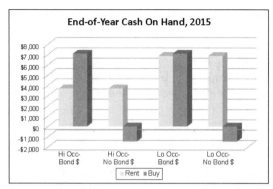

Source: Used with permission from Microsoft Corporation

FIGURE C-38   A 3-D clustered column chart created from Scenario Summary data

Printing and Submitting Your Work

Ask your instructor which worksheets need to be printed for submission. Make sure your printouts of the spreadsheet, the Scenario Manager Summary table, and the graph (if you created one) fit on one printed page apiece. Click the File tab on the Ribbon, click the Print button, and then click Page Setup at the bottom of the Print Navigation pane. When the Page Setup dialog box opens, click the Page tab if it is not already open, then click the Fit to radio button and click 1 page wide by 1 page tall. Your spreadsheet, table, and graph will be fitted to print on one page apiece.

REVIEW OF EXCEL BASICS

This section reviews some basic operations in Excel and provides some tips for good work practices. Then you will work through some cash flow calculations. Working through this section will help you complete the spreadsheet cases in the following chapters.

Save Your Work Often—and in More Than One Place

To guard against data loss in case of power outages, computer crashes, and hard drive failure, it is always a good idea to save your work to a separate storage device. Copying a file into two separate folders on the same hard disk is *not* an adequate safeguard. If you are working on your college's computer network and you have been assigned network storage, the network storage is usually "mirrored"; in other words, it has duplicate drives recording data to prevent data loss if the system goes down. However, most laptops and home computers lack this feature. An excellent way to protect your work from accidental deletion is to purchase a USB "thumb" drive and copy all of your files to it.

When you save your Excel files, Windows will usually store them in the My Documents folder unless you specify the storage location. Instead of just clicking the Save icon, a good idea is to click the File group in the Ribbon (or the Office Button in Office 2007), and then click the Save As button. A dialog box will appear with icons on the left side, as shown in Figure C-39. If you have previously saved your file to a particular location, it will appear in the Save in text box at the top of the dialog box. To save the file in the same location, click Save. If your work is stored elsewhere, you can find the location using the icons on the left side of the dialog box. If you are saving to a USB thumb drive, it will appear as a storage device when you click the My Computer icon. Click the folder where you want to save your file.

NOTE

If you are trying an operation that might damage your spreadsheet and you do not want to use the Undo command, you can use the Save As command, and then add a number or letter to the filename to save an additional copy to "play with." Your original work will be preserved.

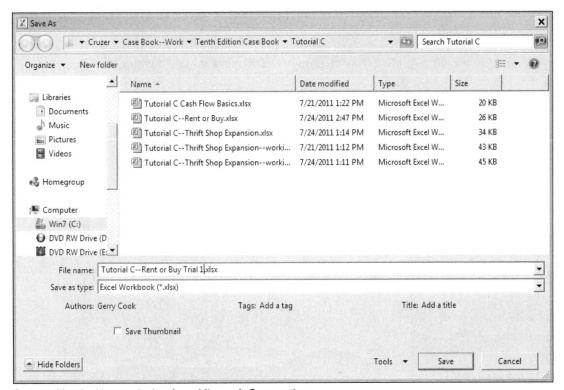

Source: Used with permission from Microsoft Corporation
FIGURE C-39   The Save As dialog box in Excel

Basic Operations

To begin, you will review the following topics: formatting cells, displaying the spreadsheet cell formulas, circular reference errors, using the AND and OR logical operators in IF statements, and using nested IF statements to produce more than two outcomes.

Formatting Cells

Cell Alignment

Headings for columns are usually centered, while numbers in cells are usually aligned to the right. To set the alignment of cell data:

1. Highlight the cell or cell range to format.
2. Select the Home tab.
3. In the Alignment group, click the button representing the horizontal alignment you want for the cell (Left Align, Center, or Right Align).
4. Also in the Alignment group, above the horizontal alignment buttons, click the vertical alignment you want (Top Align, Middle, or Bottom Align). Middle Align is the most common vertical alignment for cells.

Cell Borders

Bottom borders are common for headings, and accountants include borders and double borders to indicate subtotals and grand totals on spreadsheets. Sometimes it is also useful to put a "box" border around a table of values or a section of a spreadsheet. To create borders:

1. Highlight the cell or cell range that needs a border.
2. Select the Home tab.
3. In the Font group, click the drop-down arrow of the Border icon. A menu of border selections appears (see Figure C-40).
4. Choose the desired border for the cell or group of cells. Note that All Borders creates a box border around each cell, while Outside Borders draws a box around a group of cells.

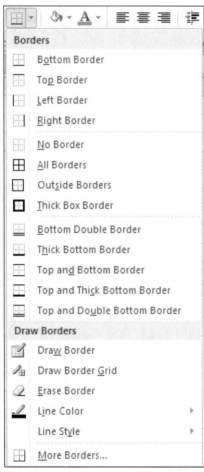

Source: Used with permission from Microsoft Corporation

FIGURE C-40   Selections in the Borders menu

Number Formats

For financial numbers, you usually use the Currency format. (Do not use the Accounting format, as it places the $ sign to the far left side of the cell). To apply the appropriate Currency format:

1. Highlight the cell or cell range to be formatted.
2. Select the Home tab.
3. In the Number group, select Currency in the Number Format drop-down list.
4. To set the desired number of decimal places, click the Increase Decimal or Decrease Decimal button in the bottom-right corner of the group (see Figure C-41).

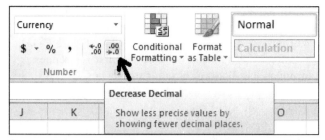

Source: Used with permission from Microsoft Corporation

FIGURE C-41 Increase Decimal and Decrease Decimal buttons

If you do not know what a button does in Office, hover your mouse pointer over the button to see a description.

Format "Painting"

If you want to copy *all* the format properties of a certain cell to other cells, use the Format Painter. First, select the cell whose format you want to copy. Then click the Format Painter button (the paintbrush icon) in the Clipboard Group under the Home tab (see Figure C-42). When you click the button, the mouse pointer turns into a paintbrush. Click the cell you want to reformat. To format multiple cells, select the cell whose format you want to copy, and then click *twice* on the Format Painter button. The mouse cursor will become a paintbrush, and the paint function will stay on so you can reformat as many cells as you want. To turn off the Format Painter, click its button again or press the Esc key.

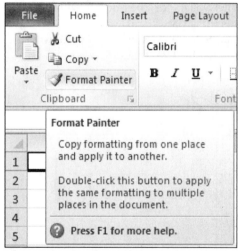

Source: Used with permission from Microsoft Corporation

FIGURE C-42 The Format Painter button

Showing the Excel Formulas in the Cells

Sometimes your instructor might want you to display or print the formulas in the spreadsheet cells. If you want the spreadsheet cells to display the actual cell formulas, follow these steps:

1. While holding down the Ctrl key, press the key in the upper-left corner of the keyboard that contains the back quote (`) and tilde (~). The spreadsheet will display the formulas in the cells. The columns may also become quite wide—if so, do not resize them.
2. The Ctrl+`~ key combination is a toggle; to restore your spreadsheet to the normal cell contents, press Ctrl+`~ again.

Understanding Circular Reference Errors

When entering formulas, you might make the mistake of referring to the cell in which you are entering the formula as part of the formula, even though it should only display the output of that formula. Referring a cell back to itself in a formula is called a *circular reference*. For example, suppose that in cell B2 of a worksheet, you enter =B2-B1. A terrible but apt analogy for a circular reference is a cannibal trying to eat himself! Fortunately, Excel informs you when you try to enter a circular reference into a formula (see Figure C-43). Excel also warns you if you try to open an existing spreadsheet that has one or more circular references. Before you can use the spreadsheet, you must fix the formulas that contain circular references.

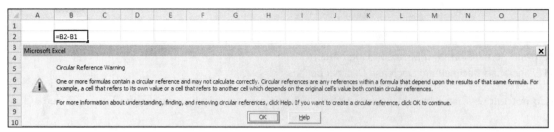

Source: Used with permission from Microsoft Corporation

FIGURE C-43   Excel circular reference warning

Using AND and OR Functions in IF Statements

Recall that the IF function has the following syntax:

> IF(test condition, result if test is True, result if test is False)

The test conditions in the previous example IF statements tested only one cell's value, but a test condition can test more than one value of a cell.

For example, look at the thrift shop tutorial again. The Total Sales Dollars for 2014 depended on the economic outlook (recession or boom). The original IF statement was =IF($C$8="R",B18*1.3,B18*1.15), as shown in Figure C-44. This function increased the baseline 2013 Total Sales Dollars by 30% if there was continued recession ("R" entered in cell C8), but only increased the total by 15% if there was a boom.

C18	f_x =IF($C$8="R",B18*1.3,B18*1.15)			
	A	B	C	D
7	**Inputs**	**2013**	**2014**	**2015**
8	Economic Outlook (R=Recession, B=Boom)	NA	R	NA
9	Inflation Outlook (H=High, L=Low)	NA	H	NA
10				
11	**Summary of Key Results**	**2013**	**2014**	**2015**
12	Net Income after Taxes (Expansion)	NA	$69,974	$73,660
13	End-of-year Cash on Hand (Expansion)	NA	$84,974	$158,634
14	Net Income after Taxes (No Expansion)	NA	$72,601	$69,936
15	End-of-year Cash on Hand (No Expansion)	NA	$87,601	$157,537
16				
17	**Calculations (Expansion)**	**2013**	**2014**	**2015**
18	Total Sales Dollars	$350,000	$455,000	$591,500
19	Cost of Goods Sold	$245,000	$337,610	$465,227
20	Cost of Goods Sold (as a percent of Sales)	70%	74%	79%
21	Interest Rate for Business Loan	5%	NA	NA

Source: Used with permission from Microsoft Corporation

FIGURE C-44   The original IF statement used to calculate the Total Sales Dollars for 2014

To take the IF argument one step further, assume that the Total Sales Dollars for 2014 depended not only on the Economic Outlook, but on the Inflation Outlook (High or Low). Suppose there are two possibilities:

- Possibility 1: If the economic outlook is for a Recession and the inflation outlook is High, the Total Sales Dollars for 2014 will be 30% higher than in 2013.
- Possibility 2: For the other three cases (Recession and Low Inflation, Boom and High Inflation, and Boom and Low Inflation), assume that the 2014 Total Sales Dollars will only be 15% higher than in 2013.

The first possibility requires two conditions to be true at the same time: C8="R" and C9="H". You can include an AND( ) function inside the IF statement to reflect the additional condition as follows:

=IF(AND($C$8="R",$C$9="H"), B18*130%,B18*115%)

When the test argument uses the AND() function, conditions "R" *and* "H" both must be present at the same time for the statement to use the true result (multiplying last year's sales by 130%). Any of the other three outcome combinations will cause the statement to use the false result (multiplying last year's sales by 115%).

You can also use an OR() function in an IF statement. For example, assume that instead of both conditions (Recession and High Inflation) having to be present, only one of the two conditions needs to be present for sales to increase by 30%. In this case, you use the OR() function in the test argument as follows:

=IF(OR($C$8="R",$C$9="H"), B18*130%,B18*115%)

In this case, if *either* of the two conditions (C8="R" or C9="H"), is true, the function will return the true argument, multiplying the 2013 sales by 130%. If *neither* of the two conditions is true, then the function will return the false argument, multiplying the 2013 sales by 115% instead.

Using IF Statements Inside IF Statements (Also Called "Nesting IFs")

By now you should be familiar with IF statements, but here is a quick review of the syntax:

=IF(test condition, result if test is True, result if test is False)

In the preceding examples, only two courses of action were possible for each of the inputs: Recession or Boom, High Inflation or Low Inflation, Rental Occupancy High or Low, Bond Money Available or No Bond Money Available. The tutorial used only two possible outcomes to keep them simple.

However, in the business world, decision support models are frequently based on three or more possible outcomes. For capital projects and new product launches, you will frequently project financial outcomes based on three possible scenarios: Most Likely, Worst Case, and Best Case. You can modify the IF statement by placing another IF statement inside the result argument if the first test is false, creating the ability to launch two more alternatives from the second IF statement. This is called "nesting" your IF statements.

Try a simple nested IF statement: In your thrift shop example, assume that three economic outlooks are possible: Recession (R), Boom (B), or Stable (S). As before, the 2014 Total Sales Dollars (cell C18) will be the 2013 Total Sales Dollars increased by some fixed percentage. In a Recession, sales will increase by 30%, in a Boom they will increase by 15%, and for a Stable Economic Outlook, sales will increase by 22%, which is roughly midway between the other two percentages. You can "nest" the IF statement in cell C18 to reflect the third outcome as follows:

=IF($C$8= "R",B18*130%,IF($C$8="B",B18*115%,B18*122%))

Note the added IF statement inside the False value argument. You can break down this statement:

- If the value in cell C8 is "R", multiply the value in cell B18 by 130%, and enter the result in cell C18.
- If the value in cell C8 is not "R", check whether the value in cell C8 is "B". If it is, multiply the value in cell B18 by 115%, and enter the result in cell C18.
- If the value in cell C8 is not "B", multiply the value in cell B18 by 122%, and enter the result in cell C18.

If you have four or more alternatives, you can keep nesting IF statements inside the false argument for the outer IF statements. (Excel 2007 and later versions have a limit of 64 levels of nesting in the IF function, which should take care of every conceivable situation.)

NOTE

The "embedded IFs" in a nested IF statement are not preceded by an equals sign. Only the first IF gets the equals sign.

Cash Flow Calculations: Borrowing and Repayments

The Scenario Manager cases that follow in this book require accounting for money that the fictional company will have to borrow or repay. This money is not like the long-term loan that the Collegetown Thrift Shop is considering for its expansion. Instead, this money is short-term borrowing that companies use to pay current obligations, such as purchasing inventory or raw materials. Such short-term borrowing is called a line of credit, and is extended to businesses by banks, much like consumers have credit cards. Lines of credit usually involve interest payments, but for simplicity's sake, focus instead on how to do short-term borrowing and repayment calculations.

To work through cash flow calculations, you must make two assumptions about a company's borrowing and repayment of short-term debt. First, you assume that the company has a desired *minimum* cash level at the end of a fiscal year (which is also its cash level at the start of the next fiscal year), to ensure that the company can cover short-term expenses and purchases. Second, assume the bank that serves the company will provide short-term loans (a line of credit) to make up the shortfall if the end-of-year cash falls below the desired minimum level.

NCP stands for Net Cash Position, which equals beginning-of-year cash plus net income after taxes for the year. NCP represents the available cash at the end of the year, *before* any borrowing or repayment. For the three examples shown in Figure C-45, set up a simple spreadsheet in Excel and determine how much the company needs to borrow to reach its minimum year-end cash level. Use the IF function to enter 0 under Amount to Borrow if the company does not need to borrow any money.

	A	B	C	D
1	Example	NCP	Minimum Cash Required	Amount to Borrow
2	1	$25,000	$10,000	?
3	2	$9,000	$10,000	?
4	3	($12,000)	$10,000	?

Source: Used with permission from Microsoft Corporation

FIGURE C-45   Examples of borrowing

You can also assume that the company will use some of its cash on hand at the end of the year to pay off as much of its outstanding debt as possible without going below its minimum cash on hand required. The "excess" cash is the company's NCP *less* the minimum cash on hand required—any cash above the minimum is available to repay any debt. In the examples shown in Figure C-46, compute the excess cash and then compute the amount to repay. In addition, compute the ending cash on hand after the debt repayment.

	A	B	C	D	E	F
9	Example	NCP	Minimum Cash Required	Beginning-of-Year Debt	Repay?	Ending Cash
10	1	$12,000	$10,000	$5,000	?	?
11	2	$13,000	$10,000	$8,000	?	?
12	3	$20,000	$10,000	$0	?	?
13	4	$60,000	$10,000	$40,000	?	?
14	5	($20,000)	$10,000	$10,000	?	?

Source: Used with permission from Microsoft Corporation

FIGURE C-46   Examples of debt repayment

In the Scenario Manager cases of the following chapters, your spreadsheets may need two bank financing sections beneath the Income and Cash Flow Statements sections. You will build the first section to calculate any needed borrowing or repayment at year's end to compute year-end cash on hand. The second section will calculate the amount of debt owed at the end of the year after any borrowing or repayment.

Return to the Collegetown Thrift Shop tutorial and assume that it includes a line of credit at a local bank for short-term cash management. The first new section extends the end-of-year cash calculation, which was shown for the thrift shop in Figure C-19. Figure C-47 shows the structure of the new section highlighted in boldface.

A	B	C	D
29 Income and Cash Flow Statements (Expansion)	2013	2014	2015
30 Beginning-of-year Cash on Hand	NA	$15,000	$0
31 Sales (Revenue)	NA	$455,000	$591,500
32 Cost of Goods Sold	NA	$337,610	$465,227
33 Business Loan Payment	NA	($12,950)	($12,950)
34 Income before Taxes	NA	$104,440	$113,323
35 Income Tax Expense	NA	$34,465	$39,663
36 Net Income after Taxes	NA	$69,974	$73,660
37 **Net Cash Position NCP Beginning-of-year Cash on Hand plus Net Income after Taxes**	NA	$84,974	$73,660
38 Line of credit borrowing from bank	NA		
39 Line of credit repayments to bank	NA		
40 End-of Year Cash on Hand	$15,000		

Source: Used with permission from Microsoft Corporation

FIGURE C-47   Calculation section for End-of-Year Cash on Hand with borrowing and repayments added

The heading in cell A36 was originally End-of-year Cash on Hand in Figure C-19, but you will add line-of-credit borrowing and repayment to the end-of-year totals. You must add the line-of-credit borrowing from the bank to the NCP and subtract the line-of-credit repayments to the bank from the NCP to obtain the End-of-Year Cash on Hand.

The second new section you add will compute the End-of year debt owed. This section is called Debt Owed, as shown in Figure C-48.

A	B	C	D
42 **Debt Owed**	2013	2014	2015
43 **Beginning-of-year debt owed**	NA		
44 **Borrowing from bank line of credit**	NA		
45 **Repayment to bank line of credit**	NA		
46 **End-of-year debt owed**	$47,000		

Source: Used with permission from Microsoft Corporation

FIGURE C-48   Debt Owed section

As you can see, the thrift shop currently owes $47,000 on its line of credit at the end of 2013. The End-of-year debt owed equals the Beginning-of-year debt owed plus any new borrowing from the bank's line of credit, minus any repayment to the bank's line of credit. Therefore, the formula in cell C46 would be:

=C43+C44-C45

Assume that the amounts for borrowing and repayment (cells C44 and C45) were calculated in the first new section (for the year 2014, the amounts would be in cells C38 and C39), and then copied into the second section. The formula for cell C44 would be =C38, and for cell C45 would be =C39. The formula for cell C43, Beginning-of-year debt owed in 2014, would simply be the End-of-year debt owed in 2013, or =B46.

Now that you have added the spreadsheet entries for borrowing and repayment, consider the logic for the borrowing and repayment formulas.

Calculation of Borrowing from the Bank Line of Credit

When using logical statements, it is sometimes easier to state the logic in plain language and then turn it into an Excel formula. For borrowing, the logic in plain language is:

If (cash on hand before financing transactions is greater than the minimum cash required,

then borrowing is not needed; else,

borrow enough to get to the minimum)

You can restate this logic as the following:

If (NCP is greater than minimum cash required,

then borrowing from bank=0;

else, borrow enough to get to the minimum)

You have not added minimum cash at the end of the year as a requirement, but you could add it to the Constants section at the top of the spreadsheet (in this case the new entry would be cell C6). Assume that you want $50,000 as the minimum cash on hand at the end of both 2014 and 2015. Assuming that the NCP is shown in cell C37, you could restate the formula for borrowing (cell C38) as the following:

IF(NCP>Minimum Cash, 0; otherwise, borrow enough to get to the minimum cash)

You have cell addresses for NCP (cell C37) and for Minimum Cash (cell $C$6). To develop the formula for cell C38, substitute the cell address for the test argument; the true argument is simply zero (0), and the false argument is the minimum cash minus the current NCP. The formula stated in Excel for cell C38 would be:

=IF(C37>=$C$6, 0, $C$6-C37)

Calculation of Repayment to the Bank Line of Credit

Simplify the statements first in plain language:

IF(beginning of year debt=0, repay 0 because nothing is owed, but

IF(NCP is less than the minimum, repay 0, because you must borrow, but

IF(extra cash equals or exceeds the debt, repay the whole debt,

ELSE (to stay above the minimum cash, repay the extra cash above the minimum)

Look at the following formula. If you assume that the repayment amount will be in cell C39, the beginning-of-year debt is in cell C43, and the minimum cash target is still in cell $C$6, the repayment formula for cell C39 with the nested IFs should look like the following:

=IF(C43=0,0,IF(C37<=$C$6,0,IF(C37-$C$6>=C43,C43,C37-$C$6)))

The new sections of the thrift shop spreadsheet would look like those in Figure C-49.

	C39	▼	f_x =IF(C43=0,0,IF(C37<=$C$6,0,IF(C37-$C$6>=C43,C43,C37-$C$6)))		
	A		B	C	D
29	**Income and Cash Flow Statements (Expansion)**		**2013**	**2014**	**2015**
30	Beginning-of-year Cash on Hand		NA	$15,000	$50,000
31	Sales (Revenue)		NA	$455,000	$591,500
32	Cost of Goods Sold		NA	$337,610	$465,227
33	*Business Loan Payment*		NA	($12,950)	($12,950)
34	Income before Taxes		NA	$104,440	$113,323
35	Income Tax Expense		NA	$34,465	$39,663
36	Net Income after Taxes		NA	$69,974	$73,660
	Net Cash Position NCP				
	Beginning-of-year Cash on Hand				
37	**plus Net Income after Taxes**		NA	$84,974	$123,660
38	**Line of credit borrowing from bank**		NA	$0	$0
39	**Line of credit repayments to bank**		NA	$34,974	$12,026
40	**End-of Year Cash on Hand**		$15,000	$50,000	$111,634
41					
42	**Debt Owed**		**2013**	**2014**	**2015**
43	Beginning-of-year debt owed		NA	$47,000	$12,026
44	Borrowing from bank line of credit		NA	$0	$0
45	Repayment to bank line of credit		NA	$34,974	$12,026
46	End-of-year debt owed		$47,000	$12,026	$0

Source: Used with permission from Microsoft Corporation

FIGURE C-49   Thrift shop spreadsheet with line-of-credit borrowing, repayments, and Debt Owed added

Answers to the Questions about Borrowing and Repayment

Figures C-50 and C-51 display solutions for the borrowing and repayment calculations.

	A	B	C	D
1	Example	NCP	Minimum Cash Required	Amount to Borrow
2	1	$25,000	$10,000	$0
3	2	$9,000	$10,000	$1,000
4	3	($12,000)	$10,000	$22,000

Source: Used with permission from Microsoft Corporation

FIGURE C-50   Answers to examples of borrowing

In Figure C-50, the formula in cell D2 for the amount to borrow is =IF(B2>=C2,0,C2-B2).

	A	B	C	D	E	F
9	Example	NCP	Minimum Cash Required	Beginning-of-Year Debt	Repay?	Ending Cash
10	1	$12,000	$10,000	$5,000	$2,000	$10,000
11	2	$13,000	$10,000	$8,000	$3,000	$10,000
12	3	$20,000	$10,000	$0	$0	$20,000
13	4	$60,000	$10,000	$40,000	$40,000	$20,000
14	5	($20,000)	$10,000	$10,000	$0	NA

Source: Used with permission from Microsoft Corporation

FIGURE C-51   Answers to examples of repayment

In Figure C-51, the formula in cell E10 for the amount to repay is

=IF(B10>=C10,IF(D10>0,MIN(B10-C10,D10),0),0).

Note the following points about the repayment calculations shown in Figure C-51.

- In Example 1, only $2,000 is available for debt repayment ($12,000 – $10,000) to avoid dropping below the Minimum Cash Required.
- In Example 2, only $3,000 is available for debt repayment.
- In Example 3, the Beginning-of-Year Debt was zero, so the Ending Cash is the same as the Net Cash Position.
- In Example 4, there was enough cash to repay the entire $40,000 debt, leaving $20,000 in Ending Cash.
- In Example 5, the company has cash problems—it cannot repay any of the Beginning-of-Year Debt of $10,000, and it will have to borrow an additional $30,000 to reach the Minimum Cash Required target of $10,000.

You should now have all the basic tools you need to tackle Scenario Manager in Cases 6 and 7. Good luck!

CRYSTAL LAKE PARK EXPANSION STRATEGY DECISION

Decision Support Using Excel

PREVIEW

Crystal Lake Park is a family-owned amusement park in Pennsylvania's Pocono Mountains. The park has been in business for more than 50 years and has grown to become the leading tourist attraction in the area. The owners, the Patterson family, have earned more than $50 million from the park over the past decade.

Sheila Patterson, the president and CEO of Crystal Lake Park, has proposed that the family invest its earnings in a park expansion. The park is adjacent to Crystal Lake and covers several hundred acres, 50 acres of which are undeveloped property. Sheila wants to build a water park on the additional land.

Sheila's brother George, vice-president of marketing, has a different proposal. Almost 1000 acres of land have been listed for sale next to the park; the land is currently zoned for commercial use. George believes that the park should purchase the property and build a drive-through safari park and zoo. George thinks that a safari park would produce more profits in the long run. He has argued that the cooler weather in the mountains does not help ticket sales to the water park.

Sheila doesn't agree. She thinks that the low cost of running a water park makes it a much safer investment alternative, even when ticket sales decrease because of cooler weather. Running a safari park and zoo requires high recurring costs, Sheila contends, and will not attract as many repeat customers as a water park.

Another factor that will affect the investment decision is the state of the economy. The economic recovery has boosted both attendance and spending per visitor at the amusement park for the past two years. The last recession affected attendance and revenues significantly, which will be important when analyzing each alternative.

PREPARATION

- Review the spreadsheet concepts discussed in class and in your textbook.
- Your instructor may assign Excel exercises to help prepare you for this case.
- Tutorial C has an excellent review of IF, nested IF, and AND statements that will help you with this case.
- Review the file-saving instructions in Tutorial C—it is always a good idea to save an extra copy of your work on a USB thumb drive.
- Review Tutorial F to brush up on your presentation skills.
- Because Crystal Lake Park is a strategic investment decision model, you will calculate the internal rate of return (IRR) in the decision model. If you are unfamiliar with the IRR function in Excel, the case includes a section that explains how to set it up.

BACKGROUND

You are an information analyst working for Crystal Lake Park. Sheila has asked you to prepare a quantitative analysis of financial, sales, and operations data to help determine which expansion path would offer the best strategic opportunity for the park. The park's managers have been asked to provide the following data from their functional areas:

- Accounting—The current cash position of the company, the cash outlays for the two investment choices, and the corporate income tax rates
- Sales—Forecasts for park attendance and concessions sales based on the current economy and projections of attendance and income for a water park and a safari park/zoo
- Operations—Materials, labor, and overhead costs for a water park and a safari park/zoo; for the latter option, these costs would include expenses for animal keepers, feed consumption, and veterinary care

The departments have given you the following data for 2014 through 2016:

- The cost to purchase the 1000 acres, convert it to a safari park, build a zoo on the existing land, and purchase the animals
- The cost to build a water park on Crystal Lake's undeveloped 50 acres
- Direct materials and labor costs per day for the water park's operating expenses and maintenance employees
- Overhead expenses per day for the water park
- Direct materials and feed costs per day for the safari park and zoo
- Direct labor costs per day for the safari park and zoo's animal keepers and maintenance employees
- Overhead expenses per day for the safari park and zoo
- Expected inflation rate
- Total operating days per year at either park
- Expected average water park attendance per day, price per ticket, and expected concession sales per ticket
- Expected average attendance per day at the safari park and zoo, price per ticket, and expected concession sales per ticket
- Corporate income tax rates

Assignment 1 also contains information you need to write the formulas for the Calculations section, Income and Cash Flow Statements section, and IRR Calculation section.

You will use Excel to see how much profit and positive cash flow each expansion alternative will generate for Crystal Lake Park for the next three years, and then you will use Excel to calculate an internal rate of return for each alternative. You will also examine the effects of the economy (Recession or Stable) and the effects of seasonal temperature (Cool or Hot) on your projected sales and profits for each alternative. In summary, your DSS will include the following inputs:

- Your decision to invest in the water park or safari park/zoo expansion
- Whether the economic outlook is for a recession or stable cycle
- Whether the seasonal temperatures are cool or hot

In a stable economy, attendance at each park would be higher than in a recession. Cool seasonal temperatures will hurt sales at the water park.

Your DSS model must account for the effects of the preceding three inputs on costs, selling prices, and other variables. If you design the model well, it will let you develop "what-if" scenarios with all the inputs, see the results, and select the expansion plan that Crystal Lake Park should adopt.

ASSIGNMENT 1: CREATING A SPREADSHEET FOR DECISION SUPPORT

In this assignment, you create a spreadsheet that models the business decision Crystal Lake Park is seeking. In Assignment 2, you write a report to the park's management team. In Assignment 3, you prepare and give a presentation of your analysis and recommendations.

To begin, you create the spreadsheet model of the company's financial and operating data. The model will cover three years of park operations and sales (2014 through 2016) for the alternative expansions considered. Assume that the park's accountants have completed preliminary research for 2013, and that the new acreage, associated buildings, and equipment are in place to begin operations and sales in 2014.

This section helps you set up each of the following spreadsheet components before entering the cell formulas:

- Constants
- Inputs
- Summary of Key Results
- Calculations
- Income and Cash Flow Statements
- Internal Rate of Return Calculation

The Internal Rate of Return Calculation section was added because Excel financial formulas such as IRR work better if the cash outflow and inflow data are arranged in a vertical column with the years in ascending order, as opposed to taking the cash flows from across the page or from nonadjacent cells.

The spreadsheet skeleton is available for you to use. To access this skeleton, go to your data files, select Case 6, and then select **Case 6—Crystal Lake Park Skeleton.xlsx**.

Constants Section

First, build the skeleton of your spreadsheet. Set up your Constants section as shown in Figure 6-1. An explanation of the line items follows the figure.

	A	B	C	D	E
1	**Crystal Lake Park--Expansion Analysis**				
2					
3	**Constants**	**2013**	**2014**	**2015**	**2016**
4	Direct Materials cost per day--Water Park	NA	$500	$515	$530
5	Direct Labor cost per day--Water Park	NA	$1,500	$1,575	$1,654
6	Overhead cost per day--Water Park	NA	$1,600	$1,648	$1,697
7	Direct Materials/Feed cost per day--Safari Park/Zoo	NA	$3,000	$3,090	$3,183
8	Direct Labor cost per day--Safari Park/Zoo	NA	$4,000	$4,200	$4,410
9	Overhead cost per day--Safari Park/Zoo	NA	$3,000	$3,090	$3,183
10	Expected Base Attendance per day--Water Park	NA	4,000	5,000	6,000
11	Expected Base Attendance per day--Safari Park/Zoo	NA	5,000	7,000	9,000
12	Average Ticket and Concession Sales per Customer--Water Park	NA	$14.00	$15.00	$16.00
13	Average Ticket and Concession Sales per Customer--Safari Park	NA	$25.00	$27.00	$29.00
14	Capital Investment for Water Park	$20,000,000	NA	NA	NA
15	Capital Investment for Safari Park/Zoo	$40,000,000	NA	NA	NA
16	Expected Inflation Rate	3%			
17	Total Park Operating Days per year	150			
18	Corporate Income Tax Rate	25%	25%	26%	27%

Source: Used with permission from Microsoft Corporation

FIGURE 6-1 Constants section

> **NOTE**
>
> The direct materials and overhead costs for 2015 and 2016 are formulas created by increasing the previous year's costs by the expected inflation rate. The direct labor costs for 2015 and 2016 were created by increasing the previous year's costs by 5 percent. Accountants frequently use such formulas in spreadsheets. However, you do not have to write any formulas in the Constants section—just enter the values shown in Figure 6-1.

- Direct Materials cost per day—Water Park—This value is the daily cost of direct materials to run the water park. Most of these materials are potable water and water treatment chemicals. If the weather is hot, these costs will increase by 5 percent due to evaporation.
- Direct Labor cost per day—Water Park—This value is the daily labor cost for water park employees and concessions employees.
- Overhead cost per day—Water Park—This value is the cost per day for maintenance employees, utilities, and repair materials.
- Direct Materials/Feed cost per day—Safari Park/Zoo—This value is the cost per day for direct materials and feed for the animals. Animals must be kept and fed all year. However, if the weather is hot during the park season, the animals will eat less, which will reduce the Direct Materials/Feed cost by 3 percent during the park season only.

- Direct Labor cost per day—Safari Park/Zoo—This value is the daily cost of animal keepers, trainers, and store employees. The full cost will be applied during park operating season (150 days). The cost will be reduced by half when the park is closed (215 days).
- Overhead cost per day—Safari Park/Zoo—This value is the cost per day for maintenance employees, utilities, and veterinary services. The full cost will be applied during park operating season (150 days). This cost will be reduced by half when the park is closed (215 days).
- Expected Base Attendance per day—Water Park—This value is a marketing estimate of the average water park attendance per day. The actual attendance will depend on the Seasonal Temperature and Economic Outlook values from the Inputs section.
- Expected Base Attendance per day—Safari Park/Zoo—This value is a marketing estimate of the average safari park attendance per day. The actual attendance will depend on the Seasonal Temperature and Economic Outlook values from the Inputs section.
- Average Ticket and Concession Sales per Customer—Water Park—This value is the average ticket and concessions sales per customer in the water park.
- Average Ticket and Concession Sales per Customer—Safari Park—This value is the average ticket and concessions sales per customer in the safari park/zoo.
- Capital Investment for Water Park—This value is the total amount of capital needed to convert the 50 undeveloped acres to a water park, including water slides, pools, and water rides.
- Capital Investment for Safari Park/Zoo—This value is the total amount of capital needed to purchase the adjoining acreage; convert it and the 50 undeveloped acres to a wildlife preserve and zoo, including all buildings, animal habitats, and equipment; and purchase the animals.
- Expected Inflation Rate—This value is the estimate of the annual inflation rate for 2015 and 2016.
- Total Park Operating Days per year—This value is an estimate of the number of days per year that either park will be open for business.
- Corporate Income Tax Rate—This value is an estimate of the corporate income tax rates that the park must pay after the expansion.

Inputs Section

Your spreadsheet model must include the following inputs that will apply for all three years, as shown in Figure 6-2.

	A	B	C	D	E
20	**Inputs**	**2013**	**2014**	**2015**	**2016**
21	Expansion Selection (W-Water Park, Z-Safari Park/Zoo)	NA		NA	NA
22	Economic Outlook (S-Stable, R-Recession)	NA		NA	NA
23	Seasonal Temperature (C-Cool, H-Hot)	NA		NA	NA

Source: Used with permission from Microsoft Corporation

FIGURE 6-2 Inputs section

- Expansion Selection—This value is the basic input for the strategic decision to expand the park. Add W for Water Park or Z for Safari Park/Zoo.
- Economic Outlook—This value is Stable (S) or Recession (R). Whether the economy is stable or in a recession affects the expected daily park attendance. Tourism is discretionary consumer spending and is affected significantly by a recession.
- Seasonal Temperature—This value is Cool (C) or Hot (H). Cool weather will decrease water park attendance, while hot weather will have a similar adverse effect on safari park/zoo attendance because of customer discomfort, animal inactivity, and customers switching to the water park attractions.

Summary of Key Results Section

This section (see Figure 6-3) contains the results data, which is of primary interest to the owners of Crystal Lake Park. This data includes income and end-of-year cash on hand information, as well as the annualized internal rate of return for a particular set of business inputs. This section summarizes the values from the Calculations, Income and Cash Flow Statements, and Internal Rate of Return Calculation sections.

	A	B	C	D	E
25	**Summary of Key Results**	**2013**	**2014**	**2015**	**2016**
26	Net Expansion Income after Taxes	NA			
27	End-of-year Cash on hand from Expansion	NA			
28	Internal Rate of Return for Investment	NA	NA	NA	

Source: Used with permission from Microsoft Corporation

FIGURE 6-3   Summary of Key Results section

For each year from 2014 to 2016, your spreadsheet should show net income after taxes and end-of-year cash on hand. The net income after taxes is also the net cash inflow for the IRR calculation. Because Crystal Lake Park is funding the capital investment from its cash on hand at the end of 2013, there is no debt to repay. However, Sheila and her management team want to know the internal rate of return for the expansion investment by the end of 2016.

Calculations Section

The Calculations section includes the calculations you need to perform to determine the operating costs and revenues for the park expansion. See Figure 6-4.

	A	B	C	D	E
30	**Calculations**	**2013**	**2014**	**2015**	**2016**
31	New Product Capital Investment (internally financed)		NA	NA	NA
32	Annual Direct Materials cost	NA			
33	Annual Direct Labor cost	NA			
34	Annual Overhead cost	NA			
35	Annual Cost of Sales	NA			
36	Annual Sales Revenues	NA			

Source: Used with permission from Microsoft Corporation

FIGURE 6-4   Calculations section

NOTE

The Calculations section includes several complex nested IF(AND) formulas for costs and revenues. If you use absolute cell references for the Total Park Operating Days per year (cell $B$17) and for the Inputs (cells $C$21, $C$22, and $C$23), you have to write the formulas only once for the 2014 column, and then you can successfully copy them to the columns for 2015 and 2016.

- New Product Capital Investment—This value is the amount of investment money spent by Crystal Lake Park at the end of 2013, depending on the Expansion Selection from the Inputs section. If the selected expansion is Water Park (W), this cell should display the capital investment amount from the Constants section for the water park (the value in cell B14). If the selected expansion is Safari Park/Zoo (Z), the cell should display the capital investment amount from the Constants section for the safari park/zoo (the value in cell B15). Note that when you see the word *if* in the text, you need to write a formula that uses the IF function in the target cell.
- Annual Direct Materials cost—The Direct Materials cost per day for both expansions are included in rows 4 and 7 of the Constants section. You must write formulas to calculate the Annual Direct Materials cost for 2014, 2015, and 2016. This cost equals the Direct Materials cost per day multiplied by the number of days in the year. For the water park, calculate a 5 percent increase in the Annual Direct Materials cost if the Seasonal Temperature is Hot, but note that the cost is incurred only during the Total Park Operating Days per year (cell B17). For the safari park/zoo, the calculation is more complex. The animals must be kept and fed all year, but if the Seasonal Temperature is Hot, the animals eat less, reducing the Annual Direct Materials cost by 3 percent. Therefore, four conditions require you to write a nested IF(AND) formula:

  - Water Park and Cool: The Annual Direct Materials cost will be calculated using the basic formula described earlier, with no modifications.
  - Water Park and Hot: The Annual Direct Materials cost will be calculated using the basic formula multiplied by 1.05 to reflect the use of additional water and chemicals due to evaporation.

- Safari Park/Zoo and Cool: The Annual Direct Materials cost will be calculated using the basic formula described earlier, except that you must calculate the cost for the entire year (365 days) instead of the park's operating days.
  - Safari Park/Zoo and Hot: The Annual Direct Materials cost will be calculated using the basic formula multiplied by 0.97 to reflect reduced animal activity for hotter operating days (the value in cell B17—150 days). The full daily cost is then multiplied by the number of days per year the park is closed (365 minus the value in cell B17, or 215 days).
- Annual Direct Labor cost—The Direct Labor cost per day for both expansions is included in rows 5 and 8 of the Constants section. You must write formulas to calculate the Annual Direct Labor cost for 2014, 2015, and 2016. This cost equals the Direct Labor cost per day multiplied by the applicable number of days in the year for either alternative. You must write the formula as an IF statement to calculate the Annual Direct Labor cost for either the water park or safari park/zoo expansion. Two conditions require you to create the following formulas:

  - Water Park: The Annual Direct Labor cost is the daily cost multiplied by the number of park operating days (the value in cell B17).
  - Safari Park/Zoo: The animals must be kept and fed, but the off-season labor requirement is only half that of the operating season. Therefore, the Annual Direct Labor cost is the daily labor cost multiplied by the number of park operating days (the value in cell B17), *plus* the daily labor cost, multiplied by the remaining days in the year (365 days minus the value in cell B17, or 215 days), multiplied by 0.5.
- Annual Overhead cost—The daily Overhead cost per day for both expansions is included in rows 6 and 9 of the Constants section. You must write formulas to calculate the Annual Overhead cost for 2014, 2015, and 2016. This cost equals the Overhead cost per day multiplied by the applicable number of days in the year for either alternative. You must write the formula as an IF statement to calculate the Annual Overhead cost for either the water park or safari park/zoo expansion. Two conditions require you to create the following formulas:

  - Water Park: The Annual Overhead cost is the daily overhead cost multiplied by the number of park operating days (the value in cell B17).
  - Safari Park/Zoo: The off-season overhead cost of keeping the animals is expected to be half that of the operating season. Therefore, the Annual Overhead cost is the daily overhead cost multiplied by the number of park operating days (the value in cell B17), *plus* the daily overhead cost, multiplied by the remaining days in the year (365 days minus the value in cell B17, or 215 days), multiplied by 0.5.
- Annual Cost of Sales—This value is the sum of the Annual Direct Materials, Annual Direct Labor, and Annual Overhead costs (rows 32 through 34).
- Annual Sales Revenues—The base formula for this calculation is the Expected Base Attendance per day for either alternative, multiplied by the Average Ticket and Concessions Sales per Customer, multiplied by the Total Park Operating Days per year. However, you must construct a complex nested IF(AND) formula that reflects the following eight combinations from the Inputs section (cells $C$21, $C$22, and $C$23):

  - Water/Stable/Cool—Multiply the base formula by 0.95 to reflect reduced water park ticket sales due to cooler weather.
  - Water/Recession/Cool—Multiply the base formula by 0.80 to reflect the effects of cooler weather and reduced business in a recession economy.
  - Water/Stable/Hot—Make no modifications to the base formula; this is the best combination of inputs for the water park.
  - Water/Recession/Hot—Multiply the base formula by 0.90 to reflect the effects of reduced business in a recession economy.
  - Zoo/Stable/Cool—Make no modifications to the base formula; this is the best combination of inputs for the safari park/zoo.
  - Zoo/Recession/Cool—Multiply the base formula by 0.90 to reflect the effects of reduced business in a recession economy.
  - Zoo/Stable/Hot—Multiply the base formula by 0.95 to reflect the effects of hot weather diverting customers to the water park instead.
  - Zoo/Recession/Hot—Multiply the base formula by 0.85 to reflect the effects of hot weather and a recession on the safari park/zoo's attendance.

The Calculations section includes several complicated formulas, and the Annual Sales Revenues calculation is the most complex—it contains seven nested IF(AND)s. You might want to write the formulas on a piece of paper and check your logic before trying to enter the formulas into the spreadsheet. Also, when you are constructing a complex nested IF(AND) formula, it is sometimes difficult to count the correct number of parentheses needed at the end of the formula. If you do not enter the correct number, Excel displays an error message and suggests the correct number of parentheses. Click OK to have Excel finish the formula for you. If you get lost while trying to write the nested IF and AND formulas, refer to Tutorial C or ask your instructor for help.

Income and Cash Flow Statements Section

The statements for income and cash flow start with the cash on hand at the beginning of the year. Because Crystal Lake Park is funding the capital investment *internally*—that is, with its own cash on hand—you must deduct the invested funds from the cash on hand at the end of 2013. Figure 6-5 and the following list show how you should structure the Income and Cash Flow Statements section.

	A	B	C	D	E
38	**Income and Cash Flow Statements**	**2013**	**2014**	**2015**	**2016**
39	Beginning-of-year Cash on Hand (in 2014 deduct Investment for 2013)	NA			
40	Annual Sales Revenues	NA			
41	less: Annual Cost of Sales	NA			
42	Expansion Profit before Income Tax	NA			
43	less: Income Tax Expense	NA			
44	Expansion Income after Taxes (Net Cash Inflow)	NA			
45	End-of-year Cash/Securities on Hand	$50,000,000			

Source: Used with permission from Microsoft Corporation

FIGURE 6-5 Income and Cash Flow Statements section

- Beginning-of-year Cash on Hand—For 2014, this value is the End-of-year Cash on Hand from 2013 minus the capital investment, depending on the expansion selection. If you choose the water park expansion, the capital investment will be $20 million (cell B14 in the Constants section). If you choose the safari park/zoo expansion, the capital investment will be $40 million (cell B15 in the Constants section). For 2015 and 2016, the Beginning-of-year Cash on Hand is the End-of-year Cash on Hand from the previous year.
- Annual Sales Revenues—This value is taken from the Annual Sales Revenues in the Calculations section for each year (cells C36, D36, and E36).
- Less: Annual Cost of Sales—This value is taken from the Annual Cost of Sales in the Calculations section for each year (cells C35, D35, and E35).
- Expansion Profit before Income Tax—This value is the Annual Sales Revenues minus the Annual Cost of Sales.
- Less: Income Tax Expense—If you make a profit (in other words, if the Expansion Profit before Income Tax is greater than zero), this value is the Expansion Profit before Income Tax multiplied by the Corporate Income Tax Rate for that year from the Constants section. If you make nothing or have a net loss, the Income Tax Expense is zero.
- Expansion Income after Taxes (Net Cash Inflow)—This value is the Expansion Profit before Income Tax minus the Income Tax Expense. From a strict accounting standpoint, the net income after taxes is not the net cash inflow because you would have to add back all noncash expenses such as depreciation or depletion to determine the true cash inflow. However, for the purposes of this case, assume that net income after taxes is equal to net cash inflow.
- End-of-year Cash/Securities on Hand—This value is the Beginning-of-year Cash on Hand plus the Expansion Income after Taxes.

Internal Rate of Return Calculation Section

This section, as shown in Figure 6-6, is set up to facilitate using the Excel built-in Internal Rate of Return function.

	A	B
47	**Internal Rate of Return Calculation**	
48	Investment (Cash Outflow)	
49	Net Cash Inflow 2014	
50	Net Cash Inflow 2015	
51	Net Cash Inflow 2016	
52	Internal Rate of Return (IRR)	

Source: Used with permission from Microsoft Corporation

FIGURE 6-6   Internal Rate of Return Calculation section

- Investment (Cash Outflow)—This value is the Capital Investment amount from cell B31 of the Calculations section, multiplied by –1. The investment value must be a *negative* number to represent it as a Cash Outflow. (Think of it as money out of your pocket.)
- Net Cash Inflow 2014—This value is the net income after tax for 2014.
- Net Cash Inflow 2015—This value is the net income after tax for 2015.
- Net Cash Inflow 2016—This value is the net income after tax for 2016.
- Internal Rate of Return (IRR)—This value is the annual rate of return that the park expansion will generate for Crystal Lake Park. To calculate the IRR, click cell B52, which is where you want to record the IRR result. Next, click the f_x symbol next to the cell-editing window below the Ribbon. The Insert Function window appears (see Figure 6-7). Type IRR in the "Search for a function" text box, and then click Go.

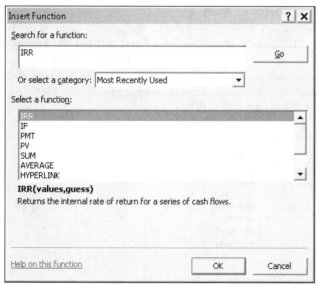

Source: Used with permission from Microsoft Corporation

FIGURE 6-7   The Insert Function window

The Function Arguments window appears to help you build the formula (see Figure 6-8). In the Values text box, enter the cells that contain all your cash outflows and inflows (B48:B51), or click and drag the mouse to select cells B48 through B51. Notice that Excel enters the formula for you in cell B52: =IRR(B48:B51). You do not have to enter a value in the Guess text box. When you click OK, Excel calculates the IRR and places the result in cell B52.

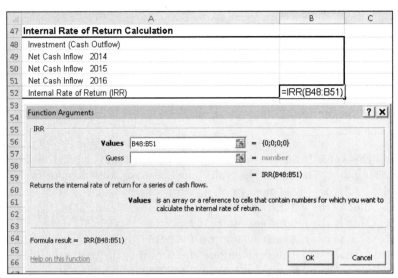

Source: Used with permission from Microsoft Corporation

FIGURE 6-8   The Function Arguments window for the IRR function

After you complete all the formulas, try testing your spreadsheet with various combinations of the three inputs. (There are eight possible combinations, as listed in the next section.) If you receive any error messages or see strange values in the cells, go back and check your formulas.

This DSS spreadsheet contains some values that represent millions of dollars. Accountants often simplify their spreadsheets by listing the outputs in multiples of thousands or millions of dollars. It is not hard to do—you simply divide the cell values by a thousand or a million, depending on the scale—but for the purposes of this case, you should keep the large numbers in the spreadsheet. If you see cell results listed as a group of "#" signs when working with large numbers (see Figure 6-9), the cell is not wide enough to display the number. Simply widen the column until the number is displayed.

	A	B	C
38	**Income and Cash Flow Statements**	**2013**	**2014**
39	Beginning-of-year Cash on Hand (in 2014 deduct Investment for 2013)	NA	########
40	Annual Sales Revenues	NA	########
41	less: Annual Cost of Sales	NA	$540,000
42	Expansion Profit before Income Tax	NA	########
43	less: Income Tax Expense	NA	########
44	Expansion Income after Taxes (Net Cash Inflow)	NA	########
45	End-of-year Cash/Securities on Hand	$50,000,000	########

Source: Used with permission from Microsoft Corporation

FIGURE 6-9   Column C is not wide enough to display the numbers

ASSIGNMENT 2: USING THE SPREADSHEET FOR DECISION SUPPORT

Next, you use the spreadsheet to gather data needed to determine the best expansion decision and to document your recommendations in a report to Sheila Patterson and the Crystal Lake Park directors.

This DSS model has eight possible financial outcomes:

1. Water Park Expansion (W)
   a. Stable Economy and Cool (W/S/C)
   b. Recession and Cool (W/R/C)
   c. Stable Economy and Hot (W/S/H)
   d. Recession and Hot (W/R/H)
2. Safari Park/Zoo Expansion (Z)
   a. Stable Economy and Cool (Z/S/C)
   b. Recession and Cool (Z/R/C)
   c. Stable Economy and Hot (Z/S/H)
   d. Recession and Hot (Z/R/H)

You are primarily interested in the park's financial position based on each of these possible outcomes. The Summary of Key Results section lets you see the Net Expansion Income after Taxes for each of the first three years of the expansion, the End-of-year Cash on Hand at the end of each of the three years, and the Internal Rate of Return for Investment. The Crystal Lake Park management team wants to make sure that the selected expansion restores the company's cash on hand to a better position by the end of the third year of operation (2016) than the cash on hand in 2013. In addition, the team prefers the expansion selection that provides a 12 percent or higher annual rate of return averaged across the four possible combinations of inputs.

Because there are eight (2^3) possible combinations of inputs when considering economic outlook, seasonal temperature, and expansion selection, you might want to run the spreadsheet model eight times, changing the inputs according to the preceding list. You should do this for two reasons:

- To ensure that no single year from 2014 through 2016 has negative income or negative end-of-year cash on hand (in other words, to make sure the park does not suffer a loss or run out of money)
- To print each spreadsheet to meet the requirements of Assignment 2A

You could then transcribe the results to a summary sheet. Next, you know that the management team is very interested in the financial data from the end of the third year of the model (2016). You can summarize the data easily using Scenario Manager.

Assignment 2A: Using Scenario Manager to Gather Data

For each of the eight situations listed earlier, you want to know the end-of-year cash on hand for the third year (2016) of the project as well as the internal rate of return generated by the three years' cash inflows.

You will run "what-if" scenarios with the eight sets of input values using Excel Scenario Manager. If necessary, review Tutorial C for tips on using Scenario Manager. In this case, the input values are stored together in one vertical group of cells (C21 through C23) in the Inputs section, as are the output cells (E27 and E28) in the Summary of Key Results section, so selecting the cells is easy. Run Scenario Manager to gather your data in a report called the Scenario Summary. Format the Scenario Summary to make it presentable, and then print it for your instructor.

If you haven't done so already, you should run the spreadsheet model for each of the eight input combinations. Save each spreadsheet in your Excel workbook with a different sheet name. Keep these names as short (but descriptive) as possible; for example, you can name the sheets WSC, WRC, WSH, WRH, ZSC, ZRC, ZSH, and ZRH. Print the spreadsheets if your instructor requires it. Make sure to save your completed Excel workbook before closing it.

Assignment 2B: Documenting Your Recommendations in a Report

Use Microsoft Word to write a brief report to the Crystal Lake Park management team. State the results of your analysis and recommend which expansion choice to make (water park or safari park/zoo). Your report must meet the following requirements:

- The first paragraph must summarize the expansion choices facing Crystal Lake Park and must state the purpose of the analysis.
- Summarize the results of your analysis and state your recommended action.
- Support your recommendation with a table outlining the Scenario Summary results. Figure 6-10 shows a recommended table format in Microsoft Word.
- If your report is well formatted, you might choose to embed an Excel object of the Scenario Summary in the body of the report. Tutorial C includes a brief description of how to copy and paste Excel objects.
- Your instructor might also ask you to provide a graph of the internal rates of return for the eight possible combinations of inputs.

Investment Selection	Seasonal Temperature	Economic Outlook	2016 Net Income ($ millions)	2016 End-of-year Cash on Hand ($ millions)	Internal Rate of Return
Water Park	Cool	Stable			
		Recession			
	Hot	Stable			
		Recession			
Safari Park/Zoo	Cool	Stable			
		Recession			
	Hot	Stable			
		Recession			

Source: © Cengage Learning 2014

FIGURE 6-10   Recommended format of a table to insert in your report

ASSIGNMENT 3: GIVING AN ORAL AND SLIDE PRESENTATION

Your instructor may ask you to summarize your analysis in an oral presentation. If so, assume that the management team at Crystal Lake Park wants you to explain your analysis and recommendations in 10 minutes or less. A well-designed PowerPoint presentation, with or without handouts, is considered appropriate in a business setting. Tutorial F provides excellent tips for preparing and delivering a presentation.

DELIVERABLES

Your completed case should include the following deliverables for the instructor:

- A printed copy of your report to management
- Printouts of your spreadsheets
- Electronic copies of all your work, including your report, PowerPoint presentation, and Excel DSS model. Ask your instructor for specific guidance on which items you should submit for grading.

CASE 7

THE POSTAL SERVICE ANALYSIS

Decision Support Using Excel

PREVIEW

The postal service of a large country has been losing money for years. Various operating reforms and financial reforms have been proposed. In this case, you will use Excel to examine whether the proposals can restore the postal service to financial health.

PREPARATION

- Review spreadsheet concepts discussed in class and in your textbook.
- Complete any exercises that your instructor assigns.
- Complete any part of Tutorial C that your instructor assigns. You may need to review the use of IF statements and the section called "Cash Flow Calculations: Borrowings and Repayments."
- Review file-saving procedures for Windows programs.
- Refer to Tutorials E and F as necessary.

BACKGROUND

A large country's postal service has been losing money for years. The postal service is officially a private entity, but the government oversees its finances and operating decisions and provides loans in the case of cash shortfalls. The postal service faces serious challenges. Proposed operating changes and financial changes need approval from appropriate unions and the legislature before they can be implemented.

Challenges

The postal service has been hurt severely by the rise of the Internet. Every e-mail sent is a letter not written and a first-class stamp that was not purchased from the postal service. Also, the postal service's shipping business has been hurt by the rise of private shipping companies, such as FedEx.

The postal service owns many trucks and airplanes. The related energy costs are significant, so increases in the cost of fuel pose a risk. Analysts estimate that a 1 percent increase in the market price of fuel adds $25 million a year to postal service costs.

Possible Operational and Financial Changes

Revenue would improve by raising the price of a first-class stamp, which costs less now in real terms than it did in 2001. But, by law, the postal service cannot raise the price of a stamp by more than the rate of inflation. This rate has been low in recent years, so the price of a stamp has increased by a penny per year at most in the last decade.

In 2006, the government passed a law requiring the postal service to prepay 75 years of pension and retiree healthcare costs by the end of 2016. In effect, the postal service was forced to prepay retirement benefits of workers who are not yet born! This requirement has increased yearly costs by $5 billion. The postal service wants the prepayment requirement lifted.

The postal service says it could save $3 billion per year by discontinuing Saturday home mail service, and surveys have shown that most citizens would be willing to give up Saturday service. However, this policy would require legislative action, and lawmakers have refused to approve it.

The postal service is prohibited by law to participate in other businesses. For example, it cannot put advertising on its delivery trucks. The postal service wants to partner with drugstore chains to deliver prescription medicines, but it is prohibited from doing so.

The postal service had 640,000 employees at the end of 2013. In the next five years, it expects to shed 20,000 employees per year by normal attrition. Management thinks it could lay off up to 20,000 additional employees per year and maintain the same level of service, but the legislature would need to approve these layoffs. So far, the postal unions have persuaded the legislature to prevent layoffs.

If the postal service loses money in a year, the government covers the shortfall with a loan. In effect, the government serves as the postal service's "bank." In recent years, the postal service has lost money, and now its loans under the line of credit total $12 billion. By law, the postal service cannot owe the government more than $15 billion. Lawmakers are now worried about the fate of the post office. They want the amounts owed to decrease, but it is becoming increasingly clear that losses and more debt are inevitable unless changes occur.

A legislative committee has asked you to create a DSS model of postal service operations with and without reforms. Your DSS model needs to account for the effects of possible operating changes. Your model will let you develop "what-if" scenarios with the inputs, see the financial results, and then help you decide if the post office can be made into a profitable operation.

ASSIGNMENT 1: CREATING A SPREADSHEET FOR DECISION SUPPORT

In this assignment, you will produce a spreadsheet that models the postal service business for the next five years. Then, in Assignment 2, you will write a memorandum that documents your analysis and conclusions. In Assignment 3, you will prepare and give an oral presentation of your analysis and conclusions to postal service management and government officials.

First, you create the spreadsheet model of the financial situation. The model covers the five years from 2014 to 2018. This section helps you set up each of the following spreadsheet components before entering cell formulas:

- Constants
- Inputs
- Summary of Key Results
- Calculations
- Income and Cash Flow Statements
- Debt Owed

A discussion of each section follows. The spreadsheet skeleton is available for you to use. To access this skeleton, go to your data files, select Case 7, and then select **PO.xlsx**.

Constants Section

Your spreadsheet should include the constants shown in Figure 7-1. An explanation of the line items follows the figure.

	A	B	C	D	E	F	G
1	**The Postal Service Analysis**						
2							
3	**Constants**	**2013**	**2014**	**2015**	**2016**	**2017**	**2018**
4	No Saturday delivery cost savings	NA	$ 3,000,000,000	$ 3,000,000,000	$ 3,000,000,000	$ 3,000,000,000	$ 3,000,000,000
5	Minimum cash needed to start year	NA	$ 2,000,000,000	$ 2,000,000,000	$ 2,000,000,000	$ 2,000,000,000	$ 2,000,000,000
6	Expected work force	NA	620,000	600,000	580,000	560,000	540,000
7	Expected increase in average compensation	NA	2%	2%	3%	3%	4%
8	Expected decrease in first class units	NA	5%	6%	7%	8%	9%
9	Pension and health costs, no pre-funding	NA	$ 3,000,000,000	$ 3,000,000,000	$ 3,000,000,000	$ 3,000,000,000	$ 3,000,000,000
10	Pension and health pre-funding cost	NA	$ 5,000,000,000	$ 5,000,000,000	$ 5,000,000,000	$ -	$ -
11	Expected other than first class revenue	NA	$ 30,000,000,000	$32,000,000,000	$ 34,000,000,000	$35,000,000,000	$ 36,000,000,000
12	Interest rate on debt owed	NA	3.5%	3.5%	3.5%	3.5%	3.5%
13	Expected other operating expense	NA	$ 9,000,000,000	$ 9,200,000,000	$ 9,400,000,000	$ 9,600,000,000	$ 9,800,000,000
14	Other Business net income	NA	$ 1,000,000,000	$ 2,000,000,000	$ 4,000,000,000	$ 6,000,000,000	$ 8,000,000,000

Source: Used with permission from Microsoft Corporation

FIGURE 7-1   Constants section

- No Saturday delivery cost savings—If Saturday delivery were discontinued, operating expenses would be $3 billion less each year.
- Minimum cash needed to start year—The postal service wants to have at least $2 billion in cash at the beginning of each year. The government will lend the amount needed at the end of a year to begin the new year with that amount.

- Expected work force—By attrition, 20,000 workers will be lost each year. This attrition will reduce compensation expenses each year.
- Expected increase in average compensation—Union agreements call for wage increases each year. Expected percentage increases in average compensation are shown.
- Expected decrease in first class units—The postal service will continue to feel the impact of the Internet on first-class mail. This field shows the expected percentage decrease in items sent first class each year.
- Pension and health costs, no pre-funding—The "normal" provision for pension and retiree health care is expected to be $3 billion a year.
- Pension and health pre-funding cost—The added cost for future pension and retiree healthcare "pre-funding" is expected to be $5 billion a year through 2016.
- Expected other than first class revenue—The postal service does more than deliver first-class mail. The postal service provides shipping and packaging services, handles "junk" mail, delivers periodicals, and provides other mail-related services. In total, these aspects of the postal service are expected to grow. Expected revenue for each year is shown.
- Interest rate on debt owed—The government acts as the postal service's banker. The average interest rate on existing debt is 3.5 percent. For planning purposes, the government would charge this interest rate for any new borrowing.
- Expected other operating expense—Other out-of-pocket expenses, such as costs that do not involve compensation or transportation, are expected to grow each year, as shown.
- Other Business net income—Profits will accrue if the postal service is allowed to branch into new businesses. The value of estimated new revenue minus added expenses will be low in early years, but it is projected to grow as time goes on, as shown.

Inputs Section

Your spreadsheet should include the following inputs for the years 2014 to 2018, as shown in Figure 7-2.

	A	B	C	D	E	F	G
16	**Inputs**	**2013**	**2014**	**2015**	**2016**	**2017**	**2018**
17	Allowed to Enter Other Business? (Y/N)	NA		NA	NA	NA	NA
18	Saturday Delivery Eliminated? (Y/N)	NA		NA	NA	NA	NA
19	Increase in Price of First Class Stamp ($.xx)	NA					
20	Remove Benefits Pre-Funding? (Y/N)	NA		NA	NA	NA	NA
21	Expected % Change in Energy Prices	NA					
22	Added attrition in year	NA					

Source: Used with permission from Microsoft Corporation

FIGURE 7-2 Inputs section

- Allowed to Enter Other Business? (Y/N)—Enter Y if the legislature will permit this policy, and N if not. The entry applies to all five years.
- Saturday Delivery Eliminated? (Y/N)—Enter Y if the legislature will permit this policy, and N if not. The entry applies to all five years.
- Increase in Price of First Class Stamp ($.xx)—Enter a value for this increase each year. For example, if no increase is expected in any year, the entry would be .00, .00, .00, .00, .00. If a three-cent increase is expected each year, the entry would be .03, .03, .03, .03, .03. Cells should be formatted for Currency.
- Remove Benefits Pre-Funding? (Y/N)—Enter Y if the legislature will permit this policy, and N if not. The entry applies to all five years.
- Expected % Change in Energy Prices—Enter the percent change expected each year. For example, if a 2 percent increase is expected each year, enter .02, .02, .02, .02, .02. If no change is expected, enter five zeroes. Negative values would indicate that energy cost decreases are expected. Cells should be formatted for Percentage.
- Added attrition in year—Enter the number of workers expected to be laid off each year, above and beyond the already expected attrition. Do not enter a number of more than 20,000 in any year. If no added layoffs are expected, enter a zero.

Summary of Key Results Section

Your spreadsheet should include the results shown in Figure 7-3.

	A	B	C	D	E	F	G
24	**Summary of Key Results**	**2013**	**2014**	**2015**	**2016**	**2017**	**2018**
25	Net Income (Loss) in Year	NA					
26	End-of-year cash on hand	NA					
27	End-of-year debt owed	NA					

Source: Used with permission from Microsoft Corporation

FIGURE 7-3   Summary of Key Results section

For each year, your spreadsheet should show net income (or loss) in the year, cash on hand at the end of the year, and debt owed to the government at the end of the year. The net income, cash, and debt cells should be formatted as currency with zero decimals. These values are computed elsewhere in the spreadsheet and should be echoed here.

Calculations Section

You should calculate intermediate results that will be used in the income and cash flow statements that follow. Calculations, as shown in Figure 7-4, may be based on expected year-end 2013 values. When called for, use absolute referencing properly. Values must be computed by cell formula; hard-code numbers in formulas only when you are told to do so. Cell formulas should not reference a cell with a value of "NA."

An explanation of each item in this section follows the figure.

	A	B	C	D	E	F	G
29	**Calculations**	**2013**	**2014**	**2015**	**2016**	**2017**	**2018**
30	Average yearly employee compensation	$ 75,000					
31	Number of employees	NA					
32	Total compensation	NA					
33	Other Business net income	NA					
34	Cost of first class stamp	$ 0.45					
35	Number of first class pieces	74,000,000,000					
36	First class mail revenue	NA					
37	Pension and health cost	NA					
38	Transportation cost	$ 6,000,000,000					
39	Other operating expense	NA					

Source: Used with permission from Microsoft Corporation

FIGURE 7-4   Calculations section

- Average yearly employee compensation—Almost all employees are unionized. The average compensation for all employees was $75,000 in 2013. The expected rate of compensation increase each year is shown in the Constants section. Format cells for currency and no decimals.
- Number of employees—This value is a function of the expected workforce (from the Constants section) and added attrition (from the Inputs section).
- Total compensation—This value is a function of the number of employees and the average yearly compensation. Both of these values are from the Calculations section.
- Other Business net income—This amount will be earned if the postal service is allowed to branch into new businesses, as indicated by the value in the Inputs section. The amount is shown in the Constants section.
- Cost of first class stamp—A stamp cost 45 cents in 2013. The increase each year is shown in the Inputs section.
- Number of first class pieces—A total of 74 billion pieces of mail were delivered in 2013. The number is expected to decrease each year by a percentage shown in the Constants section.
- First class mail revenue—This value is a function of the price of a first-class stamp and the number of first-class pieces. Both values are taken from the Calculations section. Format cells for currency and no decimals.
- Pension and health cost—The base cost and the pre-funding increment are shown in the Constants section. A value in the Inputs section indicates whether the pre-funding cost will be eliminated. Format cells for currency and no decimals.
- Transportation cost—The transportation cost in 2013 was $6 billion. The cost each year will remain at $6 billion unless energy prices change. The expected percentage change each year is

shown in the Inputs section. For each 1 percent increase in price, transportation costs will increase $25 million. For example, a +2 percent change is expected in 2014. Transportation costs in 2014 will be $6 billion + (2 × $25 million), or $6,050,000,000.

- Other operating expense—If Saturday delivery is eliminated, the expense shown in this field will be the expected other operating expenses minus the Saturday savings. Both of these values are from the Constants section.

Income and Cash Flow Statements

The forecast for net income and cash flow starts with the cash on hand at the beginning of the year. This value is followed by the income statement and the calculation of cash on hand at year's end. For readability, format cells in this section as currency with zero decimals. Values must be computed by cell formula; hard-code numbers in formulas only if you are told to do so. Cell formulas should not reference a cell with a value of "NA." Your spreadsheets should look like those in Figures 7-5 and 7-6. A discussion of each item in the section follows each figure.

	A	B	C	D	E	F	G
41	**Income and Cash Flow Statements**	**2013**	**2014**	**2015**	**2016**	**2017**	**2018**
42	Beginning-of-year cash on hand	NA					
43							
44	Revenue:	NA					
45	First class mail	NA					
46	Other mail services	NA					
47	Other business	NA					
48	Total revenue	NA					
49	Costs:						
50	Employee compensation	NA					
51	Pension and health cost	NA					
52	Transportation cost	NA					
53	Other operating expense	NA					
54	Total costs	NA					
55	Income before interest expense	NA					
56	Interest expense	NA					
57	Net Income (Loss)	NA					
58							
59	Expected capital outlays in year	NA					

Source: Used with permission from Microsoft Corporation

FIGURE 7-5   Income and Cash Flow Statements section

- Beginning-of-year cash on hand—This value is the cash on hand at the end of the prior year.
- First class mail revenue—This amount is from the Calculations section and can be echoed here.
- Other mail services revenue—This amount is from the "Expected other than first class revenue" value in the Constants section and can be echoed here.
- Other business revenue—This amount is from the Other Business net income value in the Calculations section and can be echoed here.
- Total revenue—This amount is the sum of first-class, other mail service, and other business revenues.
- Employee compensation—This amount is from the Calculations section and can be echoed here.
- Pension and health cost—This amount is from the Calculations section and can be echoed here.
- Transportation cost—This amount is from the Calculations section and can be echoed here.
- Other operating expense—This amount is from the Calculations section and can be echoed here.
- Total costs—This amount is the total of employee compensation costs, pension and health costs, transportation costs, and other operating costs.
- Income before interest expense—This amount is the difference between total revenue and total costs.
- Interest expense—This amount is the product of the debt owed to the government at the beginning of the year and the interest rate for the year. The interest rate is a value in the Constants section.
- Net Income (Loss)—This amount is the difference between income before interest expense and interest expense. The postal service does not pay income taxes, so there is no line for Net Income After Taxes.
- Expected capital outlays in year—These amounts are expected to be $2 billion in 2014, 2015, and 2016, $4 billion in 2017, and $6 billion in 2018. You can hard-code these values in the cells.

Line items for the year-end cash calculation are discussed next. In Figure 7-6, column B represents 2013, column C is for 2014, and so on. Year 2013 values are NA except for End-of-year cash on hand, which is $2 billion.

	A	B	C	D	E	F	G
61	Net cash position (NCP)	NA					
62	Borrowing from government	NA					
63	Repayment to government	NA					
64	End-of-year cash on hand	$ 2,000,000,000					

Source: Used with permission from Microsoft Corporation

FIGURE 7-6   End-of-year cash on hand section

- Net cash position (NCP)—The NCP at the end of a year equals the cash at the beginning of the year, plus the year's net income (or loss), minus the year's capital outlay.
- Borrowing from government—Assume that the government will lend enough money at the end of the year to reach the minimum cash needed to start the next year. If the NCP is less than this minimum, the postal service must borrow enough to start the next year with the minimum. Borrowing increases the cash on hand, of course.
- Repayment to government—If the NCP is more than the minimum cash needed and some debt is owed at the beginning of the year, as much debt as possible must be paid off (but do not take cash below the minimum amount required to start the next year). Repayments reduce cash on hand, of course.
- End-of-year cash on hand—This amount is the NCP plus any borrowing, and minus any repayments.

Debt Owed Section

This section shows a calculation of debt owed to the government at year's end, as shown in Figure 7-7. Year 2013 values are NA except for End-of-year debt owed, which is $12 billion. Values must be computed by cell formula; hard-code numbers in formulas only when you are told to do so. Cell formulas should not reference a cell with a value of "NA." An explanation of each item follows the figure.

	A	B	C	D	E	F	G
66	**Debt Owed**	**2013**	**2014**	**2015**	**2016**	**2017**	**2018**
67	Beginning-of-year debt owed	NA					
68	Borrowing from government	NA					
69	Repayment to government	NA					
70	End-of-year debt owed	$12,000,000,000					

Source: Used with permission from Microsoft Corporation

FIGURE 7-7   Debt Owed section

- Beginning-of-year debt owed—Debt owed at the beginning of a year equals the debt owed at the end of the prior year.
- Borrowing from government—This amount has been calculated elsewhere and can be echoed to this section. Borrowing increases the amount of debt owed.
- Repayment to government—This amount has been calculated elsewhere and can be echoed to this section. Repayments reduce the amount of debt owed.
- End-of-year debt owed—In 2014 through 2018, this is the amount owed at the beginning of the year, plus borrowing during the year, and minus repayments during the year.

ASSIGNMENT 2: USING THE SPREADSHEET FOR DECISION SUPPORT

Complete the case by (1) using the spreadsheet to gather data about possible scenarios and (2) documenting your findings in a memo.

Postal service management wants to know projected financial results in three different scenarios:

1. No operational or financial changes occur. Presumably, further losses will occur in this scenario. How bad will the debt load get?
2. Aggressive changes are made. This would require intensive negotiations with the union and the legislature. Will profitability be restored and debt eliminated?
3. Some "middling" changes are made. This would require some negotiations. Will profitability be restored and debt eliminated?

Each scenario is explained in the following sections.

Scenario #1: No changes

The first scenario is called "No Change." In this scenario:

- The postal service is not allowed to branch into other businesses.
- Saturday service is not eliminated.
- The price of a first-class stamp increases 1 cent per year in each of the years from 2014 to 2018.
- Benefits pre-funding is not eliminated; in other words, the postal service must continue to pre-fund.
- Energy prices increase only 1 percent per year in each of the years from 2014 to 2018.
- No additional attrition occurs in any of the years from 2014 to 2018.

Scenario #2: Aggressive

The second scenario is called "Aggressive." In this scenario:

- The postal service is allowed to branch into other businesses.
- Saturday service is eliminated.
- The price of a first-class stamp increases 3 cents per year in each of the years from 2014 to 2018.
- Benefits pre-funding is eliminated; in other words, the postal service does not continue to pre-fund.
- Energy prices increase only 1 percent per year in each of the years from 2014 to 2018.
- Additional layoffs (added attrition) of 20,000 occur in each of the years from 2014 to 2018.

Scenario #3: Middling

The third scenario is called "Middling." In this scenario:

- The postal service is allowed to branch into other businesses.
- Saturday service is not eliminated.
- The price of a first-class stamp increases 2 cents per year in each of the years from 2014 to 2018.
- Benefits pre-funding is eliminated; in other words, the postal service does not continue to pre-fund.
- Energy prices increase 3 percent per year in each of the years from 2014 to 2018.
- Additional layoffs (added attrition) of 5,000 occur in each of the years from 2014 to 2018.

Can adequate profitability and reasonable debt levels be achieved with less intensive negotiating? If so, the aggressive scenario need not be pursued.

In each scenario, management wants to know the projected 2018 net income and projected 2018 cash on hand. In each scenario, management also wants to know the debt owed at the end of each of the years from 2014 to 2018, and whether projected debt levels ever exceed the $15 billion threshold.

The No Change scenario results are expected to be poor, so management will want a supplemental "what-if" analysis to accompany this scenario. Manually enter the No Change scenario inputs, change the first-class stamp prices, and observe the Summary of Key Results section. Using these price changes only, what price increase would be needed in each of the five years to yield profitability in 2018 and debt levels below $15 billion in all five years? Would you need a 10-cent increase each year? Would you need a 20-cent increase?

Assignment 2A: Using the Spreadsheet to Gather Data

You have built the spreadsheet to model the business situation. For each of the three scenarios, you want to know the 2018 net income after taxes, the 2018 end-of-year cash on hand, and the end-of-year debt owed for each of the years 2014 through 2018.

You will run "what-if" scenarios with the three sets of input values using Scenario Manager. (See Tutorial C for details on using Scenario Manager.) Set up the three scenarios. Your instructor may ask you to use conditional formatting to make sure that your input values are proper. (Note that in Scenario Manager you can enter noncontiguous cell ranges, such as C19, D19, C20:F20.)

The relevant output cells are the 2018 net income and end-of-year cash on hand cells, and the end-of-year debt owed cells for 2014 through 2018 in the Summary of Key Results section. Run Scenario Manager to gather the data in a report. When you finish, print the spreadsheet with the input for any of the scenarios, print the Scenario Manager summary sheet, and then save the spreadsheet file a final time.

Finally, enter inputs manually for the No Change scenario. Develop "what-if" scenarios for stamp prices. In the Scenario Manager summary sheet, manually add a fourth scenario for the following results: No Change inputs, needed stamp price increase in each year, and resulting net income, cash, and debt levels.

Assignment 2B: Documenting Your Recommendations in a Memo

Use Microsoft Word to write a brief memo documenting your analysis and conclusions. You can address the memo to "Postal Service Management." Observe the following requirements:

- Set up your memo as described in Tutorial E.
- In the first paragraph, briefly state the business situation and the purpose of your analysis.
- Next, provide the answers to management's questions. What will happen if there are no changes? If there are no changes, what stamp price increase would be needed to restore profitability and reduced debt levels? Is an Aggressive strategy needed, or can a Middling strategy succeed?
- State your recommendation. What strategy, if any, should postal service management pursue?
- Support your statements graphically, as your instructor requires. Your instructor may ask you to return to Excel and copy the Scenario Manager summary sheet results into the memo. (See Tutorial C for details on this procedure.) Your instructor might also ask you to make a summary table in Word based on the Scenario Manager summary sheet results. (This procedure is described in Tutorial E.)

Your table should resemble the format shown in Figure 7-8.

	No Change Scenario	Aggressive Scenario	Middling Scenario	No Change – $.xx Stamp Increase
2018 Net income				
2018 Cash on hand				
2014 Debt owed				
2015 Debt owed				
2016 Debt owed				
2017 Debt owed				
2018 Debt owed				

Source: © Cengage Learning 2014

FIGURE 7-8   Format of table to insert in memo

In the far-right column heading, replace "$.xx" with the required increase. For example, if a 15-cent increase were needed each year, the header would be: "No Change – $.15 Stamp Increase."

ASSIGNMENT 3: GIVING AN ORAL PRESENTATION

Your instructor may ask you to explain your analysis and recommendations in an oral presentation. If so, assume that postal service management and key lawmakers want the presentation to last 10 minutes or less. Use visual aids or handouts that you think are appropriate. See Tutorial F for tips on preparing and giving an oral presentation.

DELIVERABLES

Your completed case should include the following deliverables for the instructor:

1. A printed copy of your report and memo
2. Printouts of your spreadsheets
3. Electronic copies of all your work, including your report and Excel DSS model. Ask your instructor which items you should submit for grading.

PART 3

DECISION SUPPORT CASES USING MICROSOFT EXCEL SOLVER

BUILDING A DECISION SUPPORT SYSTEM USING MICROSOFT EXCEL SOLVER

In Tutorial C, you learned that Decision Support Systems (DSS) are programs used to help managers solve complex business problems. Cases 6 and 7 were DSS models that used Microsoft Excel Scenario Manager to calculate and display financial outcomes given certain inputs, such as economic outlooks and mortgage interest rates. You used the outputs from Scenario Manager to see how different combinations of inputs affected cash flows and income so that you could make the best decision for expanding your business or selecting a technology to develop and market.

Many business situations require models in which the inputs are not limited to two or three choices, but include large ranges of numbers in more than three variables. For such business problems, managers want to know the best or optimal solution to the model. An optimal solution can either maximize an objective variable, such as income or revenues, or minimize the objective variable, such as operating costs. The formula or equation that represents the target income or operating cost is called an objective function. Optimizing the objective function requires the use of constraints (also called constraint equations), which are rules or conditions you must observe when solving the problem. The field of applied mathematics that addresses problem solving with objective functions and constraint equations is called linear programming. Before the advent of digital computers, linear programming required the knowledge of complex mathematical techniques. Fortunately, Excel has a tool called Solver that can compute the answers to optimization problems.

This tutorial has five sections:

1. **Adding Solver to the Ribbon**—Solver is not installed by default with Excel 2010; you must add it to the application. You may need to use Excel Options to add Solver to the Ribbon.
2. **Using Solver**—This section explains how to use Solver. You will start by determining the best mix of vehicles for shipping exercise equipment to stores throughout the country.
3. **Extending the example**—This section tests your knowledge of Solver as you modify the transportation mix to accommodate changes: additional stores to supply and redesign of the product to reduce shipping volume.
4. **Using Solver on a new problem**—In this section, you will use Solver on a new problem: maximizing the profits for a mix of products.
5. **Troubleshooting Solver**—Because Solver is a complex tool, you will sometimes have problems using it. This section explains how to recognize and overcome such problems.

NOTE

If you need a refresher, Tutorial C offers guidance on basic Excel concepts such as formatting cells and using the =IF( ) and AND( ) functions.

ADDING SOLVER TO THE RIBBON

Before you can use Solver, you must determine whether it is installed in Excel. Start Excel and then click the Data tab on the Ribbon. If you see a group on the right side named Analysis that contains Solver, you do not need to install Solver (see Figure D-1).

Source: Used with permission from Microsoft Corporation

FIGURE D-1   Analysis group with Solver installed

If the Analysis group or Solver is not shown on the Data tab of the Ribbon, do the following:

1. Click the File tab.
2. Click Options (see Figure D-2).
3. Click Add-Ins (see Figure D-3) to display the available add-ins in the right pane.
4. Click Go at the bottom of the right pane. The window shown in Figure D-4 appears.
5. Click the Solver Add-in box as well as the Analysis ToolPak and Analysis ToolPak-VBA boxes. (You will need the latter options in a subsequent case, so install them now with Solver.)
6. Click OK to close the window and return to the Ribbon. If you click the Data tab again, you should see the Analysis group with Data Analysis and Solver on the right.

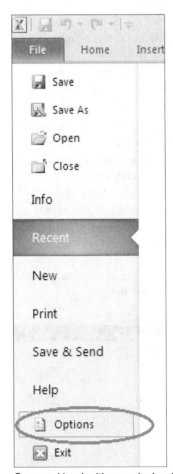

Source: Used with permission from Microsoft Corporation

FIGURE D-2   Excel Options selection

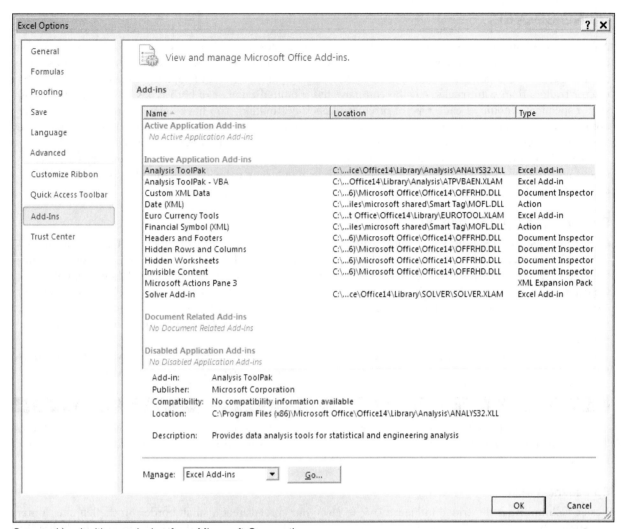

Source: Used with permission from Microsoft Corporation

FIGURE D-3   Add-Ins pane

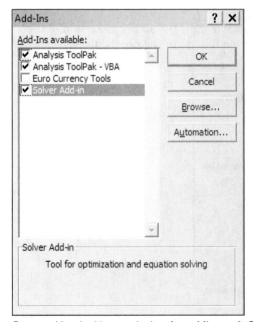

Source: Used with permission from Microsoft Corporation

FIGURE D-4   Add-Ins window with Solver, Analysis ToolPak, and Analysis ToolPak VBA selected

USING SOLVER

A fictional company called CV Fitness builds exercise machines in its plant in Memphis, Tennessee and ships them to its stores across the country. The company has a small fleet of trucks and tractor-trailers to ship its products from the factory to its stores. It costs less money per cubic foot of capacity to ship products with tractor-trailers than with trucks, but the company has a limited number of both types of vehicles and must ship a specified amount of each type of product to each destination. You have been asked to determine the optimal mix of trucks and tractor-trailers to send merchandise to each store. The optimal mix will have the lowest total shipping cost while ensuring that the required quantity of products is shipped to each store.

To use Solver, you must set up a model of the problem, including the factors that can vary (the mix of trucks and tractor-trailers) and the constraints on how much they can vary (the number of each vehicle available). Your goal is to minimize the shipping cost.

Setting Up a Spreadsheet Skeleton

CV Fitness makes three fitness machines: exercise bikes (EB), elliptical cross-trainers (CT), and treadmills (TM). When packaged for shipment, their shipping volumes are 12, 15, and 22 cubic feet, respectively. The finished machines are shipped via ground transportation to five stores in Philadelphia, Atlanta, Miami, Chicago, and Los Angeles. Your vehicle fleet consists of 12 trucks and six tractor-trailers. Each truck has a capacity of 1500 cubic feet, and each tractor-trailer has a capacity of 2350 cubic feet. The spreadsheet includes the road distances from your plant in Memphis to each store, along with each store's demand for the three fitness machines.

What is the best mix of trucks and tractor-trailers to send to each destination? You will learn how to use Solver to determine the answer. The spreadsheet components are discussed in the following sections.

AT THE KEYBOARD

Start by saving your blank spreadsheet. Use a descriptive filename so you can find it easily later—**CV Fitness Trucking Problem.xlsx** should work well. Then enter the skeleton and formulas as directed in the following sections.

Spreadsheet Title

Resize Column A, as illustrated in Figure D-5, to give your spreadsheet a small border on the left side. Enter the spreadsheet title in cell B1. Merge and center cells B1 through F1 using the Merge and Center button in the Alignment group of the Home tab.

Constants Section

Your spreadsheet should have a section for values that will not change. Figure D-5 shows a skeleton of the Constants section and the values you should enter. A discussion of the line items follows the figure.

A	B	C	D	E	F
1	CV Fitness, Inc. Truck Load Management Problem				
2					
3	Constants Section:				
4		Volume Cu. Ft.	Operating Cost per mi.	Operating Cost per mi-cu. Ft.	Available Fleet
5	Truck	1500	$1.00	$0.000667	12
6	Tractor Trailer	2350	$1.30	$0.000553	6
7					
8	Exercise Bike (EB)	12			
9	Elliptical Crosstrainer (CT)	15			
10	Treadmill (TM)	22			

Source: Used with permission from Microsoft Corporation

FIGURE D-5   Spreadsheet title and Constants section

- In column C, enter the Volume Cu. Ft., which is the cubic-foot capacity of the vehicles as well as the shipping volume for each item of exercise equipment.
- In column D, enter the Operating Cost per mi., which is the cost per mile driven for each type of vehicle.
- In column E, enter the Operating Cost per mi.-cu.ft. This value is actually a formula: the operating cost per mile divided by the vehicle volume in cubic feet. Normally you do not put formulas in the Constants section, but in this case it lets you see the relative cost efficiencies of each vehicle. Assuming that both types of vehicles can be filled to capacity, the tractor-trailer is the preferred vehicle for shipping cost efficiency.
- In column F, enter the values for the Available Fleet, which is the number of each type of vehicle your company owns or leases.

You can update the Constants section as the company adds more products to its offerings or adds vehicles to its fleet.

N O T E

The column headings in the Constants section contain two or three lines to keep the columns from becoming too wide. To create a new line in a cell, hold down the Alt key and press Enter.

Now is a good time to save your workbook again. Keep the name you assigned earlier.

Calculations and Results Section

The structure and format of your Calculations and Results section will vary greatly depending on the nature of the problem you need to solve. In some Solver models, you might need to maximize income, which means you might also have an Income Statement section. In other Solver models, you may want to have a separate Changing Cells section that contains cells Solver will manipulate to obtain a solution. In this tutorial, you want to minimize shipping costs while meeting the product demand of your stores. You can accomplish this task by building a single unified table that includes the distances to the stores, the product demand for each store, and the shipping alternatives and costs.

A unified Calculations and Results section makes sense in this model for several reasons. First, it simplifies writing and copying the formulas for the needed shipping volumes, the vehicle capacity totals, and the shipping costs to each destination. Second, a well-organized table allows you to easily identify the changing cells, which Solver will manipulate to optimize the solution, as well as the total cost (or optimization cell). Finally, a unified table allows your management team to visualize both the problem and its solution.

When creating a complex table, it is often a good idea to sketch the table's structure first to see how you want to organize the data. Format the table structure, then enter the data you are given for the problem. Write the cells that contain the formulas last, starting with all the formulas in the first row. If you do a good job structuring your table, you will be able to copy the first-row formulas to the other rows.

Build the blank table shown in Figure D-6. A discussion of the rows and columns follows the figure.

N O T E

Leave rows 11 and 12 blank between the Constants section and the Calculations and Results section. You then will have room to add an extra product to your Constants section later.

	Distance/Demand Table		Store Demand			Vehicle Loading							Cost
						Volume Required	Trucks	Volume for Trucks	Tractor-Trailers	Volume for Tractor-Trailers	Total Vehicle Capacity	% of Vehicle Capacity Utilized	Shipping Cost
Distance Table (from Memphis Plant)		Miles	EB	CT	TM								
Philadelphia Store		1010	140	96	86								
Atlanta Store		380	76	81	63								
Miami Store		1000	56	64	52								
Chicago Store		540	115	130	150								
Los Angeles Store		1810	150	135	180								
					Totals:								
Fill Legend:				Changing Cells									Total Cost
				Optimization Cell									

Calculations and Results Section: (row 13)

Source: Used with permission from Microsoft Corporation

FIGURE D-6 Blank table for Calculations and Results section

- In row 13, enter "Calculations and Results Section:" as the title of the table.
- In row 14, columns B and C, enter "Distance/Demand Table" as a column heading. Merge and center the heading in the two columns.
- In row 14, columns D, E, and F, enter "Store Demand" as a column heading. Merge and center the heading in the three columns.
- In row 14, columns G through M, enter "Vehicle Loading" as a column heading. Merge and center the heading across the columns.
- In row 14, column N, enter "Cost" as a centered column heading.
- In row 15, column B, enter "Distance Table (from Memphis Plant)" as a centered column heading.
- In row 15, column C, enter "Miles" as a centered column heading.
- In row 15, columns D, E, and F, enter "EB," "CT," and "TM," respectively, as equipment headings.
- In row 15, columns G through N, enter "Volume Required," "Trucks," "Volume for Trucks," "Tractor-Trailers," "Volume for Tractor-Trailers," "Total Vehicle Capacity," "% of Vehicle Capacity Utilized," and "Shipping Cost," respectively, as column headings.
- In rows 16 through 20, column B, enter the destination store locations.
- In rows 16 through 20, column C, enter the number of miles to the destination store locations.
- In rows 16 through 20, columns D through F, enter the number of exercise bikes (EB), cross-trainers (CT), and treadmills (TM) to be shipped to each store location.
- Rows 16 through 20, columns G through N, will contain formulas or "seed values" later. Leave them blank for now, but fill cells H16 through H20 and cells J16 through J20 with a light color to indicate that they are the changing cells for Solver. To fill a cell, use the Fill Color button in the Font group.
- In cell F21, enter "Totals:" to label the following cells in the row.
- Cells G21 through N21 will be used for column totals. Fill cell N21 with a slightly darker shade than you used for the changing cells. Cell N21 is your optimization cell.
- In cell B22, enter "Fill Legend:" as a label.
- Fill cell C22 with the fill color you selected for the changing cells.
- In cells D22 and E22, enter "Changing Cells" as the label for the fill color. Merge and center the label in the cells.
- In cell N22, enter "Total Cost" as the label for the value in cell N21.
- Fill cell C23 with the fill color you selected for the optimization cell.
- In cells D23 and E23, enter "Optimization Cell" as the label for the fill color. Merge and center the label in the cells.

Figure D-7 illustrates a magnified section of the Distance/Demand table in case the numbers in Figure D-6 are difficult to read.

	Distance Table (from Memphis Plant)	Miles	EB	CT	TM
	Calculations and Results Section:				
	Distance/Demand Table		**Store Demand**		
16	Philadelphia Store	1010	140	96	86
17	Atlanta Store	380	76	81	63
18	Miami Store	1000	56	64	52
19	Chicago Store	540	115	130	150
20	Los Angeles Store	1810	150	135	180
21					Totals:
22	Fill Legend:		Changing Cells		
23			Optimization Cell		

Source: Used with permission from Microsoft Corporation

FIGURE D-7   Magnified view of the Distance/Demand table

Use the Borders menu in the Font group to select and place appropriate borders around parts of the Calculations and Results section (see Figure D-8). The All Borders and Outside Borders selections are the most useful borders for your table.

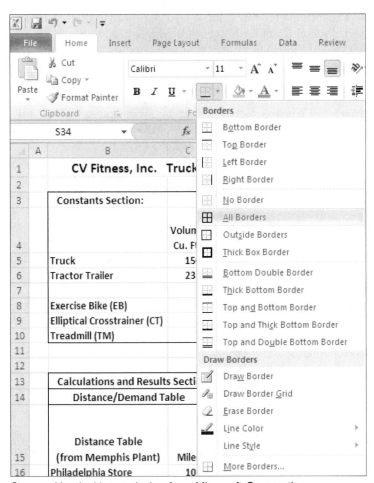

Source: Used with permission from Microsoft Corporation

FIGURE D-8   Borders menu

Next, you write the formulas for the volume and cost calculations. Figure D-9 shows a magnified view of the Vehicle Loading and Cost sections. A discussion of the formulas required for the cells follows the figure.

	G	H	I	J	K	L	M	N
13								
14			Vehicle Loading					Cost
15	Volume Required	Trucks	Volume for Trucks	Tractor-Trailers	Volume for Tractor-Trailers	Total Vehicle Capacity	% of Vehicle Capacity Utilized	Shipping Cost
16								
17								
18								
19								
20								
21								
22								Total Cost

Source: Used with permission from Microsoft Corporation

FIGURE D-9   Vehicle Loading and Cost sections

For illustration purposes, the cell numbers in the following list refer to values for the Philadelphia store.

- Volume Required—Cell G16 contains the total shipping volume of the three types of equipment shipped to the Philadelphia store. The formula for this cell is =D16*$C$8+E16*$C$9+F16*$C$10. Cells D16, E16, and F16 are the quantities of each item to be shipped, and cells $C$8, $C$9, and $C$10 are the shipping volumes for the exercise bike, cross-trainer, and treadmill, respectively. When taking values from the Constants section to calculate formulas, you almost always should use absolute cell references ($) because you will copy the formulas down the columns.
- Trucks—Cell H16 contains the number of trucks selected to ship the merchandise. Cell H16 is a changing cell, which means Solver will determine the best number of trucks to use and place the number in this cell. For now, you should "seed" the cell with a value of 1.
- Volume for Trucks—Cell I16 contains the number of trucks selected, multiplied by the capacity of a truck. The capacity value is taken from the Constants section. The formula for this cell is =H16*$C$5. Cell H16 is the number of trucks selected, and cell $C$5 is the volume capacity of the truck in cubic feet.
- Tractor-Trailers—Cell J16 contains the number of tractor-trailers selected to ship the merchandise. Cell J16 is a changing cell, which means Solver will determine the best number of tractor-trailers to use and place the number in this cell. For now, you should "seed" the cell with a value of 1.
- Volume for Tractor-Trailers—Cell K16 contains the number of tractor-trailers selected, multiplied by the capacity of a tractor-trailer. The capacity value is taken from the Constants section. The formula for this cell is =J16*$C$6. Cell J16 is the number of tractor-trailers selected, and cell $C$6 is the cubic feet capacity of the tractor-trailer.
- Total Vehicle Capacity—Cell L16 contains the sum of the Volume for Trucks and the Volume for Tractor-Trailers. The formula for this cell is =I16+K16. You need to know the Total Vehicle Capacity to make sure that you have enough capacity to ship the Volume Required. This value will be one of your constraints in Solver.
- % of Vehicle Capacity Utilized—Cell M16 contains the Volume Required divided by the Total Vehicle Capacity. The formula for this cell is =G16/L16; after entering the formula, format it as a percentage using the % button in the Number group. Although this information is not required to minimize shipping costs, it is useful for managers to know how much space was filled in the selected vehicles. Alternatively, you could run Solver to determine the highest space utilization on the vehicles rather than the lowest cost. Note that you cannot use more than 100% of the available space on the vehicles.

- Shipping Cost—Cell N16 contains the following calculation:

  Mileage to destination store × Number of trucks selected × Cost per mile for trucks + Mileage to destination store × Number of tractor-trailers selected × Cost per mile for tractor-trailers

  The formula for this cell is =H16*C16*$D$5+J16*C16*$D$6. Note that absolute cell references for the cost-per-mile values are taken from the Constants section.

If you entered the formulas correctly in row 16, your table should look like Figure D-10.

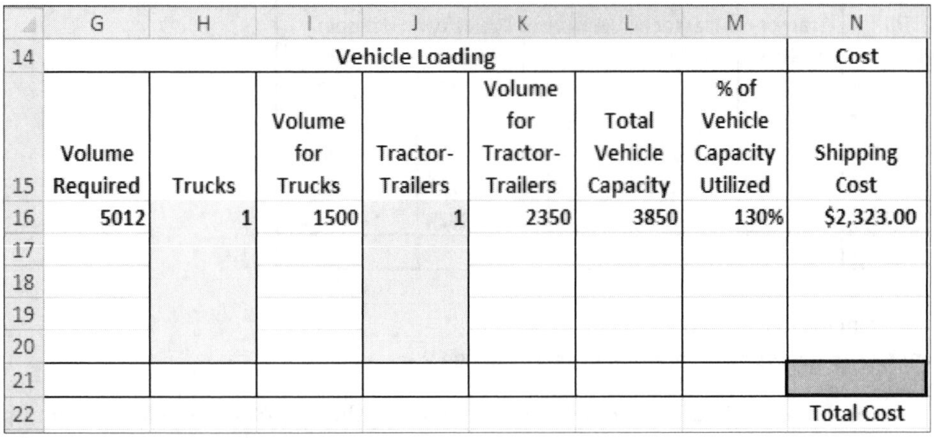

	G	H	I	J	K	L	M	N
14	Vehicle Loading							Cost
15	Volume Required	Trucks	Volume for Trucks	Tractor-Trailers	Volume for Tractor-Trailers	Total Vehicle Capacity	% of Vehicle Capacity Utilized	Shipping Cost
16	5012	1	1500	1	2350	3850	130%	$2,323.00
17								
18								
19								
20								
21								
22								Total Cost

Source: Used with permission from Microsoft Corporation

FIGURE D-10 Vehicle Loading and Cost sections with formulas entered in the first row

To complete the empty cells in rows 17 through 20, you can copy the formulas from cells G16 through N16 to the rest of the rows. Click and drag to select cells G16 through N16, then right-click and select Copy from the menu (see Figure D-11).

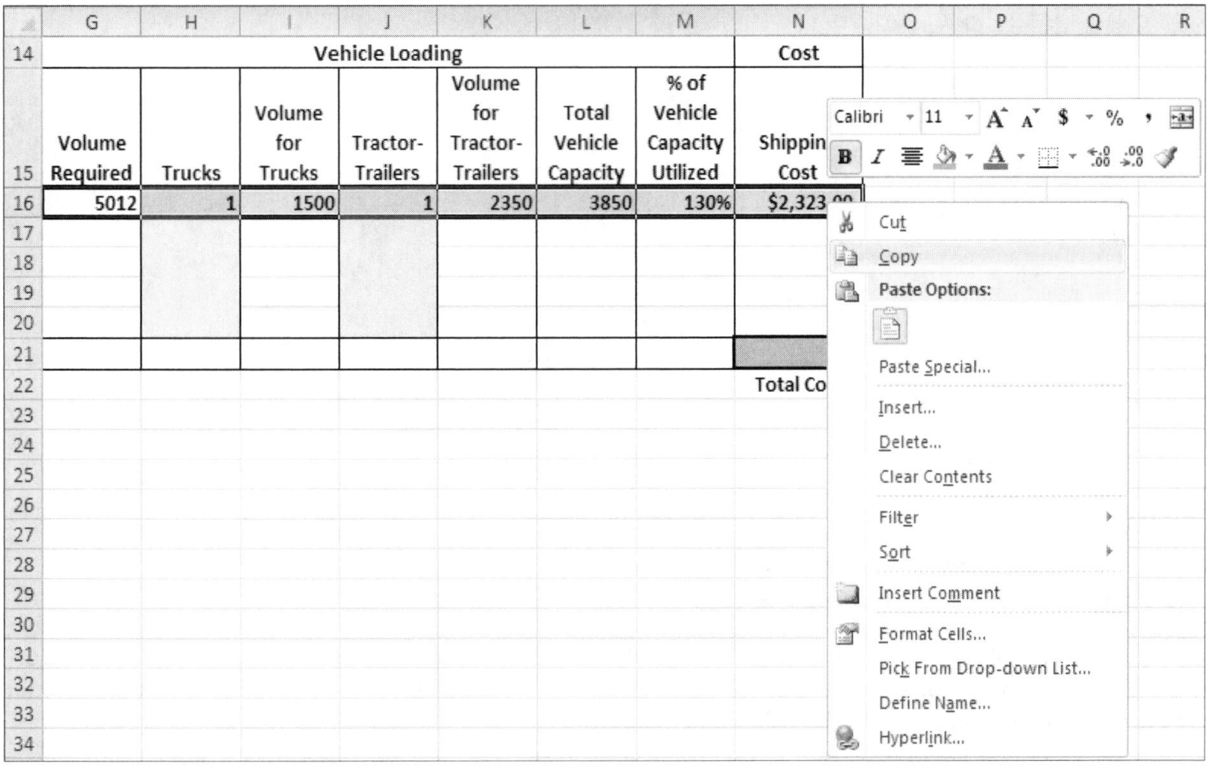

Source: Used with permission from Microsoft Corporation

FIGURE D-11 Copying formulas

Next, select cells G17 through N20, which are in the four rows beneath row 16. Either press Enter or click Paste in the Clipboard group. The formulas from row 16 should be copied to the rest of the destination cities (see Figure D-12).

	G	H	I	J	K	L	M	N
14	Vehicle Loading							Cost
15	Volume Required	Trucks	Volume for Trucks	Tractor-Trailers	Volume for Tractor-Trailers	Total Vehicle Capacity	% of Vehicle Capacity Utilized	Shipping Cost
16	5012	1	1500	1	2350	3850	130%	$2,323.00
17	3513	1	1500	1	2350	3850	91%	$874.00
18	2776	1	1500	1	2350	3850	72%	$2,300.00
19	6630	1	1500	1	2350	3850	172%	$1,242.00
20	7785	1	1500	1	2350	3850	202%	$4,163.00
21								
22								Total Cost

Source: Used with permission from Microsoft Corporation

FIGURE D-12   Formulas from row 16 successfully copied to rows 17 through 20

You have one row of formulas to complete: the Totals row. You will use the AutoSum function to sum up one column, and then copy the formula to the rest of the columns *except* cell M21. This cell is not actually a total, but an overall capacity utilization rate.

To enter the sum of cells G16 through G20 in cell G21, select cells G16 through G21, then click AutoSum in the Editing group on the Home tab of the Ribbon (see Figure D-13).

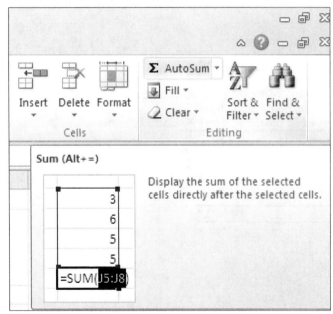

Source: Used with permission from Microsoft Corporation

FIGURE D-13   AutoSum button in the Editing group

Cell G21 should now contain the formula =SUM(G16:G20), and the displayed answer should be 25716. Now you can copy cell G21 to cells H21, I21, J21, K21, L21, and N21. When you have completed this section of the table, it should have the values shown in Figure D-14.

	G	H	I	J	K	L	M	N
14			Vehicle Loading					Cost
15	Volume Required	Trucks	Volume for Trucks	Tractor-Trailers	Volume for Tractor-Trailers	Total Vehicle Capacity	% of Vehicle Capacity Utilized	Shipping Cost
16	5012	1	1500	1	2350	3850	130%	$2,323.00
17	3513	1	1500	1	2350	3850	91%	$874.00
18	2776	1	1500	1	2350	3850	72%	$2,300.00
19	6630	1	1500	1	2350	3850	172%	$1,242.00
20	7785	1	1500	1	2350	3850	202%	$4,163.00
21	25716	5	7500	5	11750	19250		$10,902.00
22								Total Cost

Source: Used with permission from Microsoft Corporation

FIGURE D-14 Totals cells completed

The last formula to enter is for cell M21. This is not a total, but an overall percentage of Vehicle Capacity Utilized for all the vehicles used. This calculation uses the same formula as the cell above it, so you can simply copy cell M20 to cell M21. The formula for this cell is =G21/L21, which is Volume Required divided by Total Vehicle Capacity, expressed as a percentage. Your completed spreadsheet should look like Figure D-15.

	A	B	C	D	E	F	G	H	I	J	K	L	M	N
14		Distance/Demand Table			Store Demand						Vehicle Loading			Cost
15		Distance Table (from Memphis Plant)	Miles	EB	CT	TM	Volume Required	Trucks	Volume for Trucks	Tractor-Trailers	Volume for Tractor-Trailers	Total Vehicle Capacity	% of Vehicle Capacity Utilized	Shipping Cost
16		Philadelphia Store	1010	140	96	86	5012	1	1500	1	2350	3850	130%	$2,323.00
17		Atlanta Store	380	76	81	63	3513	1	1500	1	2350	3850	91%	$874.00
18		Miami Store	1000	56	64	52	2776	1	1500	1	2350	3850	72%	$2,300.00
19		Chicago Store	540	115	130	150	6630	1	1500	1	2350	3850	172%	$1,242.00
20		Los Angeles Store	1810	150	135	180	7785	1	1500	1	2350	3850	202%	$4,163.00
21						Totals:	25716	5	7500	5	11750	19250	134%	$10,902.00
22		Fill Legend:			Changing Cells									Total Cost
23					Optimization Cell									

Source: Used with permission from Microsoft Corporation

FIGURE D-15 Completed Calculations and Results section

Working the Model Manually

Now that you have a working model, you could manipulate the number of trucks and tractor-trailers manually to obtain a solution to the shipping problem. You would need to observe the following rules (or constraints):

1. Assign enough Total Vehicle Capacity to meet the Volume Required for each destination. (In other words, you cannot exceed 100% of Vehicle Capacity Utilized.)
2. The total number of trucks and tractor-trailers you assign cannot exceed the number available in your fleet.

Try to assign your trucks and tractor-trailers to meet your shipping requirements, and note the total shipping costs—you may get lucky and come up with an optimal solution. The tractor-trailers are more cost efficient than the trucks, but the problem is complicated by the fact that you want to achieve the best capacity utilization as well. In some instances, the trucks may be a better fit. Figure D-16 shows a sample solution determined from working the problem manually.

Source: Used with permission from Microsoft Corporation

FIGURE D-16   Manual attempt to solve the vehicle loading problem optimally

This probably looks like a good solution—after all, you have not violated any of your constraints, and you have a 94% average vehicle capacity utilization. But is it the most cost-effective solution for your company? This is where Solver comes in.

Setting Up Solver Using the Solver Parameters Window

To access the Solver pane, click the Data tab on the Ribbon, then click Solver in the Analysis group on the far right side of the Ribbon. The Solver Parameters window appears (see Figure D-17).

NOTE

Solver in Excel 2010 has changed significantly from earlier versions of Excel. It allows three different calculation methods, and it allows you to specify an amount of time and number of iterations to perform before Excel ends the calculation. Refer to Microsoft Help for more information.

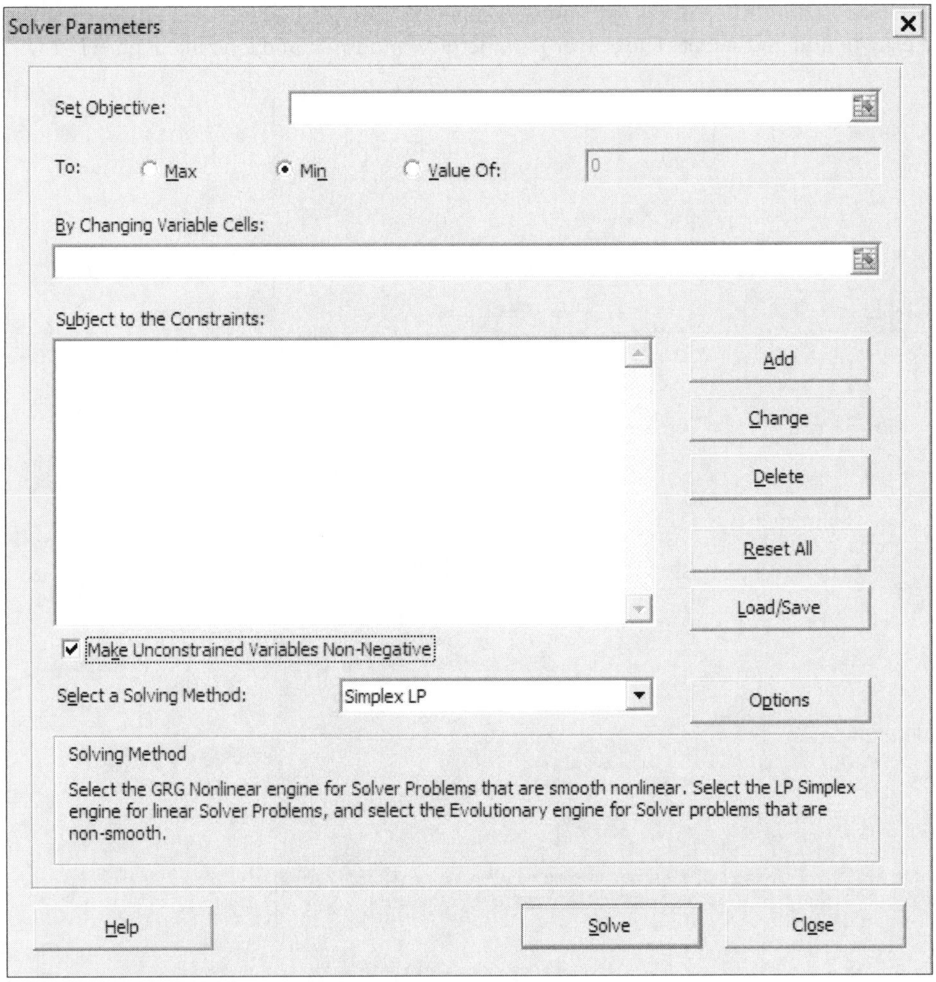

Source: Used with permission from Microsoft Corporation
FIGURE D-17   Solver Parameters window

The Solver Parameters window in Excel 2010 looks intimidating at first. However, to solve linear optimization problems, you have to satisfy only three sets of conditions by filling in the following fields:

- Set Objective—Specify the optimization cell.
- By Changing Variable Cells—Specify the changing cells in your worksheet.
- Subject to the Constraints—Define all of the conditions and limitations that must be met when seeking the optimal solution.

The following sections explain these fields in detail. You may also need to click the Options button and select one or more options for solving the problem. Most of the cases in this book are linear problems, so you can set the solving method to Simplex LP, as shown in Figure D-17. If this method does not work in later cases, you can select the GRG Nonlinear or Evolutionary method to try to solve the problem. Note that the GRG Nonlinear and Evolutionary solving methods are available only in Excel 2010.

Optimization Cell and Changing Cells

To use Solver successfully, you must first specify the cell you want to optimize—in this case, the total shipping cost, or cell N21. To fill the Set Objective field, click the button at the right edge of the field, and then click cell N21 in the spreadsheet. You could also type the cell address in the window, but selecting the cell in the spreadsheet reduces your chance of entering the wrong cell address. Next, specify whether you want Solver to seek the maximum or minimum value for cell N21. Because you want to minimize the total shipping cost, click the radio button next to Min.

Next, tell Solver which cell values it will change to determine the optimal solution. Use the By Changing Variable Cells field to specify the range of cells that you want Solver to manipulate. Again, click the button at the right edge of the field, select the cells that contain the numbers of trucks (H16 to H20), and then hold down the

Ctrl key and select the cells that contain the numbers of tractor-trailers (J16 to J20). If you used a fill color for the changing cells, they will be easy to find and select. The Solver Parameters window should look like Figure D-18.

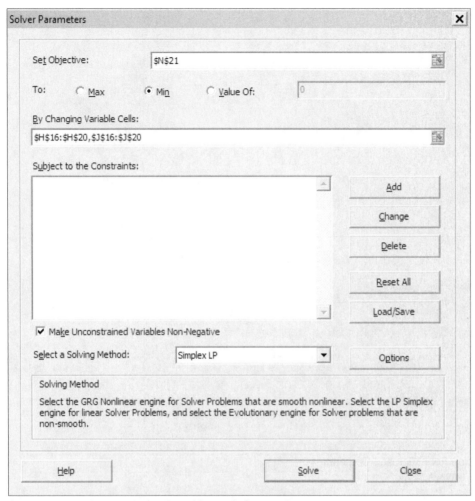

Source: Used with permission from Microsoft Corporation

FIGURE D-18   Solver Parameters window with the objective cell and changing cells entered

Note that Solver has added absolute cell references (the $ signs before the column and row designators) for the cells you have specified. Solver will also add these references to the constraints you define. Solver adds the references to preserve the links to the cells in case you revise the worksheet in the future. In fact, you will make changes to the worksheet later in the tutorial.

Defining and Entering Constraints

For Solver to successfully determine the optimum solution for the shipping problem, you need to specify what constraints or rules it must observe to calculate the solution. Without constraints, Solver theoretically might calculate that the best solution is not to ship anything, resulting in a cost of zero. Furthermore, if you failed to define variables as positive numbers, Solver would select "negative trucks" to maximize "negative costs." Finally, the vehicles are indivisible units—you cannot assign a fraction of a vehicle for a fraction of the cost, so you must define your changing cells as integers to satisfy this constraint.

Aside from the preceding logical constraints, you have operational constraints as well. You cannot assign more vehicles than you have in your fleet, and the vehicles you assign must have at least as much total capacity as your shipping volume.

Before entering the constraints in the Solver Parameters window, it is a good idea to write them down in regular language. You must enter the following constraints for this model:

- All trucks and tractor-trailers in the changing cells must be integers greater than or equal to zero.
- The sums of trucks and tractor-trailers assigned (cells H21 and J21) must be less than or equal to the available trucks and tractor-trailers (cells F5 and F6, respectively).

- The Total Vehicle Capacity for the vehicles assigned to each store (cells L16 to L20) must be greater than or equal to the Volume Required to be shipped to each store (cells G16 to G20, respectively).

You are ready to enter the constraints as equations or inequalities in the Add Constraint window. To begin, click the Add button in the Solver Parameters window. In the window that appears (see Figure D-19), click the button at the right edge of the Cell Reference box, select cells H16 to H20, and then click the button again. Next, click the drop-down menu in the middle field and select > =. Then go to the Constraint field and type 0. Finally, click Add; otherwise, the constraint you defined will not be added to the list defined in the Solver Parameters window.

Source: Used with permission from Microsoft Corporation

FIGURE D-19   Add Constraint window

You can continue to add constraints in the Add Constraint window. For this example, enter the constraints shown in the completed Solver Parameters window in Figure D-20. When you finish, click Add to save the last constraint, then click Cancel in the Add Constraint window to return to the Solver Parameters window.

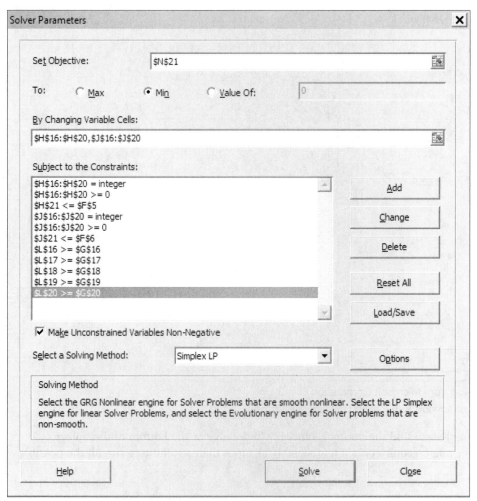

Source: Used with permission from Microsoft Corporation

FIGURE D-20   Completed Solver Parameters window

If you have difficulty reading the constraints listed in Figure D-20, use the following list instead:

- $H$16:$H$20 = integer
- $H$16:$H$20 >= 0
- $H$21 <= $F$5
- $J$16:$J$20 = integer
- $J$16:$J$20 >= 0
- $J$21 <= $F$6
- $L$16 >= $G$16
- $L$17 >= $G$17
- $L$18 >= $G$18
- $L$19 >= $G$19
- $L$20 >= $G$20

You should also click the Options button in the Solver Parameters window and check the Options window shown in Figure D-21. You can use this window to set the maximum amount of time and iterations you want Solver to run before stopping. Leave both options at 100 for now, but remember that Solver may need more time and iterations for more complex problems. To get the best solution, you should set the Integer Optimality (%) to zero. Click OK to close the window.

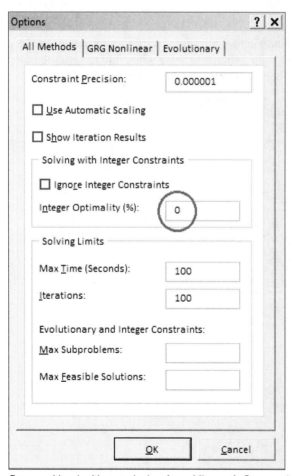

Source: Used with permission from Microsoft Corporation

FIGURE D-21   Solver Options window with Integer Optimality set to zero

You are ready to run Solver to find the optimal solution. Click Solve at the bottom of the Solver Parameters window. Solver might require only a few seconds or more than a minute to run all the possible iterations—the status bar at the bottom of the Excel window displays iterations and possible solutions continuously until Solver finds an optimal solution or runs out of time (see Figure D-22).

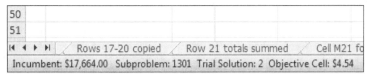

Source: Used with permission from Microsoft Corporation

FIGURE D-22   Excel status bar showing Solver running through possible solutions

A new window will appear eventually, indicating that Solver has found an optimal solution to the problem (see Figure D-23). The portion of the spreadsheet that displays the assigned vehicles and shipping cost should be visible below the Solver Results window. Solver has assigned nine of the 12 trucks and all six tractor-trailers, for a total shipping cost of $17,398. The earlier manual attempt to solve the problem (see Figure D-16) assigned all 12 trucks and four tractor-trailers, for a total shipping cost of $18,122. Using Solver in this situation saved your company $724.

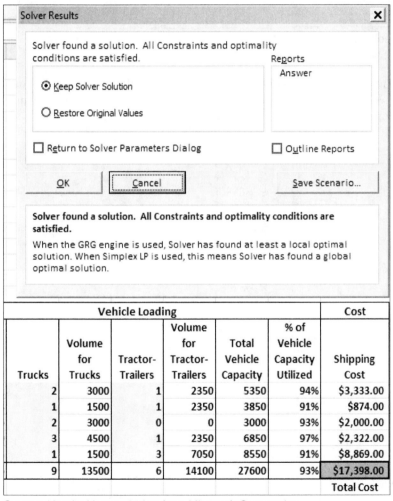

Source: Used with permission from Microsoft Corporation

FIGURE D-23   Solver Results window

If the Solver Results window does not report an optimal solution to the problem, it will report that the problem could not be solved given the changing cells and constraints you specified. For instance, if you had not had enough vehicles in your fleet to carry the required shipping volume to all the destinations, the

Solver Results window might have looked like Figure D-24. In the figure, your vehicle fleet was reduced to 10 trucks and five tractor-trailers, so Solver could not find a solution that satisfied the shipping volume constraints.

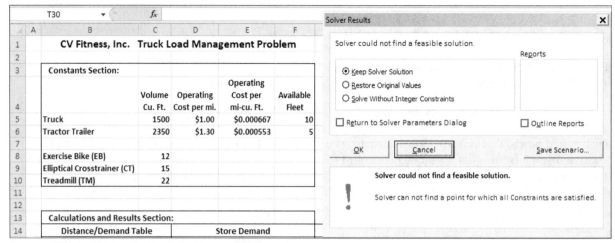

Source: Used with permission from Microsoft Corporation

FIGURE D-24   Solver could not find a feasible solution with a reduced vehicle fleet

Fortunately, Solver did find an optimal solution. To update the spreadsheet with the new optimal values for the changing cells and optimization cell, click OK in the Solver Results window. You can also create an Answer Report by clicking the Answer option in the Solver Results window (see Figure D-25) and then clicking OK.

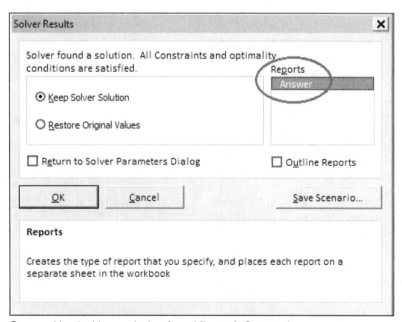

Source: Used with permission from Microsoft Corporation

FIGURE D-25   Creating an Answer Report

Excel will create a report in a separate sheet called Answer Report 1. The Answer Report is shown in Figures D-26 and D-27.

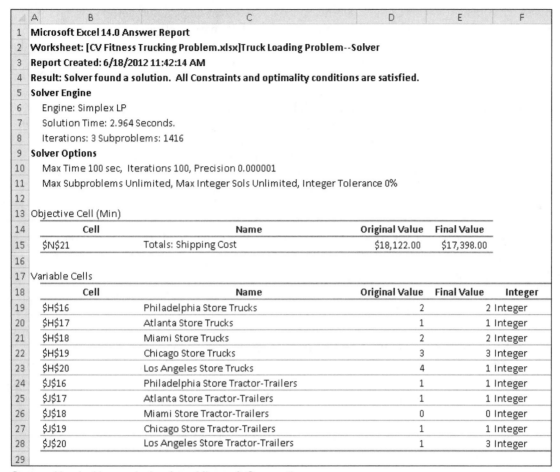

	A	B	C	D	E	F
1	**Microsoft Excel 14.0 Answer Report**					
2	**Worksheet: [CV Fitness Trucking Problem.xlsx]Truck Loading Problem--Solver**					
3	**Report Created: 6/18/2012 11:42:14 AM**					
4	**Result: Solver found a solution. All Constraints and optimality conditions are satisfied.**					
5	**Solver Engine**					
6	Engine: Simplex LP					
7	Solution Time: 2.964 Seconds.					
8	Iterations: 3 Subproblems: 1416					
9	**Solver Options**					
10	Max Time 100 sec, Iterations 100, Precision 0.000001					
11	Max Subproblems Unlimited, Max Integer Sols Unlimited, Integer Tolerance 0%					
12						
13	Objective Cell (Min)					
14		**Cell**	**Name**	**Original Value**	**Final Value**	
15	$N$21	Totals: Shipping Cost	$18,122.00	$17,398.00	
16						
17	Variable Cells					
18		**Cell**	**Name**	**Original Value**	**Final Value**	**Integer**
19	$H$16	Philadelphia Store Trucks	2	2	Integer
20	$H$17	Atlanta Store Trucks	1	1	Integer
21	$H$18	Miami Store Trucks	2	2	Integer
22	$H$19	Chicago Store Trucks	3	3	Integer
23	$H$20	Los Angeles Store Trucks	4	1	Integer
24	$J$16	Philadelphia Store Tractor-Trailers	1	1	Integer
25	$J$17	Atlanta Store Tractor-Trailers	1	1	Integer
26	$J$18	Miami Store Tractor-Trailers	0	0	Integer
27	$J$19	Chicago Store Tractor-Trailers	1	1	Integer
28	$J$20	Los Angeles Store Tractor-Trailers	1	3	Integer
29						

Source: Used with permission from Microsoft Corporation

FIGURE D-26   Top portion of the Answer Report

	A	B	C	D	E	F	G
30	Constraints						
31		**Cell**	**Name**	**Cell Value**	**Formula**	**Status**	**Slack**
32	$L$19	Chicago Store Total Vehicle Capacity	6850	$L$19>=$G$19	Not Binding	220
33	$L$18	Miami Store Total Vehicle Capacity	3000	$L$18>=$G$18	Not Binding	224
34	$L$20	Los Angeles Store Total Vehicle Capacity	8550	$L$20>=$G$20	Not Binding	765
35	$L$16	Philadelphia Store Total Vehicle Capacity	5350	$L$16>=$G$16	Not Binding	338
36	$L$17	Atlanta Store Total Vehicle Capacity	3850	$L$17>=$G$17	Not Binding	337
37	$H$21	Totals: Trucks	9	$H$21<=$F$5	Not Binding	3
38	$J$21	Totals: Tractor-	6	$J$21<=$F$6	Binding	0
39	$H$16	Philadelphia Store Trucks	2	$H$16>=0	Binding	0
40	$H$17	Atlanta Store Trucks	1	$H$17>=0	Binding	0
41	$H$18	Miami Store Trucks	2	$H$18>=0	Binding	0
42	$H$19	Chicago Store Trucks	3	$H$19>=0	Binding	0
43	$H$20	Los Angeles Store Trucks	1	$H$20>=0	Binding	0
44	$J$16	Philadelphia Store Tractor-Trailers	1	$J$16>=0	Binding	0
45	$J$17	Atlanta Store Tractor-Trailers	1	$J$17>=0	Binding	0
46	$J$18	Miami Store Tractor-Trailers	0	$J$18>=0	Binding	0
47	$J$19	Chicago Store Tractor-Trailers	1	$J$19>=0	Binding	0
48	$J$20	Los Angeles Store Tractor-Trailers	3	$J$20>=0	Binding	0
49	$H$16:$H$20=Integer						
50	$J$16:$J$20=Integer						
51							

Source: Used with permission from Microsoft Corporation

FIGURE D-27   Bottom portion of the Answer Report—note the new tab created by Solver

The Answer Report gives you a wealth of information about the solution. The top portion displays the original and final values of the Objective cell. The second part of the report displays the original and final values of the changing cells. The last part of the report lists the constraints. Binding constraints are those that reached their maximum or minimum value; nonbinding constraints did not.

Perhaps a savings of $724 does not seem significant—however, this problem does not have a specified time frame. The example probably represents one week of shipments for CV Fitness. The store demands will change from week to week, but you could use Solver each time to optimize the truck assignments. In a 50-week business year, the savings that Solver helps you find in shipping costs could be well over $30,000!

Go to the File tab to print the worksheets you created. Save the Excel file as **CV Fitness Trucking Problem.xlsx**, then select the Save As command in the File tab to create a new file called **CV Fitness Trucking Problem 2.xlsx**. You will use the new file in the next section.

EXTENDING THE EXAMPLE

Like all successful companies, CV Fitness looks for ways to grow its business and optimize its costs. Your management team is considering two changes:

- Opening two new stores and expanding the vehicle fleet if necessary
- Improving product design and packaging to reduce the shipping volume of the treadmill from 22 cubic feet to 17 cubic feet

You have been asked to modify your model to see the new requirements for each change separately. The two new stores would be in Denver and Phoenix, and they are 1,040 and 1,470 miles from the Memphis plant, respectively. If necessary, open the CV Fitness Trucking Problem 2.xlsx file, then right-click row 21 at the left worksheet border. Click Insert to enter a new row between rows 20 and 21. Repeat the steps to insert a second new row. Your spreadsheet should look like Figure D-28. Do not worry about the borders for now— you can fix them later.

	A	B	C	D	E	F	G	H	I	J	K	L	M	N
13		Calculations and Results Section:												
14		Distance/Demand Table			Store Demand					Vehicle Loading				Cost
15		Distance Table (from Memphis Plant)	Miles	EB	CT	TM	Volume Required	Trucks	Volume for Trucks	Tractor-Trailers	Volume for Tractor-Trailers	Total Vehicle Capacity	% of Vehicle Capacity Utilized	Shipping Cost
16		Philadelphia Store	1010	140	96	86	5012	2	3000	1	2350	5350	94%	$3,333.00
17		Atlanta Store	380	76	81	63	3513	1	1500	1	2350	3850	91%	$874.00
18		Miami Store	1000	56	64	52	2776	2	3000	0	0	3000	93%	$2,000.00
19		Chicago Store	540	115	130	150	6630	3	4500	1	2350	6850	97%	$2,322.00
20		Los Angeles Store	1810	150	135	180	7785	1	1500	3	7050	8550	91%	$8,869.00
21														
22														
23						Totals:	25716	9	13500	6	14100	27600	93%	$17,398.00
24		Fill Legend:			Changing Cells									Total Cost
25					Optimization Cell									

Source: Used with permission from Microsoft Corporation

FIGURE D-28   Distance/Demand table with two blank rows inserted for the new stores

Enter the two new stores in cells B21 and B22, enter their distances in cells C21 and C22, and enter the Store Demands in cells D21 through F22, as shown in Figure D-29. When you complete this part of the table, fix the borders to include the two new stores. Select the area in the table you want to fix, click the No Borders button to clear the old borders, highlight the area to which you want to add the border, and then click the Outside Borders button.

	A	B	C	D	E	F
13		Calculations and Results Section:				
14		Distance/Demand Table			Store Demand	
15		Distance Table (from Memphis Plant)	Miles	EB	CT	TM
16		Philadelphia Store	1010	140	96	86
17		Atlanta Store	380	76	81	63
18		Miami Store	1000	56	64	52
19		Chicago Store	540	115	130	150
20		Los Angeles Store	1810	150	135	180
21		Denver Store	1040	74	67	43
22		Phoenix Store	1470	41	28	37
23						Totals:
24		Fill Legend:			Changing Cells	
25					Optimization Cell	

Source: Used with permission from Microsoft Corporation

FIGURE D-29 Distance/Demand table with new store locations and demands entered

Next, copy the formulas from cells G20 to N20 to the two new rows in the Vehicle Loading and Cost sections of the table. Select cells G20 to N20, right-click, and click Copy on the menu. Then select cells G21 to N22 and click Paste in the Clipboard group. Your table should look like Figure D-30.

	G	H	I	J	K	L	M	N
13								
14				Vehicle Loading				Cost
15	Volume Required	Trucks	Volume for Trucks	Tractor-Trailers	Volume for Tractor-Trailers	Total Vehicle Capacity	% of Vehicle Capacity Utilized	Shipping Cost
16	5012	2	3000	1	2350	5350	94%	$3,333.00
17	3513	1	1500	1	2350	3850	91%	$874.00
18	2776	2	3000	0	0	3000	93%	$2,000.00
19	6630	3	4500	1	2350	6850	97%	$2,322.00
20	7785	1	1500	3	7050	8550	91%	$8,869.00
21	2839	1	1500	3	7050	8550	33%	$5,096.00
22	1726	1	1500	3	7050	8550	20%	$7,203.00
23	30281	9	13500	6	14100	27600	110%	$17,398.00
24								Total Cost

Source: Used with permission from Microsoft Corporation

FIGURE D-30 Formulas from row 20 copied into rows 21 and 22

Note that most cells in the Totals row have not changed—their formulas need to be updated to include the values in rows 21 and 22. To quickly check which cells you need to update, display the formulas in the Totals row. Hold down the Ctrl key and press the ~ key (on most keyboards, this key is next to the "1" key). The Vehicle Loading and Cost sections now display formulas in the cells (see Figure D-31).

	G	H	I	J	K	L	M	N
13								
14			Vehicle Loading					Cost
15	Volume Required	Trucks	Volume for Trucks	Tractor-Trailers	Volume for Tractor-Trailers	Total Vehicle Capacity	% of Vehicle Capacity Utilized	Shipping Cost
16	=D16*$C$8+E16*$C$9+F16*$C$10	2	=H16*$C$5	1	=J16*$C$6	=I16+K16	=G16/L16	=H16*C16*$D$5+J16*C16*$D$6
17	=D17*$C$8+E17*$C$9+F17*$C$10	1	=H17*$C$5	1	=J17*$C$6	=I17+K17	=G17/L17	=H17*C17*$D$5+J17*C17*$D$6
18	=D18*$C$8+E18*$C$9+F18*$C$10	2	=H18*$C$5	0	=J18*$C$6	=I18+K18	=G18/L18	=H18*C18*$D$5+J18*C18*$D$6
19	=D19*$C$8+E19*$C$9+F19*$C$10	3	=H19*$C$5	1	=J19*$C$6	=I19+K19	=G19/L19	=H19*C19*$D$5+J19*C19*$D$6
20	=D20*$C$8+E20*$C$9+F20*$C$10	1	=H20*$C$5	3	=J20*$C$6	=I20+K20	=G20/L20	=H20*C20*$D$5+J20*C20*$D$6
21	=D21*$C$8+E21*$C$9+F21*$C$10	1	=H21*$C$5	3	=J21*$C$6	=I21+K21	=G21/L21	=H21*C21*$D$5+J21*C21*$D$6
22	=D22*$C$8+E22*$C$9+F22*$C$10	1	=H22*$C$5	3	=J22*$C$6	=I22+K22	=G22/L22	=H22*C22*$D$5+J22*C22*$D$6
23	=SUM(G16:G22)	=SUM(H16:H20)	=SUM(I16:I20)	=SUM(J16:J20)	=SUM(K16:K20)	=SUM(L16:L20)	=G23/L23	=SUM(N16:N20)
24								Total Cost

Source: Used with permission from Microsoft Corporation

FIGURE D-31   Vehicle Loading and Cost sections with formulas displayed in the cells

You must update any Totals cells that do not include the contents of rows 21 and 22. For example, you need to update the Totals cells H23 through L23 and cell N23. Cell M23 is not really a total; it is a cumulative ratio formula, so you do not need to update the cell. Use the following formulas to revise the Totals cells:

- Cell H23: =SUM(H16:H22)
- Cell I23: =SUM(I16:I22)
- Cell J23: =SUM(J16:J22)
- Cell K23: =SUM(K16:K22)
- Cell L23: =SUM(L16:L22)
- Cell N23: =SUM(N16:N22)

The updated sections should look like Figure D-32.

	G	H	I	J	K	L	M	N
14			Vehicle Loading					Cost
15	Volume Required	Trucks	Volume for Trucks	Tractor-Trailers	Volume for Tractor-Trailers	Total Vehicle Capacity	% of Vehicle Capacity Utilized	Shipping Cost
16	5012	2	3000	1	2350	5350	94%	$3,333.00
17	3513	1	1500	1	2350	3850	91%	$874.00
18	2776	2	3000	0	0	3000	93%	$2,000.00
19	6630	3	4500	1	2350	6850	97%	$2,322.00
20	7785	1	1500	3	7050	8550	91%	$8,869.00
21	2839	1	1500	3	7050	8550	33%	$5,096.00
22	1726	1	1500	3	7050	8550	20%	$7,203.00
23	30281	11	16500	12	28200	44700	68%	$29,697.00
24								Total Cost

Source: Used with permission from Microsoft Corporation

FIGURE D-32   Vehicle Loading and Cost sections with the formulas updated

You are ready to use Solver to determine the optimal vehicle assignment. Click Solver in the Analysis group of the Data tab. You should notice immediately that you must revise the changing cells to include the two new stores; you must also change some of the constraints and add others. Solver has already updated the Objective cell from $N$21 to $N$23 and has updated the $H$23<=$F$5 and $J$23<=$F$6 constraints for vehicle fleet size. To update the changing cells, click the button to the right of the By Changing Variable Cells field and select the cells again, or edit the formula in the window by changing cell address $H$20 to $H$22 and cell address $J$20 to $J$22.

To change a constraint, select the one you want to change, and then click Change (see Figure D-33).

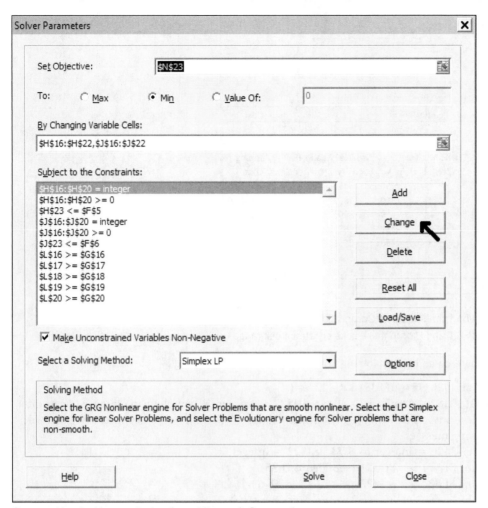

Source: Used with permission from Microsoft Corporation

FIGURE D-33   Selecting a constraint to change

When you click Change, the Change Constraint window appears. Click the Cell Reference button; the selected cells will appear on the spreadsheet with a moving marquee around them (see Figure D-34). Highlight the new group of cells; when the new range appears in the Cell Reference field, click OK. The Solver Parameters window appears with the constraint changed.

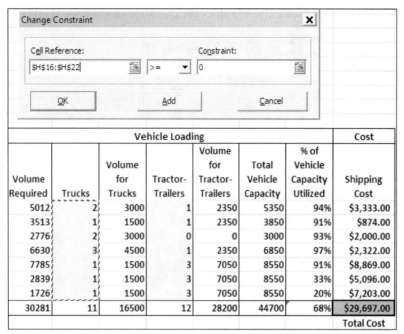

Source: Used with permission from Microsoft Corporation

FIGURE D-34   Adding cells H21 and H22 to the Trucks constraint cell range

You also need to update or add the following constraints:

- Update $J$16:$J$20 >=0 to $J$16:$J$22 >=0.
- Update $H$16:$H$20 = integer to $H$16:$H$22 = integer. When changing integer constraints, you must click "int" in the middle field of the Change Constraint window; otherwise, you will receive an error message.
- Update $J$16:$J$20 = integer to $J$16:$J$22 = integer.
- Add constraint $L$21 >= $G$21 (see Figure D-35).
- Add constraint $L$22 >= $G$22.

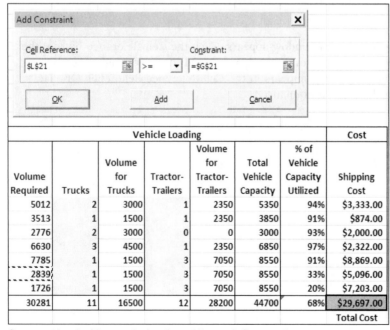

Source: Used with permission from Microsoft Corporation

FIGURE D-35   Adding a constraint using the Add Constraint window

You are ready to solve the shipping problem to include the new stores in Denver and Phoenix. Figure D-36 shows the updated Solver Parameters window.

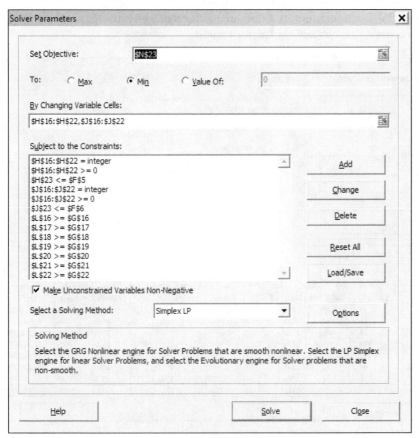

Source: Used with permission from Microsoft Corporation

FIGURE D-36   Solver parameters updated for shipping to seven stores

Before you run Solver again, you might want to attempt to assign the vehicles manually, because your fleet may not be large enough to handle two more stores. In this case, you will quickly realize that the vehicle fleet is at least one truck or tractor-trailer short of the minimum required to ship the needed volume. You can confirm this by running Solver (see Figure D-37).

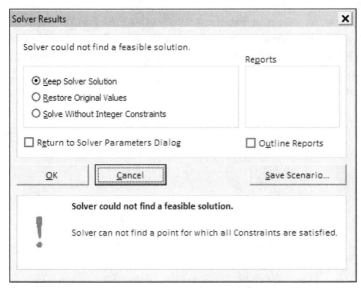

Source: Used with permission from Microsoft Corporation

FIGURE D-37   Vehicle fleet does not meet minimum requirements

The Solver Results window confirms that your truck fleet is too small, so change the value in cell F5 from 12 to 13 to add another truck to your fleet, and then run Solver again. As you add more stores and vehicles to make the problem more complex, Solver will take longer to run, especially on older computers. You may have to wait a minute or more for Solver to finish its iterations and find an answer (see Figure D-38). In this example, Solver recommends that you use 13 trucks and six tractor-trailers.

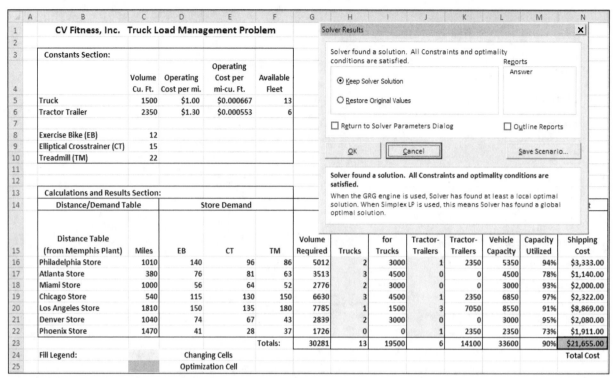

Source: Used with permission from Microsoft Corporation

FIGURE D-38   Solver's solution

Select Answer in the Reports list to add an Answer Report to the workbook, and then click OK. You can keep or delete the old Answer Report 1 tab from the earlier workbook. The new Answer Report is in a new worksheet named Answer Report 2.

You can meet the shipping requirements by adding one more truck, but is it really the most cost-effective solution? What if you add a tractor-trailer instead? Set the number of trucks back to 12, and add a tractor-trailer by entering 7 instead of 6 in cell F6. Run Solver again.

This time Solver finds a less expensive solution, as shown in Figures D-39 and D-40. At first it does not make sense—how can adding a more expensive vehicle (a tractor-trailer) reduce the overall expense? In fact, the additional tractor-trailer has replaced two trucks. With seven tractor-trailers, you only need 11 trucks instead of the original 13.

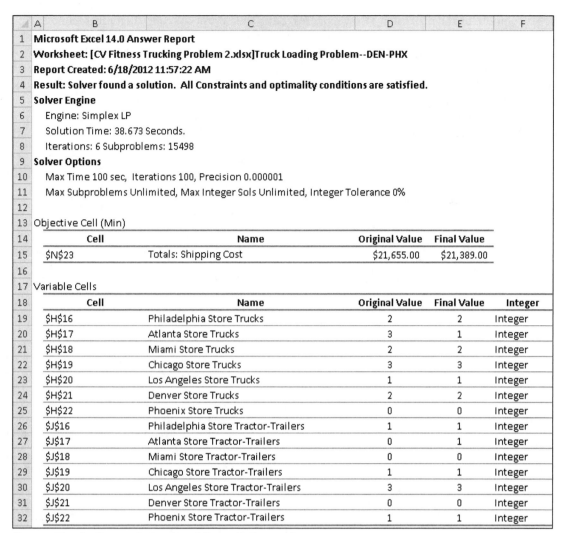

	A	B	C	D	E	F
1	**Microsoft Excel 14.0 Answer Report**					
2	**Worksheet: [CV Fitness Trucking Problem 2.xlsx]Truck Loading Problem--DEN-PHX**					
3	**Report Created: 6/18/2012 11:57:22 AM**					
4	**Result: Solver found a solution. All Constraints and optimality conditions are satisfied.**					
5	**Solver Engine**					
6	Engine: Simplex LP					
7	Solution Time: 38.673 Seconds.					
8	Iterations: 6 Subproblems: 15498					
9	**Solver Options**					
10	Max Time 100 sec, Iterations 100, Precision 0.000001					
11	Max Subproblems Unlimited, Max Integer Sols Unlimited, Integer Tolerance 0%					
12						
13	Objective Cell (Min)					
14	**Cell**		**Name**	**Original Value**	**Final Value**	
15	$N$23		Totals: Shipping Cost	$21,655.00	$21,389.00	
16						
17	Variable Cells					
18	**Cell**		**Name**	**Original Value**	**Final Value**	**Integer**
19	$H$16		Philadelphia Store Trucks	2	2	Integer
20	$H$17		Atlanta Store Trucks	3	1	Integer
21	$H$18		Miami Store Trucks	2	2	Integer
22	$H$19		Chicago Store Trucks	3	3	Integer
23	$H$20		Los Angeles Store Trucks	1	1	Integer
24	$H$21		Denver Store Trucks	2	2	Integer
25	$H$22		Phoenix Store Trucks	0	0	Integer
26	$J$16		Philadelphia Store Tractor-Trailers	1	1	Integer
27	$J$17		Atlanta Store Tractor-Trailers	0	1	Integer
28	$J$18		Miami Store Tractor-Trailers	0	0	Integer
29	$J$19		Chicago Store Tractor-Trailers	1	1	Integer
30	$J$20		Los Angeles Store Tractor-Trailers	3	3	Integer
31	$J$21		Denver Store Tractor-Trailers	0	0	Integer
32	$J$22		Phoenix Store Tractor-Trailers	1	1	Integer

Source: Used with permission from Microsoft Corporation

FIGURE D-39 Answer Report 3 displays a more cost-effective solution

	G	H	I	J	K	L	M	N
13								
14				Vehicle Loading				Cost
15	Volume Required	Trucks	Volume for Trucks	Tractor-Trailers	Volume for Tractor-Trailers	Total Vehicle Capacity	% of Vehicle Capacity Utilized	Shipping Cost
16	5012	2	3000	1	2350	5350	94%	$3,333.00
17	3513	1	1500	1	2350	3850	91%	$874.00
18	2776	2	3000	0	0	3000	93%	$2,000.00
19	6630	3	4500	1	2350	6850	97%	$2,322.00
20	7785	1	1500	3	7050	8550	91%	$8,869.00
21	2839	2	3000	0	0	3000	95%	$2,080.00
22	1726	0	0	1	2350	2350	73%	$1,911.00
23	30281	11	16500	7	16450	32950	92%	$21,389.00
24								Total Cost

Source: Used with permission from Microsoft Corporation

FIGURE D-40 Seven tractor-trailers and 11 trucks are the optimal mix

You have a solution for the expansion to seven stores. Save your workbook, and then create a new workbook using the Save As command. Name the new workbook **CV Fitness Trucking Problem 3.xlsx**.

Next, evaluate the potential cost savings if the company redesigns its treadmill product and packaging to reduce the shipping volume from 22 cubic feet to 17 cubic feet. Your engineers report that the redesign will cost approximately $10,000. If you can save at least $500 per shipment, the project will pay for itself in less than six months (20 weekly shipments).

Go to cell C10 on the worksheet, replace 22 with 17, and run Solver again. When Solver finds the solution, select Answer to create another Answer Report, and then click OK. See Figure D-41.

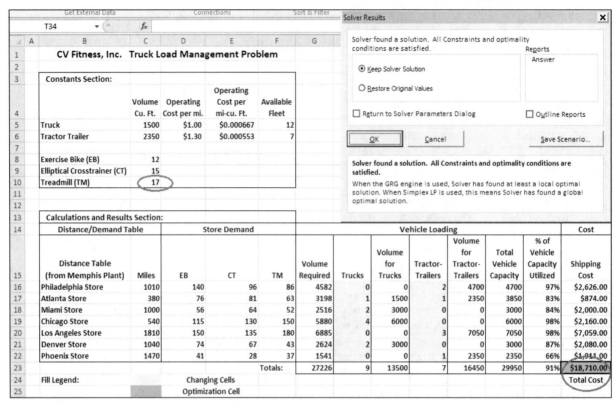

Source: Used with permission from Microsoft Corporation

FIGURE D-41   Solver solution with redesigned treadmill and packaging

Check the Answer Report to see the cost difference between shipping the old treadmills and the redesigned models (see Figure D-42). The cost savings for one shipment is $2,679, which is more than five times the minimum savings you needed. You should go ahead with the project.

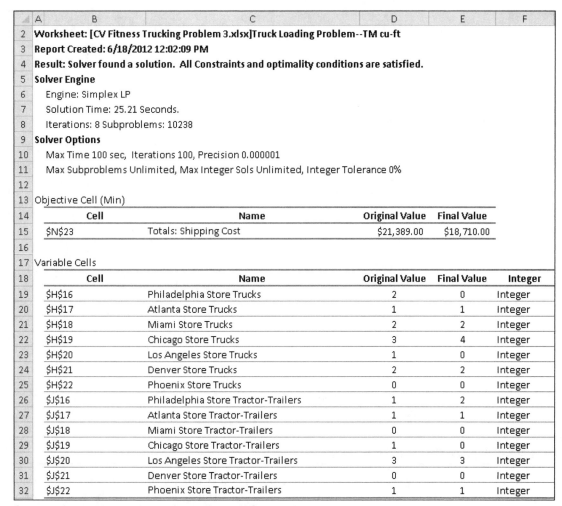

	A	B	C	D	E	F
2	**Worksheet: [CV Fitness Trucking Problem 3.xlsx]Truck Loading Problem--TM cu-ft**					
3	**Report Created: 6/18/2012 12:02:09 PM**					
4	**Result: Solver found a solution. All Constraints and optimality conditions are satisfied.**					
5	**Solver Engine**					
6	Engine: Simplex LP					
7	Solution Time: 25.21 Seconds.					
8	Iterations: 8 Subproblems: 10238					
9	**Solver Options**					
10	Max Time 100 sec, Iterations 100, Precision 0.000001					
11	Max Subproblems Unlimited, Max Integer Sols Unlimited, Integer Tolerance 0%					
12						
13	Objective Cell (Min)					
14		**Cell**	**Name**	**Original Value**	**Final Value**	
15		$N$23	Totals: Shipping Cost	$21,389.00	$18,710.00	
16						
17	Variable Cells					
18		**Cell**	**Name**	**Original Value**	**Final Value**	**Integer**
19		$H$16	Philadelphia Store Trucks	2	0	Integer
20		$H$17	Atlanta Store Trucks	1	1	Integer
21		$H$18	Miami Store Trucks	2	2	Integer
22		$H$19	Chicago Store Trucks	3	4	Integer
23		$H$20	Los Angeles Store Trucks	1	0	Integer
24		$H$21	Denver Store Trucks	2	2	Integer
25		$H$22	Phoenix Store Trucks	0	0	Integer
26		$J$16	Philadelphia Store Tractor-Trailers	1	2	Integer
27		$J$17	Atlanta Store Tractor-Trailers	1	1	Integer
28		$J$18	Miami Store Tractor-Trailers	0	0	Integer
29		$J$19	Chicago Store Tractor-Trailers	1	0	Integer
30		$J$20	Los Angeles Store Tractor-Trailers	3	3	Integer
31		$J$21	Denver Store Tractor-Trailers	0	0	Integer
32		$J$22	Phoenix Store Tractor-Trailers	1	1	Integer

Source: Used with permission from Microsoft Corporation

FIGURE D-42   Answer Report for the treadmill redesign

When you finish examining the Answer Report, save your file and then close it. To close the workbook, click the File tab and then click Close (see Figure D-43).

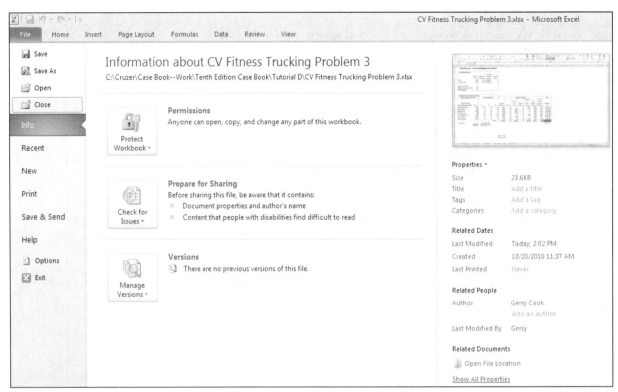

Source: Used with permission from Microsoft Corporation
FIGURE D-43   Closing the Excel workbook

USING SOLVER ON A NEW PROBLEM

A common problem in manufacturing businesses is deciding on a product mix for different items in the same product family. Sensuous Scents Inc. makes a premium collection of perfume, cologne, and body spray for sale in large department stores and boutiques. The primary ingredient is ambergris, a valuable digestive excretion from whales that is harvested without harming the animals. Ambergris costs more than $9,000 per pound and is very difficult to obtain in large quantities; Sensuous Scents can obtain only about 20 pounds of ambergris each year. The other ingredients—deionized water, ethanol, and various additives—are available in unlimited quantities for a reasonable cost.

You have been asked to create a spreadsheet model for Solver to determine the optimal product mix that maximizes Sensuous Scents' net income after taxes.

Setting up the Spreadsheet

The sections in this spreadsheet are different from those in the preceding trucking problem. You will create a Constants section, a Bill of Materials section for the three products, a Quantity Manufactured section that contains the changing cells, a Calculations section (to calculate ambergris usage, manufacturing costs, and sales revenue per product line), and an Income Statement section to determine the net income after taxes, which will be the optimization cell.

AT THE KEYBOARD

Start a new file called **Sensuous Scents Inc.xlsx** and set up the spreadsheet.

Spreadsheet Title and Constants Section

Your spreadsheet title and Constants section should look like Figure D-44. A discussion of the section entries follows the figure.

	Sensuous Scents Inc. Product Mix			Body Spray	Cologne	Perfume
1						
2						
3	Constants:			Body Spray	Cologne	Perfume
4	Sales Price per bottle			$11.95	$21.00	$53.00
5	Conversion Cost per Unit (Direct Labor plus Manufacturing Overhead)			$2.60	$6.50	$13.00
6	Minimum Sales Demand			60000	25000	12000
7	Income Tax Rate		0.32			
8	Sales, General and Administrative Expenses per Dollar Revenue		0.30			
9	Available Ambergris (lbs)		20			
10	Cost per lb, Deionized Water		$0.50			
11	Cost per lb, Ethanol		$1.00			
12	Cost per lb, other Additives		$182.00			
13	Cost per lb, Ambergris		$9,072.00			

Source: Used with permission from Microsoft Corporation

FIGURE D-44   Spreadsheet title and Constants section for Sensuous Scents Inc.

- Sales Price per bottle—These values are the sales prices for each of the three products.
- Conversion Cost per Unit—These values are the direct labor costs plus the manufacturing overhead costs budgeted per unit manufactured. A conversion cost is often used in industries that manufacture liquid products.
- Minimum Sales Demand—These values reflect the forecast minimum sales demand that you must supply to your customers. These values will be used later as constraints.
- Income Tax Rate—The rate is 32% of your pretax income. No taxes are paid on losses.
- Sales, General and Administrative Expenses per Dollar Revenue—This value is an estimate of the non-manufacturing costs that Sensuous Scents will incur per dollar of sales revenue. These expenses are subtracted from the Gross Profit value in the Income Statement section to obtain Net Income before taxes.
- Available Ambergris (lbs.)—This value is the amount of ambergris that Sensuous Scents obtained this year for production.
- Cost per lb., Deionized Water—This value is the current cost per pound of deionized water.
- Cost per lb., Ethanol—This value is the current cost per pound of ethanol.
- Cost per lb., other Additives—Scent products contain other additives and fixatives to enhance or preserve the fragrance. This value is the cost per pound of the other additives.
- Cost per lb., Ambergris—This value is the current market price per pound of naturally harvested ambergris. Again, no whales are harmed to obtain the ambergris.

The rest of the cells are filled with a gray background to indicate that you will not use their values or formulas. The section is arranged this way to maintain one column per product all the way down the spreadsheet, which will simplify writing the formulas later.

Bill of Materials Section

Your spreadsheet should contain a Bill of Materials section, as shown in Figure D-45. The section entries are explained after the figure. A bill of materials is a list of raw materials and ingredients required to make one unit of a product.

	A	B	C	D	E	F
14						
15		Bill of Materials:		Body Spray	Cologne	Perfume
16		Deionized Water (lb)		0.4	0.1	0.05
17		Ethanol (lb)		0.1	0.02	0.01
18		Other Additives (lb)		0.01	0.001	0.0001
19		Ambergris (lb)		0.0001	0.00018	0.00055

Source: Used with permission from Microsoft Corporation

FIGURE D-45   Bill of Materials section

- Deionized Water (lb.)—The amount of deionized water required to make one unit of each product
- Ethanol (lb.)—The amount of ethanol required to make one unit of each product
- Other Additives (lb.)—The amount of other additives required to make one unit of each product
- Ambergris (lb.)—The amount of ambergris required to make one unit of each product

Extremely small quantities of ambergris and other additives are required to make one bottle of each product. Also, each product requires a different amount of ambergris. Check the values to make sure you entered the correct number of decimal places.

Quantity Manufactured (Changing Cells) Section

This model contains a separate Changing Cells section called Quantity Manufactured, as shown in Figure D-46. This section contains the cells that you want Solver to manipulate to achieve the highest net income after taxes.

	A	B	C	D	E	F
20						
21		Quantity Manufactured (Changing Cells)		Body Spray	Cologne	Perfume
22		Units Produced		60000	25000	12000

Source: Used with permission from Microsoft Corporation

FIGURE D-46   Quantity Manufactured (changing cells) section

Cells D22, E22, and F22 are yellow to indicate that Solver will change them to reach an optimal solution. To begin, enter the minimum sales demand in these cells, which will remind you to specify the minimum demand constraints from the Constants section in the Solver Parameters window.

Calculations Section

Your model should contain the Calculations section shown in Figure D-47.

	A	B	C	D	E	F	G
23							
24		Calculations:		Body Spray	Cologne	Perfume	Totals
25		Lbs of Ambergris Used					
26		Manufacturing Cost per Unit (Materials Costs plus Conversion Cost)					
27		Total Manufacturing Costs per Product Line					
28		Sales Revenues per Product Line					

Source: Used with permission from Microsoft Corporation

FIGURE D-47   Calculations section

The section contains the following calculations:

- Lbs. of Ambergris Used—This value is the pounds of ambergris per unit from the Bill of Materials section, multiplied by Units Produced from the Quantity Manufactured section for each of the three products. The Totals cell (G25) is the sum of cells D25, E25, and F25. Use the value in this cell to specify the constraint that you have only 20 pounds of ambergris available to use for raw materials (Constants section, cell C9).
- Manufacturing Cost per Unit (Materials Costs plus Conversion Cost)—To get this value, write a formula that multiplies the unit cost for each of the four product ingredients by the amount per unit specified in the bill of materials, multiplied by Units Produced. The total materials costs for the four ingredients are added together, and then the Conversion Cost per Unit is added from the Constants section to obtain the Manufacturing Cost per Unit. Enter the following formula for the Body Spray Manufacturing Cost per Unit in cell D26:

    =$C$10*D16+$C$11*D17+$C$12*D18+$C$13*D19+D5

    Use absolute cell references for the cells that hold values for costs per pound ($C$10, $C$11, $C$12, and $C$13). By doing so, you can copy the body spray formula to the Manufacturing Cost per Unit cells for the cologne and perfume values (cells E26 and F26). The Totals cell (G26) is not used in this row—you can fill the cell in gray to indicate that it is not used.
- Total Manufacturing Costs per Product Line—This value is the Manufacturing Cost per Unit multiplied by Units Produced from the Quantity Manufactured section. The Totals cell (G27) is the sum of cells D27, E27, and F27. You will use the value in the Totals cell in the Income Statement section.

- Sales Revenues per Product Line—This value is the Sales Price per bottle from the Constants section multiplied by Units Produced from the Quantity Manufactured section. The Totals cell (G28) is the sum of cells D28, E28, and F28. You will use the value in this cell in the Income Statement section.

Income Statement Section

The last section you need to construct is the Income Statement, as shown in Figure D-48. An explanation of the needed formulas follows the figure.

	A	B	C	D	E
29					
30		**Income Statement:**			
31		Sales Revenues			
32		Less: Manufacturing Cost			
33		Gross Profit			
34		Less: Sales, General, and Administrative Expenses			
35		Net Income before taxes			
36		Less: Income Tax Expense			
37		Net Income after taxes			
38					
39		Fill Legend:		Changing Cells	
40				Optimization Cell	

Source: Used with permission from Microsoft Corporation

FIGURE D-48 Income Statement section with fill legend

- Sales Revenues—This value is the total sales revenues from the Calculations section (cell G28).
- Less: Manufacturing Cost—This value is the total manufacturing costs from the Calculations section (cell G27).
- Gross Profit—This value is the Sales Revenues minus the Manufacturing Cost.
- Less: Sales, General, and Administrative Expenses—This value is the Sales Revenues multiplied by the Sales, General, and Administrative Expenses per Dollar Revenue from the Constants section (cell C8).
- Net Income before taxes—This value is the Gross Profit minus the Sales, General, and Administrative Expenses.
- Less: Income Tax Expense—If the Net Income before taxes is greater than zero, this value is the Net Income before taxes multiplied by the Income Tax Rate in the Constants section. If Net Income before taxes is zero or less, the Income Tax Expense is zero.
- Net Income after taxes—This value is the Net Income before taxes minus the Income Tax Expense. You will use this value as your optimization cell because you want to maximize Net Income after taxes.

Setting up Solver

You need to satisfy the following conditions when running Solver:

- Your objective is to maximize Net Income after taxes (cell C37).
- Your changing cells are the Units Produced (cells D22, E22, and F22).
- Observe the following constraints:
  - You must produce at least the Minimum Sales Demand for each product (cells D6, E6, and F6).
  - Your total Lbs. of Ambergris Used (cell G25) cannot exceed the Available Ambergris (cell C9).
  - You cannot produce negative units of any product (enter constraints for the changing cells to be greater than or equal to zero).
  - You can produce only whole units of any product (enter constraints for the changing cells to be integers).

Run Solver and create an Answer Report when Solver finds the solution. When you complete the program, print your spreadsheet with the Solver solution, and print the Answer Report. Save your work and close Excel.

TROUBLESHOOTING SOLVER

Solver is a fairly complex software program. This section helps you address common problems you may encounter when attempting to run Solver.

Using Whole Numbers in Changing Cells

Before you run your first Solver model or rerun a previous model, always enter a positive whole number in each of the changing cells. If you have not already defined maximum and minimum constraints for the values in the changing cells, enter 1 in each cell before running Solver.

Getting Negative or Fractional Answers

If you receive negative or fractional answers when running Solver, you may have neglected to specify one or more of the changing cells as non-negative integers. Alternatively, if you are working on a cost minimization problem and you fail to specify the optimization cell as non-negative, you may receive a negative answer for the cost. Sometimes Solver will also warn you that you have one or more unbounded constraints (see Figure D-49).

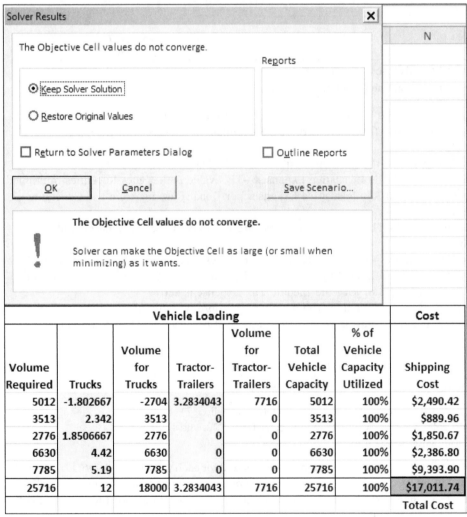

	Vehicle Loading						Cost
Volume Required	Trucks	Volume for Trucks	Tractor-Trailers	Volume for Tractor-Trailers	Total Vehicle Capacity	% of Vehicle Capacity Utilized	Shipping Cost
5012	-1.802667	-2704	3.2834043	7716	5012	100%	$2,490.42
3513	2.342	3513	0	0	3513	100%	$889.96
2776	1.8506667	2776	0	0	2776	100%	$1,850.67
6630	4.42	6630	0	0	6630	100%	$2,386.80
7785	5.19	7785	0	0	7785	100%	$9,393.90
25716	12	18000	3.2834043	7716	25716	100%	$17,011.74
							Total Cost

Source: Used with permission from Microsoft Corporation

FIGURE D-49   Solver has an "unbounded" objective function because you did not specify non-negative integer constraints

Creating Overconstrained Models

If Solver cannot find a solution because it cannot meet the constraints you defined, you will receive an error message. When this happens, Solver may even violate the integer constraints you defined in an attempt to find an answer, as shown in Figure D-50.

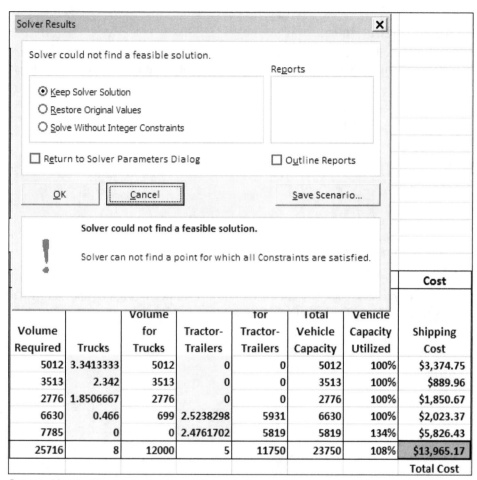

							Cost
Volume Required	Trucks	Volume for Trucks	Tractor-Trailers	for Tractor-Trailers	Total Vehicle Capacity	Vehicle Capacity Utilized	Shipping Cost
5012	3.3413333	5012	0	0	5012	100%	$3,374.75
3513	2.342	3513	0	0	3513	100%	$889.96
2776	1.8506667	2776	0	0	2776	100%	$1,850.67
6630	0.466	699	2.5238298	5931	6630	100%	$2,023.37
7785	0	0	2.4761702	5819	5819	134%	$5,826.43
25716	8	12000	5	11750	23750	108%	$13,965.17
							Total Cost

Source: Used with permission from Microsoft Corporation

FIGURE D-50   Solver could not find a feasible solution because not enough vehicles were available

Setting a Constraint to a Single Amount

Sometimes you may want to enter an exact amount into a constraint, as opposed to a number in a range. For example, if you wanted to assign exactly 11 trucks in the CV Fitness problem instead of a maximum of 12, you would select the equals (=) operator in the Change Constraint window, as shown in Figure D-51.

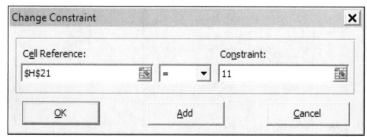

Source: Used with permission from Microsoft Corporation

FIGURE D-51   Constraining a value to a specific amount

Setting Changing Cells to Integers

Throughout the tutorial, you were directed to set the changing cells to integers in the Solver constraints. In many business situations, there is a logical reason for demanding integer solutions, but this approach does have disadvantages. Forcing integers can sometimes increase the amount of time Solver needs to find a feasible solution. In addition, Solver sometimes can find a solution using real numbers in the changing cells instead of integers. If Solver cannot find a feasible solution or reports that it has reached its calculation time limit, consider removing the integer constraints from the changing cells and rerunning Solver to see if it finds an optimal solution that makes sense.

Restarting Solver with New Constraints

Suppose you want to start over with a completely new set of constraints. In the Solver Parameters window, click Reset All. You will be asked to confirm that you want to reset all the Solver options and cell selections (see Figure D-52).

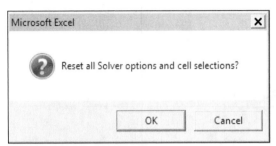

Source: Used with permission from Microsoft Corporation

FIGURE D-52   Reset options warning

If you want to clear all the Solver settings, click OK. An empty Solver Parameters window appears with all the former entries deleted, as shown in Figure D-53. You can then set up a new model.

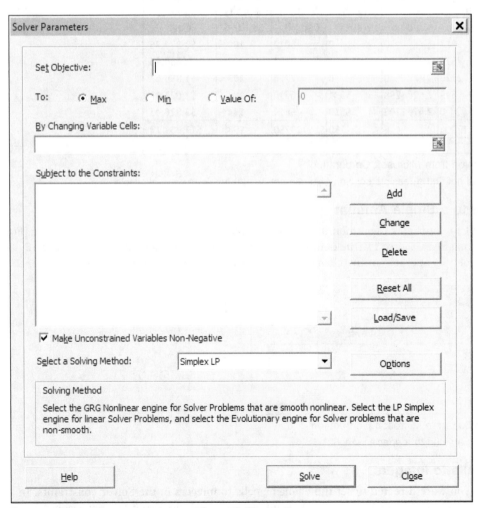

Source: Used with permission from Microsoft Corporation

FIGURE D-53   Solver Parameters window after selecting Reset All

Using the Solver Options Window

Solver has several internal settings that govern its search for an optimal answer. Click the Options button in the Solver Parameters window to see the default selections for these settings, as shown in Figure D-54.

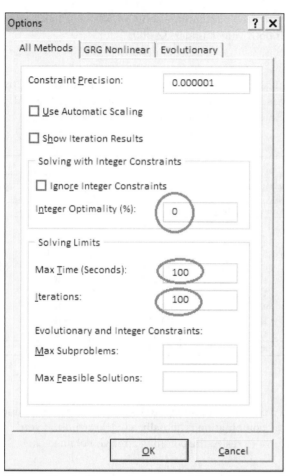

Source: Used with permission from Microsoft Corporation
FIGURE D-54   Solver Options window

You should not need to change the settings in the Options window except for the default value of 5% for Integer Optimality. When it is set at 5%, Solver will get within 5% of the optimal answer, but this setting might not give you the lowest cost or highest income. Change the setting to 0 and click OK.

In more complex problems that have a dozen or more constraints, Solver may not find the optimal solution within the default 100 seconds or 100 iterations. If so, a window will prompt you to continue or stop (see Figure D-55). If you have time, click Continue and let Solver keep working toward the best possible solution. If Solver works for several minutes and still does not find the optimal solution, you can stop by pressing the Ctrl and Break keys together. Click Stop in the resulting window.

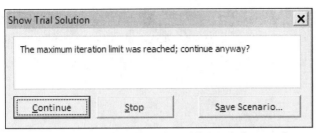

Source: Used with permission from Microsoft Corporation

FIGURE D-55   Prompt that appears when Solver reaches its maximum iteration limit

If you think that Solver needs more time and iterations to reach an optimal solution, you can increase the Max Time and Iterations, but you should probably keep both values under 32,000.

Printing Cell Formulas in Excel

Earlier in the tutorial, you learned how to display cell formulas in your spreadsheet cells. Hold down the Ctrl key and then press the ~ key (on most keyboards, this key is next to the "1" key). You can change the cell widths to see the entire formula by clicking and dragging the column by the dividing lines between the column letters. See Figure D-56.

	R27 ▾	f_x Width: 14.14 (208 pixels)				
	G	H	I	J	K	L
13						
14			Vehicle Loading			
15	Volume Required	Trucks	Volume for Trucks	Tractor-Trailers	Volume for Tractor-Trailers	Total Vehicle Capacity
16	=D16*$C$8+E16*$C$9+F16*$C$10	2	=H16*$C$5	1	=J16*$C$6	=I16+K16
17	=D17*$C$8+E17*$C$9+F17*$C$10	1	=H17*$C$5	1	=J17*$C$6	=I17+K17
18	=D18*$C$8+E18*$C$9+F18*$C$10	2	=H18*$C$5	0	=J18*$C$6	=I18+K18
19	=D19*$C$8+E19*$C$9+F19*$C$10	3	=H19*$C$5	1	=J19*$C$6	=I19+K19
20	=D20*$C$8+E20*$C$9+F20*$C$10	4	=H20*$C$5	1	=J20*$C$6	=I20+K20
21	=SUM(G16:G20)	=SUM(H16:H20)	=SUM(I16:I20)	=SUM(J16:J20)	=SUM(K16:K20)	=SUM(L16:L20)

Source: Used with permission from Microsoft Corporation

FIGURE D-56   Spreadsheet with formulas displayed in the cells

To print the formulas, click the File tab and select Print. To restore the screen to its normal appearance and display values instead of formulas, press Ctrl+~ again; the key combination is actually a toggle switch. If you changed the column widths in the formula view, you might have to resize the columns after you change back.

"Fatal" Errors in Solver

When you run Solver, you might sometimes receive a message like the one shown in Figure D-57.

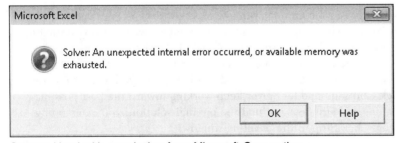

Source: Used with permission from Microsoft Corporation

FIGURE D-57   Fatal error in Solver

Solver usually attempts to find a solution or reports why it cannot. When Solver reports a fatal error, the root cause is difficult to troubleshoot. Possible causes include merged cells on the spreadsheet or printing multiple Answer Reports after running Solver multiple times. A common solution to this error has been to remove the Solver add-in, close Excel, reopen it, and then reinstall Solver. If you encounter a fatal error when using this book, check with your instructor.

Sometimes Solver will generate strange results. Even when your cell formulas and constraints match the ones your instructor has created, Solver's answers might not match the "book" answers. You might have entered your constraints into Solver in a different order, you may have changed some of the options in Solver, or you may have specified real numbers instead of integers for the constraints (or vice versa). Also, the solving method you selected and the amount of time you gave Solver to work can affect the final answer. If your solution is close to the one posted by your instructor, but not exactly the same, show the instructor your setup in the Solver Parameters window. Solver is a powerful tool, but it is not infallible—ask your instructor for guidance if necessary.

GLOBAL ELECTRONICS ASSIGNMENT PROBLEM

Decision Support Using Microsoft Excel Solver

PREVIEW

Global Electronics is a multinational manufacturer and distributor of portable electronic devices, such as cell phones, tablet computers, and game consoles. Global recently introduced its newest product, the Hawk 5G smartphone. The product rollout has been a huge international success, largely because the Hawk has a unique feature—it can automatically translate text messages from other Hawk phones into any language the user wants. The Hawk has been rolled out on five continents from distribution centers in Atlanta, Nigeria, Brazil, Japan, and Germany. The smartphone is being manufactured at four different facilities in Phoenix, China, Mexico, and Belgium.

Tommy Wong, the president and CEO of Global Electronics, is pleased with the success of the Hawk's introduction, but is concerned that Global might not be optimizing the assignment of production from the manufacturing sites to the distribution centers. Currently, his planners are assigning production manually based on the distances between manufacturing and distribution. He thinks that transportation costs from the parts suppliers and the manufacturing costs at each site are not being taken into account using the current method.

You have been hired as an information systems analyst to develop a production assignment DSS model for Global Electronics. Your completed model will be used to assign Hawk smartphone production from the four manufacturing sites to the five distribution centers at a minimal cost. You will also be asked to modify your model to examine the effect of a manufacturing site improvement on total operating cost.

PREPARATION

- Review the spreadsheet concepts discussed in class and in your textbook.
- Your instructor may assign Excel exercises to help prepare you for this case.
- Tutorial D explains how to set up and use Solver for maximization and minimization problems. The CV Fitness exercise in the tutorial should be particularly helpful.
- Review the file-saving instructions in Tutorial C—saving an extra copy of your work on a USB thumb drive is always a good idea.
- Review Tutorial F to brush up on your presentation skills.

BACKGROUND

You will use your Excel skills to build a decision model and determine how many smartphones should be produced at each manufacturing site and sent to each distribution center. The model requires the following data, which the management team has compiled:

- Direct materials cost for the Hawk smartphone, which has the following components:
    - Circuit board
    - Case
    - Battery
    - Touch screen

- Air freight shipping costs per mile per 10,000-unit container for the components and finished Hawk smartphones
- Specific manufacturing data for each site:
  - Direct labor cost per unit
  - Overhead cost per unit
  - Plant capacity in units per month
- Monthly sales demand at each of the five distribution centers
- Air freight tables for the distances from the four component supplier sites to the four manufacturing sites and from the four manufacturing sites to the five distribution centers

Your DSS model will assign the monthly production from each manufacturing site to the five distribution centers. The model will also calculate the total manufacturing and "freight out" costs to each of the distribution centers. You will first use the model to manually assign production to each distribution center. Next, you will run Solver to minimize the total operating cost. Then, you will modify the model to examine a proposed improvement to the Brussels manufacturing site and determine the resulting cost savings to the total operating cost.

ASSIGNMENT 1: CREATING SPREADSHEET MODELS FOR DECISION SUPPORT

In this assignment, you create spreadsheets that model the business decision Global Electronics is seeking. In Assignment 1A, you create a spreadsheet to assign the production from each manufacturing site to each distribution center and attempt to assign the production manually to minimize the total operating cost. In Assignment 1B, you copy the spreadsheet to a new worksheet and then set up and run Solver to minimize the total operating cost. In Assignment 1C, you copy the Solver solution to a new worksheet and modify it for a proposed upgrade to the Brussels manufacturing site. You then rerun Solver to determine if the proposed upgrade can further improve the total operating cost. In Assignment 1D, you modify the Solver solution from Assignment 1C to examine a proposal to manufacture the smartphone case onsite at each facility, instead of buying and shipping it to each manufacturing site from the supplier in Mexico City.

This section helps you set up each of the following spreadsheet components before entering the cell formulas:

- Constants
- Manufacturing Cost by site
- Production Assigned and total costs

The Production Assigned and total costs section is the heart of the decision model. You will set up sections of units assigned from each manufacturing site. Within these sections, you will set up columns to calculate subtotals by distribution center, total manufacturing cost, total freight-out cost, and total cost by distribution center. The rows in this section represent the distribution centers. The Production Assigned section will be the range of changing cells for Solver to manipulate. The Total Operating Cost cell will serve as the optimization cell for all the assignments. In Assignment 1C, you add a section to calculate cost savings for a proposed manufacturing site improvement. In Assignment 1D, you modify the section to calculate the cost savings for a proposal to produce the phone's case component onsite.

Assignment 1A: Creating the Spreadsheet—Base Case

A discussion of each spreadsheet section follows. This information helps you set up each section of the model and learn the logic of the formulas in the spreadsheet. If you choose to enter the data directly, follow the cell structure shown in the figures. *You can also download the spreadsheet skeleton if you prefer.* To access the base spreadsheet skeleton, select **Case 8** from your data files, and then select **Global Electronics Skeleton.xlsx**.

Constants Section

First, build the skeleton of your spreadsheet. Set up the spreadsheet title and Constants section, as shown in Figure 8-1. An explanation of the column items follows the figure.

	A	B	C	D	E	F
1		Global Electronics, Inc. Manufacturing Planning Problem				
2						
3		**Constants**				
4		Direct Materials Cost:				
5		Circuit Board	$8.00			
6		Case	$1.00			
7		Battery	$3.00			
8		Touchscreen	$7.00			
9						
10		Air Freight Shipping Cost:	(cost per mile for a 10,000 unit container)			
11		Freight-In (Parts)	$1.00			
12		Freight-Out (Finished Goods)	$1.20			
13						
14		Other Costs by Mfg Site:	Phoenix	Shanghai	Monterrey	Brussels
15		Direct Labor Cost per unit	$1.00	$0.40	$0.50	$1.20
16		Overhead Cost per unit	$3.00	$1.00	$2.00	$4.00
17		Site Capacity (units	500,000	200,000	400,000	250,000
18		per month)				
19		Monthly Demand by Distribution Center (finished units):				
20		Atlanta, U.S.	220,000			
21		Lagos, Nigeria	180,000			
22		Rio De Janeiro, Brazil	210,000			
23		Osaka, Japan	400,000			
24		Frankfurt, Germany	200,000			
25						
26		Air Freight Distance Tables	To/From: Manufacturing Site			
27		(in miles)	Phoenix	Shanghai	Monterrey	Brussels
28		From: Supplier Site				
29		Beijing (Circuit Board)	6,515	665	7,340	4,960
30		Mexico City (Case)	1,255	8,035	440	5,755
31		Singapore (Battery)	9,100	2,360	9,980	6,565
32		Nagoya (Touchscreen)	5,940	935	6,825	5,845
33		To: Distribution Center				
34		Atlanta, U.S.	1,600	7,655	965	4,400
35		Lagos, Nigeria	7,410	7,610	6,875	3,060
36		Rio De Janeiro, Brazil	5,970	11,340	5,080	5,850
37		Osaka, Japan	6,030	850	6,910	5,840
38		Frankfurt, Germany	5,640	5,495	5,660	200

Source: Used with permission from Microsoft Corporation

FIGURE 8-1   Spreadsheet title and Constants section

- Spreadsheet title—Enter the spreadsheet title in cell B1, and then merge and center the title across cells B1 through F1.
- Constants—Enter the heading for this section in cell B3 and center it.
- Direct Materials Cost table—Enter the heading and the four component names shown in cells B4 through B8, and enter the costs of the components in cells C5 through C8.
- Air Freight Shipping Cost table—Enter the heading text shown in cells B10 and C10, enter the text shown in cells B11 and B12, and then enter the costs shown in cells C11 and C12.
- Other Costs by Mfg Site table—Enter the headings shown in cells B14 through F14, the text shown in cells B15 through B17, and the values shown in cells C15 through F17. All costs are shown in U.S. dollars, and the site capacities are in units per month.
- Monthly Demand by Distribution Center (finished units)—Merge and center cells B19 through E19, and enter the title. Enter the locations of the five distribution centers in cells B20 through B24, and enter the corresponding demand quantities in cells C20 through C24.
- Air Freight Distance Tables (in miles)—Enter the title text in cells B26 and B27, merge and center cells C26 through F26, and then enter the To/From heading text. Enter the four manufacturing sites in cells C27 through F27, and then enter the Supplier Site heading in cell B28. In cells B29 through B32, enter the four supplier locations with the supplied components in parentheses. Then, in cells C29 through F32, enter the mileages shown in Figure 8-1. Enter the Distribution Center heading in cell B33 and the locations of the five distribution centers in cells B34 through B38. Enter the mileages shown in Figure 8-1 in cells C34 through F38. Check your mileage entries for both tables to make sure they are correct.

Manufacturing Cost by Site Section

You use the Manufacturing Cost by site section to calculate the manufacturing cost per unit for each of the four manufacturing sites (see Figure 8-2). The purpose of the section is twofold: It provides a quick display of the total manufacturing cost per unit for each of Global's four manufacturing sites, and it helps simplify the cost calculations in the Production Assigned and total costs section. An explanation of the column items follows the figure.

	A	B	C	D	E	F
39						
40		**Manufacturing Cost**	**Manufacturing Site**			
41		**by site (per unit):**	**Phoenix**	**Shanghai**	**Monterrey**	**Brussels**
42		Direct Materials Cost				
43		Air Freight In				
44		Direct Labor Cost				
45		Overhead Cost				
46		Total Mfg Cost per Unit:				

Source: Used with permission from Microsoft Corporation

FIGURE 8-2   Manufacturing Cost by site section

- Headings—Type the heading text shown in Figure 8-2 into cells B40 and B41. Merge and center cells C40 through F40 and enter the heading "Manufacturing Site." Enter the four manufacturing sites shown in cells C41 through F41.
- Cost descriptions and Total Mfg Cost per Unit—Enter the text shown in cells B42 through B46.
- Direct Materials Cost—Cells C42 through F42 hold the sums of values in cells C5 through C8; these values represent the components required to build one Hawk phone. Because all the sites buy components from the same suppliers, the Direct Materials Cost is the same for all the sites. Use *absolute* cell references for component costs for the formula in cell C42, and then copy the formula to cells D42, E42, and F42.
- Air Freight In—Cells C43 through F43 hold the values from calculating the air freight costs of the components from each site. This calculation is relatively simple because one smartphone uses only one component from each site. Just add the mileages from the four supplier sites, multiply the sum by the Freight-In cost per mile from cell C11, and then divide by 10,000 to account for the freight cost of a 10,000-unit container. The air freight distance tables are laid out so that if you use the correct *absolute* and *relative* cell references in the formula for Phoenix (cell C43), you can copy the formula for the other manufacturing sites (cells D43, E43, and F43).
- Direct Labor Cost—Cells C44 through F44 hold the direct labor costs for each manufacturing site. These values are copied from cells C15 through F15.
- Overhead Cost—Cells C45 through F45 hold the overhead costs for each manufacturing site. These values are copied from cells C16 through F16.
- Total Mfg Cost per Unit—Cells C46 through F46 hold the sums of the Direct Materials Cost, Air Freight In, Direct Labor Cost, and Overhead Cost for each site.

If you wrote your formulas correctly, the Manufacturing Cost by site section should look like Figure 8-3.

	A	B	C	D	E	F
39						
40		**Manufacturing Cost**	**Manufacturing Site**			
41		**by site (per unit):**	**Phoenix**	**Shanghai**	**Monterrey**	**Brussels**
42		Direct Materials Cost	$19.00	$19.00	$19.00	$19.00
43		Air Freight In	$2.28	$1.20	$2.46	$2.31
44		Direct Labor Cost	$1.00	$0.40	$0.50	$1.20
45		Overhead Cost	$3.00	$1.00	$2.00	$4.00
46		Total Mfg Cost per Unit:	$25.28	$21.60	$23.96	$26.51

Source: Used with permission from Microsoft Corporation

FIGURE 8-3   Completed Manufacturing Cost by site section

Production Assigned and Total Costs Section

This section is the heart of the DSS model. It contains a table with cells for assigning production for every combination of manufacturing site and distribution center. These cells will be the changing cells defined in Solver. The table also contains column and row subtotals that you need for defining demand and capacity constraints. You will use the right side of the section to calculate Total Manufacturing Cost by distribution center, Total Freight Out Cost to each distribution center, Total Cost by Destination, and Total Operating Cost. The Total Operating Cost is the optimization cell for this case. A description of the cell entries follows Figure 8-4.

	A	B	C	D	E	F	G	H	I	J
47										
48		**Production Assigned**		**From: Manufacturing Site**				Total	Total Freight	Total Cost
49		**To: Distribution Center:**	Phoenix	Shanghai	Monterrey	Brussels	Subtotal	Mfg Cost	Out Cost	by Destination
50		Atlanta, U.S.								
51		Lagos, Nigeria								
52		Rio De Janeiro, Brazil								
53		Osaka, Japan								
54		Frankfurt, Germany								
55		Subtotal								
56										Total
57		Legend:		Changing Cells						Operating Cost
58				Optimization Cell						

Source: Used with permission from Microsoft Corporation

FIGURE 8-4 Production Assigned and total costs section

- Title and column headings—In cells B48 through J49, you enter the title and column headings for the section. Enter "Production Assigned" in cell B48. Merge and center cells C48 through F48 and enter "From: Manufacturing Site." Enter "To: Distribution Center" in cell B49, and then enter the locations of the four manufacturing sites listed in cells C49 through F49. Complete the heading entries in cells G49 through J49.

- Distribution centers—In cells B50 through B54, enter the locations of the five distribution centers in the same order shown in Figure 8-4. You can copy this information from the Constants section.

- Production assigned—Cells C50 through F54 are the changing cells for Solver to manipulate. For now, enter "1" in each cell, and select a fill color for them to remind you that they are the changing cells.

- Column subtotals—Cells C55 through F55 hold the column subtotals. Enter "Subtotal" in cell B55. Cell C55 is the sum of cells C50 through C54. Write the formula for cell C55, and then copy it to cells D55 through F55. You will need these subtotals when defining capacity constraints for the four manufacturing sites.

- Row subtotals—Cells G50 through G54 hold the row subtotals. Cell G50 is the sum of cells C50 through F50. Write the formula for cell G50, and then copy it to cells G51 through G54. You will need these subtotals when defining demand constraints for the five distribution centers.

- Total Mfg Cost—Cells H50 through H54 hold the total manufacturing cost for all units sent to one distribution center from any or all of the manufacturing sites, depending on the number assigned. To calculate this cost for Atlanta in cell H50, multiply the Total Mfg Cost per Unit for each manufacturing site (cells C46 through F46) by the corresponding number of units assigned from each site to the distribution center, and then take the sum of the products. If you have already placed "1" in each of the cells C50 through F50, the calculated value in cell H50 should be the sum of cells C46 through F46. If you use *absolute* cell references for cells C46 through F46 in your formula, you can easily copy the formula from cell H50 to cells H51 through H54.

- Total Freight Out Cost—Cells I50 through I54 hold the cost of shipping the finished units from the manufacturing sites to each distribution center, depending on the number assigned. To calculate this cost for cell I50, multiply the number of units assigned from each manufacturing site (cells C50 through F50) by the applicable air freight distance between the manufacturing site and the distribution center (cells C34 to F34). Next, take the sum of those products, multiply the total by the Freight-Out cost per mile for finished goods (cell C12), and divide the result by 10,000 to account for the freight charge per 10,000-unit container. Use an *absolute* cell reference for cell C12 in your formula so you can copy the formula from cell I50 to cells I51 through I54.

- Total Cost by Destination—Cells J50 through J54 hold the sums of the Total Mfg Cost and Total Freight Out Cost for each distribution center.
- Total Operating Cost—Cell J55 holds the sum of cells J50 through J54. This value represents the total operating cost for Global Electronics for one month's production and distribution. This value is the optimization cell for Solver. You should fill this cell with a color of your choice and indicate the color in the legend.
- Legend—In cells B57, C57, C58, D57, and D58, enter the cell colors and text descriptions for your changing cells and optimization cell. This information will make it easier to locate the appropriate cells and cell ranges when using Solver later in this case.

If you entered your formulas correctly, the Production Assigned and total costs section will appear, as shown in Figure 8-5.

	A	B	C	D	E	F	G	H	I	J
47										
48		**Production Assigned**	**From: Manufacturing Site**					**Total**	**Total Freight**	**Total Cost**
49		To: Distribution Center:	Phoenix	Shanghai	Monterrey	Brussels	Subtotal	Mfg Cost	Out Cost	by Destination
50		Atlanta, U.S.	1	1	1	1	4	$97	$2	$99
51		Lagos, Nigeria	1	1	1	1	4	$97	$3	$100
52		Rio De Janeiro, Brazil	1	1	1	1	4	$97	$3	$101
53		Osaka, Japan	1	1	1	1	4	$97	$2	$100
54		Frankfurt, Germany	1	1	1	1	4	$97	$2	$99
55		Subtotal	5	5	5	5		$487	$13	$499
56										Total
57		Legend:		Changing Cells						Operating Cost
58				Optimization Cell						

Source: Used with permission from Microsoft Corporation

FIGURE 8-5 Production Assigned and total costs section completed correctly

Attempting a Manual Solution

First, attempt to assign your production manually. You have several good reasons for doing this. First, you can make sure that your model is working correctly before you set up Solver to run. Second, assigning the production manually demonstrates which constraints you must meet to solve the problem. For instance, the total production you assign from one manufacturing site cannot exceed its monthly capacity, which is a "less than or equal to" constraint. Similarly, you must meet the total monthly demand of each distribution center, which is a "greater than or equal to" constraint. However, do you have enough total capacity to meet all the demand? You can find out by taking the sum of your site capacities, taking the sum of your distribution center demands, and subtracting the total demand from the total capacity. Fortunately, Global Electronics has more total capacity than demand. However, is the current total operating cost the least expensive solution? Running the problem manually provides an initial total operating cost for you to compare to your Solver solution later. The Solver optimization tool should provide a less expensive solution than assigning production manually.

When attempting to assign production manually in the assignment section (cells C50 through F54), you can use the same logic as Global's planners. In other words, you can assign the production from the manufacturing sites that are closest to the distribution centers. To do this, look at the second air freight distance table in cells C34 through F38. For example, you would assign most of the Brussels production to the Frankfurt distribution center because they are only 200 miles apart. Similarly, the Shanghai manufacturing facility is closest to the distribution center in Osaka, Japan. However, notice that Shanghai can supply only about half of Osaka's demand. You must assign additional production to Osaka from one or more other manufacturing sites that are farther away.

You can also use the lowest manufacturing cost as a criterion by assigning production from the least expensive sites first, while also considering sites in order of distance from the distribution center. Will this approach give you a less expensive solution? You can see that the complexity of the problem makes it an excellent candidate for using Solver to find an optimal solution.

When you have reached a manual solution that satisfies the preceding constraints, save your workbook. Next, name the worksheet **Global Electronics Guess**, and then right-click the worksheet name tab (see Figure 8-6). Click Move or Copy, and then copy the worksheet, as shown in Figure 8-7. Rename the new copy **Global Electronics Solver**. You will use the new worksheet to complete the next part of the assignment.

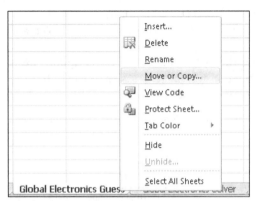

Source: Used with permission from Microsoft Corporation
FIGURE 8-6   Right-clicking the worksheet name tab

Source: Used with permission from Microsoft Corporation
FIGURE 8-7   Move or Copy menu

Assignment 1B: Setting Up and Running Solver

Before using the Solver Parameters window, you should jot down the parameters you must define and their cell addresses. Here is a suggested list:

- The cell you want to minimize (Total Operating Cost, cell J55)
- The cells you want Solver to manipulate to obtain the optimal solution (Production Assigned, cells C50 through F54)
- The constraints you must define:
  - All the production assignment cells are non-negative integers.
  - The total production assigned from each manufacturing site (cells C55 through F55) cannot exceed the production capacity of that site, as shown in the corresponding cells C17 through F17.
  - The total production assigned to each distribution center (cells G50 through G54) must at least equal the monthly demand for that distribution center, as shown in the corresponding cells C20 through C24.

Next, set up your problem. In the Analysis group on the Data tab, click Solver; the Solver Parameters window appears, as shown in Figure 8-8. Enter "Total Operating Cost" in the Set Objective text box. Click the Min button to minimize the cost, designate your Changing Cells (cells C50 through F54), and add the constraints from the preceding list. Use the default Simplex LP solving method. If you need help defining your constraints, refer to Tutorial D.

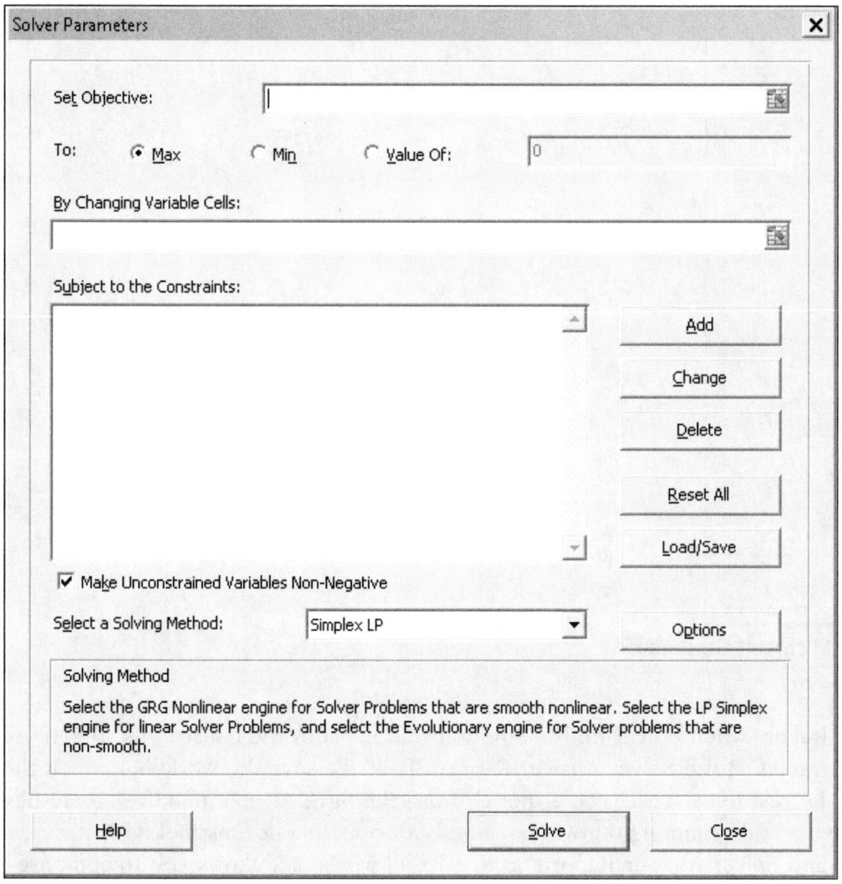

Source: Used with permission from Microsoft Corporation

FIGURE 8-8   The Solver Parameters window

Next, you should click the Options button and check the Options window that appears (see Figure 8-9). The default Integer Optimality is 5%; change it to 1% to get a better answer. Make sure that the Constraint Precision is set to the default value of .000001 and that Use Automatic Scaling is checked. When you finish setting the options, click OK to return to the Solver Parameters window.

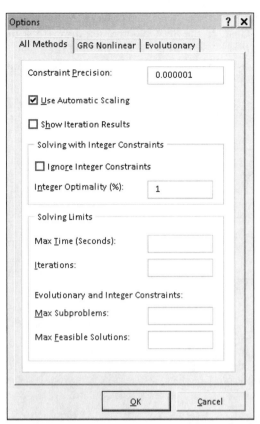

Source: Used with permission from Microsoft Corporation
FIGURE 8-9   The Solver Options window

Run Solver and click Answer Report when Solver finds a solution that satisfies the constraints. When you finish, print the entire workbook, including the Solver Answer Report Sheet. To save the workbook, click the File tab and then click Save. For the rest of the case, you either can use the Save As command to create new Excel workbooks or continue copying and renaming the worksheets. Both options offer distinct advantages, but having all of your worksheets and Solver Answer Reports in one Excel workbook allows you to compare different solutions easily as well as prepare summary reports.

Before continuing, examine the production assignments that Solver chose for minimizing the total operating cost. If you set up Solver correctly, you should see a reduction in total cost from your manual assignment. You should also see that Solver assigned less production from the costly Brussels manufacturing site. This could lead management to try to make the Brussels site more cost efficient through manufacturing improvements or innovations, or to increase the capacity of more cost-efficient manufacturing sites and close the Brussels facility.

Assignment 1C: Proposed Improvements to the Brussels Site

Using the DSS model you created in Assignment 1A, Global Electronics can also determine the cost benefit of improving manufacturing costs at the Brussels site. The engineers in Brussels have proposed some equipment and method changes in the assembly process for the Hawk smartphone. These changes will significantly reduce the direct labor and overhead costs and increase the monthly production capacity. The proposed changes can be made fairly quickly, but they will cost approximately $3 million to implement. You have been asked to modify your Solver model to determine whether the proposed changes at the Brussels site will reduce the total operating cost enough to recover the investment within one year. You will add a small analysis section to the worksheet to determine the investment savings, change the direct labor and overhead costs for Brussels in the Constants section, and then rerun Solver.

Copy your Global Electronics Solver worksheet and rename the copy **Global Electronics Solver 2**. Before running Solver again, do the following:

1. Enter new values for direct labor, overhead cost, and site capacity at the Brussels site (cells F15 through F17).
2. Build a section at the end of the worksheet to determine if the annual savings in total operating costs exceeds the capital cost of the Brussels site improvements.

The proposed improvements require you to change the following values in your worksheet for the Brussels manufacturing site, as shown in Figure 8-10.

1. The Direct Labor Cost per unit drops from $1.20 to $.90.
2. The Overhead Cost per unit drops from $4.00 to $3.00.
3. The Site Capacity in units per month increases from 250,000 to 400,000 units.

	A	B	C	D	E	F
1		Global Electronics, Inc. Manufacturing Planning Problem				
2						
3		Constants				
4		Direct Materials Cost:				
5		Circuit Board	$8.00			
6		Case	$1.00			
7		Battery	$3.00			
8		Touchscreen	$7.00			
9						
10		Air Freight Shipping Cost:	(cost per mile for a 10,000 unit container)			
11		Freight-In (Parts)	$1.00			
12		Freight-Out (Finished Goods)	$1.20			
13						
14		Other Costs by Mfg Site:	Phoenix	Shanghai	Monterrey	Brussels
15		Direct Labor Cost per unit	$1.00	$0.40	$0.50	$0.90
16		Overhead Cost per unit	$3.00	$1.00	$2.00	$3.00
17		Site Capacity (units	500,000	200,000	400,000	400,000
18		per month)				

Source: Used with permission from Microsoft Corporation

FIGURE 8-10   Changes to cells F15 through F17

Next, you must set up a financial analysis section at the end of the worksheet, as shown in Figure 8-11.

	A	B	C	D	E	F	G	H	I	J
47										
48		Production Assigned	From: Manufacturing Site					Total	Total Freight	Total Cost
49		To: Distribution Center:	Phoenix	Shanghai	Monterrey	Brussels	Subtotal	Mfg Cost	Out Cost	by Destination
50		Atlanta, U.S.	1	1	1	1	4	$97	$2	$99
51		Lagos, Nigeria	1	1	1	1	4	$97	$3	$100
52		Rio De Janeiro, Brazil	1	1	1	1	4	$97	$3	$101
53		Osaka, Japan	1	1	1	1	4	$97	$2	$100
54		Frankfurt, Germany	1	1	1	1	4	$97	$2	$99
55		Subtotal	5	5	5	5		$487	$13	$499
56										Total
57		Legend:		Changing Cells						Operating Cost
58				Optimization Cell						
59										
60						Proposed Brussels Site Upgrade				
61						Capital Cost of Upgrade:				($3,000,000)
62						Total Operating Cost (monthly) before Upgrade:				
63						Total Operating Cost (monthly) after Upgrade:				
64						Annual Cost Savings:				
65						First Year Net Cash Flow:				

Source: Used with permission from Microsoft Corporation

FIGURE 8-11   Financial analysis section—proposed Brussels site upgrade

Set up the financial analysis section as shown. You will find it easiest to merge columns F, G, H, and I in each of rows 60 through 65 before entering the text labels. Enter $3,000,000 as a negative number in cell J61 to represent a cash outlay by Global. The other cells in the table are as follows:

*   Total Operating Cost (monthly) before Upgrade—Cell J62 holds the total operating cost from your first Solver worksheet. You should link the contents of the cell from cell J55 of the first Solver worksheet. Ask your instructor for help if you do not know how to link values between Excel worksheets.
*   Total Operating Cost (monthly) after Upgrade—The value in cell J63 is taken directly from cell J55 after Solver has been rerun for the Brussels site upgrade.

- Annual Cost Savings—Try to determine the value for cell J64. Remember that your total operating costs are monthly.
- First Year Net Cash Flow—Cell J65 holds the sum of the Annual Cost Savings and the Capital Cost of Upgrade (a negative number). If the net cash flow is positive, the Annual Cost Savings paid for the Brussels site upgrade.

When you finish the financial analysis section, rerun Solver and create another Answer Report. The results may surprise you. What effects did changes at the Brussels site have on the assignment table and the total operating cost? Did the proposed site improvement pay for itself in one year?

Assignment 1D: Manufacturing the Smartphone Case Onsite

The DSS model in Assignment 1A can also be used to determine the economic feasibility of manufacturing the smartphone case at each site rather than purchasing and shipping it from one supplier. The case is a clamshell assembly of injected molded plastic. Global's corporate engineering staff has proposed that an injection molding machine be installed at each of the four manufacturing sites, which would eliminate the need to purchase cases from the supplier in Mexico City. The changes can be made quickly, but they will cost approximately $2 million to implement globally. Because of the change, the cost of the case will increase from $1.00 to $1.20: Fewer cases will be produced at each facility, startup costs must be absorbed, and new workers must learn the injection molding process.

You have been asked to modify your Solver model to determine whether the proposed case manufacturing changes will reduce the total operating cost enough to recover the investment within one year. You must modify some of the values in the Constants section and the Air Freight In formulas in the Manufacturing Cost section. Finally, you must modify the investment analysis section of the worksheet to determine the investment savings.

Copy the Global Electronics Solver 2 worksheet and rename the copy **Global Electronics Solver 3—** assume that Global Electronics has upgraded the Brussels manufacturing site. You do not have to run Solver again because the changes are the same across all four manufacturing sites. To determine the effect of manufacturing the cases onsite, you must take the following steps:

1. Enter the new cost for the smartphone case in cell C6.
2. Modify the Air Freight In formulas in the Manufacturing Cost by site section to remove the Mexico City air freight.
3. Modify the financial analysis section and rename it as "Proposal—Manufacturing the Case Onsite."

The proposed improvements require you to change a value in the Constants section of the worksheet. Enter "$1.20" in cell C6, as shown in Figure 8-12.

	A	B	C	D	E	F
1		Global Electronics, Inc. Manufacturing Planning Problem				
2						
3		Constants				
4		Direct Materials Cost:				
5		Circuit Board	$8.00			
6		Case	$1.20			
7		Battery	$3.00			
8		Touchscreen	$7.00			
9						

Source: Used with permission from Microsoft Corporation

FIGURE 8-12   New cost for the case produced onsite

Next, modify the Air Freight In formulas in cells C43 through F43 of the Manufacturing Cost by site section to remove the calculations for the Mexico City air shipments. See if you can figure this out for yourself. The values in cells C43 through F43 should match those in Figure 8-13.

	Manufacturing Cost by site (per unit):	Manufacturing Site			
		Phoenix	Shanghai	Monterrey	Brussels
42	Direct Materials Cost	$19.20	$19.20	$19.20	$19.20
43	Air Freight In	$2.16	$0.40	$2.41	$1.74
44	Direct Labor Cost	$1.00	$0.40	$0.50	$0.90
45	Overhead Cost	$3.00	$1.00	$2.00	$3.00
46	Total Mfg Cost per Unit:	$25.36	$21.00	$24.11	$24.84

Source: Used with permission from Microsoft Corporation

FIGURE 8-13   Modified Manufacturing Cost by site section

Finally, modify the financial analysis section, as shown in Figure 8-14. Make sure to enter a negative value for the cash outlay of the proposal. You should be able to determine the values or formulas to enter into the remaining cells. For the Total Operating Cost (monthly) before Upgrade value in cell J62, use the Total Operating Cost (monthly) value from cell J55 of the Global Electronics Solver 2 worksheet. In other words, assume that the Brussels upgrade has been implemented, and link cell J55 from that worksheet into cell J62 of your new worksheet.

	Proposal--Manufacturing the Case Onsite			
61	Capital Cost of Upgrade:			($2,000,000)
62	Total Operating Cost (monthly) before Upgrade:			
63	Total Operating Cost (monthly) after Upgrade:			
64	Annual Cost Savings:			
65	First Year Net Cash Flow:			

Source: Used with permission from Microsoft Corporation

FIGURE 8-14   Financial analysis section—Manufacturing the case onsite

When you finish the financial analysis section, it should display the results you are seeking. You can run Solver again and create another Answer Report, but because the costs were modified in the same manner across all manufacturing sites, the Solver 3 production assignments will be the same as the Solver 2 assignments. What is the effect of manufacturing the cases onsite on the Total Operating Cost? Does the proposed change pay for itself in one year?

ASSIGNMENT 2: USING THE WORKBOOK FOR DECISION SUPPORT

You have built worksheets to determine the best assignment solution, the feasibility of upgrading the Brussels manufacturing site, and the feasibility of manufacturing the smartphone cases onsite by eliminating the Mexico City supplier. You now complete the case by using your solutions and Answer Reports to make your recommendations in a memorandum.

Use Microsoft Word to write a memo to the management team at Global Electronics. State the results of your analysis, recommend whether to use Solver's assignment of current production or assign the production manually, tell the team whether you think the proposed Brussels upgrade is economically desirable, and state whether you think Global should manufacture the smartphone cases onsite.

- Set up your memo as described in Tutorial E.
- In the first paragraph, briefly describe the situation and state the purpose of your analysis.
- Next, summarize the results of your analysis and state your recommendations.
- Support your recommendation with appropriate graphics or Excel objects from the Excel workbook. (Tutorial C describes how to copy and paste Excel objects.)
- In a future case, you might suggest revisiting the Solver analysis to determine the profitability of relocating certain distribution centers or identifying alternate suppliers for other smartphone components.

ASSIGNMENT 3: GIVING AN ORAL PRESENTATION

Your instructor may request that you summarize your analysis and recommendations in an oral presentation. If so, prepare a presentation for the CEO and other managers that lasts 10 minutes or less. When preparing your presentation, use PowerPoint slides or handouts that you think are appropriate. Tutorial F explains how to prepare and give an effective oral presentation.

DELIVERABLES

Prepare the following deliverables for your instructor:

1. A printout of the memo
2. Printouts of your worksheets and Answer Reports
3. Your Word document, Excel workbook, and PowerPoint presentation on electronic media or sent to your course site as directed by your instructor

Staple the printouts together with the memo on top. If you have more than one Excel workbook file for your case, write your instructor a note that describes the different files.

CASE **9**

APPLIANCE WORLD WAREHOUSE INVENTORY MANAGEMENT

Decision Support Using Microsoft Excel Solver

PREVIEW

Appliance World is a national retailer of household appliances, with warehouses and stores across the country. The company's warehouse in Louisville, Kentucky, stocks eight families of appliances: ovens and ranges, compact refrigerators, standard refrigerators, washing machines, clothes dryers, dishwashers, microwave ovens, and chest freezers.

The Louisville warehouse has been operating for several years, but the new logistics manager, Harry Murvin, is concerned that the traditional operations model used for managing the inventory, known as EOQ (Economic Order Quantity), is not the most cost-efficient way to run the warehouse. The EOQ model fails to consider real-world constraints such as storage capacity and asset management. Moreover, in a recent meeting with marketing vice president Susan Barnes, Harry learned that the Louisville operation has an order service level of only 50 percent because the warehouse does not carry safety stock to deal with variation in sales demand. Susan is concerned that Appliance World is losing customers to the large home-improvement chains, which have been selling household appliances for the past several years.

Harry would like to develop an inventory management DSS model that addresses these concerns while optimizing the warehouse operating cost. He also wants to identify possible improvements to the operation.

You are the corporate MIS manager for Appliance World. You have traveled to the Louisville warehouse to meet with Harry and develop an optimized inventory management program in Microsoft Excel.

PREPARATION

- Review the spreadsheet concepts discussed in class and in your textbook.
- Your instructor may assign Excel exercises to help prepare you for this case.
- Tutorial D explains how to set up and use Solver for maximization and minimization problems.
- Review the file-saving instructions in Tutorial C—saving an extra copy of your work on a USB thumb drive is always a good idea.
- Review Tutorial F to brush up on your presentation skills.

BACKGROUND

You will use your Excel skills to build a decision model that determines the best reorder quantities for each of the eight product groups in the Louisville warehouse while meeting constraints for customer service level, storage space, and maximum inventory value.

Before building the DSS model, however, you need to familiarize yourself with some basic concepts in inventory control. Inventory control is a field of study in operations management that addresses how to meet varying customer demand for a product, given the constraints of time, space, and money. Inventory policy is a set of guidelines an organization uses to determine the quantities of each item it orders or makes and how frequently it orders each item. Inventory policy also determines the average inventory level and the maximum quantities of each item that require storage space in the warehouse.

The following list of inventory control terms and their definitions are important to addressing the problem at the Appliance World warehouse:

- Standard Accounting Period—Many businesses, especially manufacturers, distributors, and retailers, do *not* use calendar months for planning purposes. Instead, they use a period of 28 days so that budgets, market projections, sales, and costs can be standardized and compared. A standard accounting period allows a firm to plan its operations over 13 periods of equal length instead of 12 months of variable length.

- Period Demand—This value is the amount of an item demanded by customers during a period. Demand is always variable and often unpredictable, which makes inventory control particularly challenging. In this case, period demand represents an average of actual customer demand calculated from a number of previous periods.

- Standard Deviation of Demand—As a firm gathers historical data about the demand for an item over time, it can apply statistics to the data to determine the mean and standard deviation of the demand and to determine the type of probability distribution that characterizes the demand. Standard deviation of demand helps an organization determine the probability of not having an item in stock when a customer asks for it.

- Holding Cost—Also called *storage cost*, this value represents the cost to an organization of keeping an item in storage for a certain period of time. The biggest component of holding cost is the time value of the money invested by the firm in making or buying the item. Other components include the cost of storage space and warehouse operations, taxes, insurance, shelf life, depreciation, product obsolescence, and damage or pilferage. Holding cost can be expressed as a percentage of the purchase price or manufacturing cost of an item per period, or simply as a dollar value per unit per period. Annual holding costs can run as high as 30 percent or more of an item's value, but typically the costs fall between 12 and 25 percent.

- Ordering Cost—Also called *setup cost* in manufacturing, ordering cost is the administrative cost that an organization incurs each time it reorders a quantity of a given item. This cost includes IT expense and extraordinary freight consolidation, routing, and expediting charges. A popular misconception about ordering cost is that it includes a freight-in (or shipping) cost for the items. Freight-in cost is usually added to the purchase price of an item for its inventory valuation and is *not* included in the ordering cost.

- Order Lead Time, Days—This value is the amount of time between reordering an item and having it delivered. Before the creation of supply chain management and integrated software solutions in the 1990s, order lead times were longer and more variable, depending on the reliability of the supplier and the transportation mode used. The increasing sophistication of global logistics and supply chain solutions has removed most of this variability.

- Stockout—When a customer tries to order an item that is not currently available, a stockout occurs. Naturally, an organization wants to avoid stockouts because customers might switch to a competitor's product as a result.

- Desired Service Level—Service level, expressed as a percentage, is the probability that an item's supply will meet all customer demand during an inventory cycle, particularly when the item's inventory is being depleted before a reorder has been received. The probability of a stockout is equal to 1 minus the desired service level.

- Stockout Penalty Cost—This value is an estimate by marketing and operations staff of the cost of a stockout to the organization. A stockout penalty cost includes the opportunity cost of lost sales as well as an expected loss in future sales from a customer switching to a competitor. Stockout penalty costs are difficult to quantify accurately, mostly because customers behave unpredictably. If a company has strong brand recognition and high customer loyalty, stockout penalty cost is probably low. The Marketing department at Appliance World has researched the stockout penalty costs for all company products and has supplied the data for the inventory model.

- Order Quantity—This value is the amount of an item that is ordered or reordered each time the inventory must be replenished.

- Economic Order Quantity—Also known by its acronym (EOQ), economic order quantity is one of the oldest inventory control models. It states that the total cost of inventory is minimized at the order quantity at which the ordering cost and holding cost are equal. You can calculate EOQ using the following formula:

$$\mathrm{EOQ} = \sqrt{2 \times Demand \times Ordering\ Cost \div Holding\ Cost}$$

The EOQ involves several assumptions about demand, storage, and replenishment lead time, some of which are not valid in this case. These assumptions are discussed later.

- Average Inventory—Because inventory is depleted by demand until it is replenished, the average inventory of an item is the order quantity divided by two. As with EOQ, this definition assumes that the rate of demand is known and spread evenly throughout the period. In actual practice, demand rates are variable and frequently unpredictable.
- Reorder Point—The reorder point is the level of inventory at which an item must be reordered. It is the daily average demand multiplied by the order lead time, and it represents the remaining inventory that must satisfy demand while waiting for replenishment. Reorder points are not needed to build the inventory model.
- Safety Stock—This value is the quantity of an item carried in addition to normal inventory to handle variation in demand. Currently, the Louisville warehouse does not carry safety stock, so Harry and Susan want to address this issue. They want to institute a safety stock policy to minimize the penalty cost of stockouts, and they want you to figure out how to incorporate this policy in your DSS model.
- Average Inventory Value—For an item in the warehouse, the average inventory value is the item's average inventory level plus safety stock, multiplied by its purchase price per unit. Most organizations define a ceiling or maximum average inventory value for their stored products.

The Logistics and Marketing departments at Appliance World have provided data you will use to populate your DSS model.

ASSIGNMENT 1: CREATING A SPREADSHEET FOR DECISION SUPPORT

In this assignment, you create a spreadsheet that models the inventory management plan for Appliance World's Louisville warehouse, with an emphasis on minimizing the inventory operating cost. In Assignment 1A, you create a spreadsheet to model the inventory plan and attempt to assign the order quantities manually. Then you set up Solver to minimize the inventory operating cost. In Assignment 1B, you copy your model to a new worksheet and modify it to include service levels, add safety stock calculations, and examine the effects of stockout penalty cost on the inventory operating cost. In Assignment 1C, you modify the model and determine whether leasing additional storage space results in a net saving of operating costs. In Assignments 2 and 3, you use data from the spreadsheet models to summarize the results and provide a financial analysis and recommendation for the proposed leasing of storage space. You might also be asked to prepare and give an oral presentation to support your analysis and recommendations.

The Appliance World spreadsheet model contains two sections: a Constants section and a Calculations and Results section. The spreadsheet model for the lease proposal also includes a small Lease Cost Savings Analysis section.

Assignment 1A: Creating the Inventory Management Spreadsheet

A discussion of each spreadsheet section follows. This information helps you set up each section of the model and learn the logic of the formulas. If you choose to enter the data directly, follow the cell structure shown in the figures. To save time, it is highly recommended that you use the spreadsheet skeleton. To access the spreadsheet skeleton, select **Case 9** in your data files, and then select **Appliance World Warehouse Skeleton.xlsx**.

Constants Section

First, build the skeleton of your spreadsheet. Set up the spreadsheet title and Constants section as shown in Figure 9-1. An explanation of the line items follows the figure.

	Range/ Oven	Compact Refrigerator	Standard Refrigerator	Washing Machine	Clothes Dryer	Dishwasher	Microwave Oven	Chest Freezer
Appliance World Warehouse--Inventory Management								
Constants								
Warehouse Capacity (cubic feet)	100,000			Legend:		Changing Cells		
Average Inventory Budget	$800,000					Optimization Cell		
Standard Accounting Period (days)	28					Not Used		
Period Demand (units)	2250	5500	4500	6150	6150	3200	6250	1500
Standard Deviation of Daily Demand	10.2	14.4	13.5	9.2	9.3	16.8	18.3	9.7
Purchase Price per Unit	$495	$200	$550	$350	$325	$425	$125	$400
Holding Cost (per unit, period)	$10	$4	$11	$7	$7	$9	$3	$8
Ordering Cost (per order)	$200	$450	$550	$270	$270	$250	$180	$200
Order Lead Time, Days	10	5	10	10	10	15	5	15
Desired Service Level	95%	95%	99%	99%	99%	95%	95%	90%
Stockout Penalty Cost (per stockout)	$5,000	$2,500	$6,000	$6,000	$6,000	$4,000	$2,500	$3,000
Storage Space Required (cubic feet per unit)	28	12	55	22	22	15	5	35

Source: Used with permission from Microsoft Corporation

FIGURE 9-1   Spreadsheet title and Constants section

- Spreadsheet title—Enter the spreadsheet title in cell B1, and then merge and center the title across cells B1 through J1.
- Warehouse Capacity (cubic feet)—Cell C5 holds the maximum storage capacity of the Louisville warehouse.
- Average Inventory Budget—Cell C6 holds the amount budgeted by Appliance World for the maximum value of the inventory kept in the Louisville warehouse.
- Standard Accounting Period (days)—Cell C7 holds the number of days in one accounting period. This value is needed for some of the calculations in the Calculations and Results section.
- Legend (fill colors)—In cells G5 through H7, place the fill colors for the changing cells, the optimization cell, and the Not Used cells. Enter labels for these fill colors in cells H5 through H7.
- Product headings—In cells C9 through J9, enter the headings for the eight different products, as shown in Figure 9-1.
- Period Demand (units)—Cells C10 through J10 hold the historical averages of the period sales demand for the eight items. Note that these values are *historical* averages. Customer demand has the highest variability of any inventory control factor, which is reflected in the next line.
- Standard Deviation of Daily Demand—Cells C11 through J11 hold the calculated standard deviations for daily demand using historical data.
- Purchase Price per Unit—Cells C12 through J12 hold the prices that Appliance World paid its suppliers for each item, including freight-in charges. Remember that Appliance World is a retailer; it does not manufacture its own appliances.
- Holding Cost (per unit, period)—Cells C13 through J13 hold the costs that Appliance World incurs to keep each item in storage for a specified period.
- Ordering Cost (per order)—Cells C14 through J14 hold the costs that Appliance World incurs each time it reorders more stock of each item.
- Order Lead Time, Days—Cells C15 through J15 hold the number of days needed to receive a product reorder from the day the order is placed.
- Desired Service Level—Cells C16 through J16 hold the service level that the Marketing department has determined is needed to maintain brand loyalty for each item. Service level is equal to 1 minus the probability of stocking out; therefore, an item with a 99 percent service level has a 1 percent probability of stocking out during an inventory reorder cycle. Desired Service Level is a needed parameter when calculating Safety Stock levels.

- Stockout Penalty Cost (per stockout)—This cost is reported in cells C17 through J17. After considerable research, the Marketing department has determined how much money Appliance World loses in terms of sales, customers, and expediting expenses each time an item stockout occurs. These values represent the cost to Appliance World for a stockout of each item.
- Storage Space Required (cubic feet per unit)—The values in cells C18 through J18 represent the amount of warehouse storage space each packaged appliance occupies.

When you have completed the Constants section, check your values and format them as shown in Figure 9-1. Place a thick border around the entire section (cells B3 through J18).

Calculations and Results Section

This section is the heart of your decision model. It contains the optimized order quantity for each item (the changing cells), the inventory operating cost (the optimization cell), various calculations, and totals needed for defining constraints.

Again, it is highly recommended that you use the skeleton file for this case. If you choose to set up the model yourself, refer to Figure 9-2. An explanation of the line items follows the figure. Note that cells filled with yellow are the changing cells for this model, the optimization cell is filled in orange, and cells in the Totals column that are *not* used are light gray.

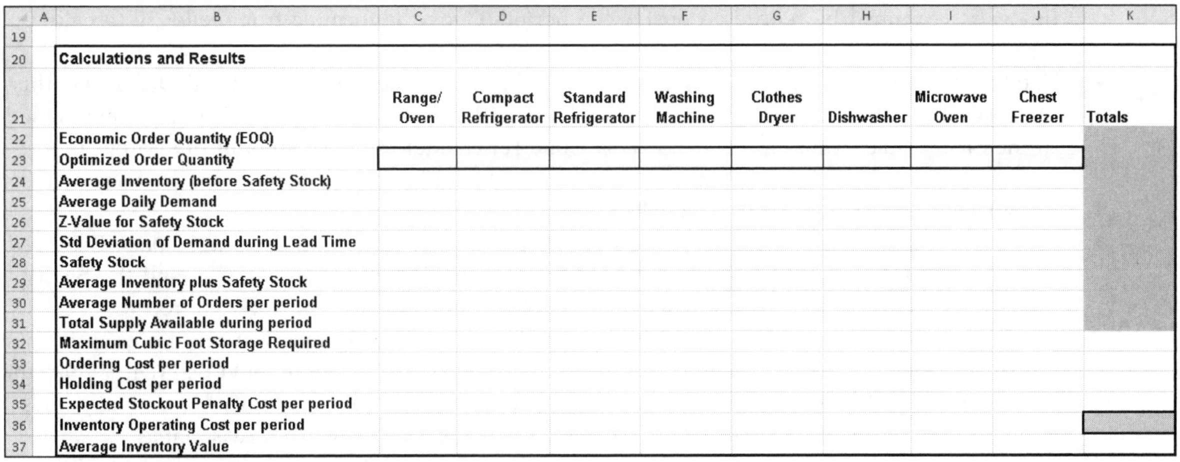

	A	B	C	D	E	F	G	H	I	J	K
19											
20		Calculations and Results									
21			Range/ Oven	Compact Refrigerator	Standard Refrigerator	Washing Machine	Clothes Dryer	Dishwasher	Microwave Oven	Chest Freezer	Totals
22		Economic Order Quantity (EOQ)									
23		Optimized Order Quantity									
24		Average Inventory (before Safety Stock)									
25		Average Daily Demand									
26		Z-Value for Safety Stock									
27		Std Deviation of Demand during Lead Time									
28		Safety Stock									
29		Average Inventory plus Safety Stock									
30		Average Number of Orders per period									
31		Total Supply Available during period									
32		Maximum Cubic Foot Storage Required									
33		Ordering Cost per period									
34		Holding Cost per period									
35		Expected Stockout Penalty Cost per period									
36		Inventory Operating Cost per period									
37		Average Inventory Value									

Source: Used with permission from Microsoft Corporation

FIGURE 9-2   Calculations and Results section

- Economic Order Quantity (EOQ)—The EOQ values in cells C22 through J22 reflect the current method used to determine Appliance World's reorder quantity for each item. The values follow the EOQ formula that was explained in the Background section. For cell C22, EOQ equals the square root of the following expression:

  2 × Period Demand (cell C10) × Ordering Cost (cell C14)/Holding Cost (cell C13)

  Because there are no absolute cell references, simply copy your completed formula from cell C22 to cells D22 through J22. Format the answers as integers because only whole units can be ordered. In other words, round up the decimals to whole numbers.
- Optimized Order Quantity—Cells C23 through J23 will be the "answer" cells when Solver is used to determine the best inventory policy. Copy the EOQ values as whole numbers from cells C22 through J22 into cells C23 through J23 for now. Copy only the values, *not* the formulas.
- Average Inventory (before Safety Stock)—Cells C24 through J24 hold the average inventory. For cell C24, the average inventory is the Optimized Order Quantity in cell C23 divided by 2. Based on the assumption that daily demand is even throughout the inventory cycle, average inventory represents the halfway point between the highest inventory (when an order is received) and when the inventory falls to zero. Copy the formula from cell C24 to cells D24 through J24.
- Average Daily Demand—Cells C25 through J25 hold the average daily demand. For cell C25, this value for an item is the Period Demand in cell C10 divided by the Standard Accounting Period in cell C7, which is 28 days. When writing the formula for cell C25, remember to use an *absolute* cell reference for cell C7 ($C$7) to facilitate copying the formula to cells D25 through J25.

Note that the formula will result in a decimal value. Your answer should contain two decimal values. A decimal answer is acceptable when you are working with averages for discrete items. In other words, you do not have to express calculated averages as integers.

- Z-Value for Safety Stock—The values in cells C26 through J26 represent the number of standard deviations to the right of the mean on a standard normal distribution curve. In this case, the area under the curve up to the Z-value represents the desired service level. Z-values are explained in depth in Assignment 1B. For now, leave these cells blank.
- Std Deviation of Demand during Lead Time—The values in cells C27 through J27 represent the standard deviation of the demand during the order lead time. These values are needed for the Safety Stock calculations in Assignment 1B. For now, leave these cells blank.
- Safety Stock—The values in cells C28 through J28 represent the amount of safety stock (or extra inventory) needed to meet the demand variation and avoid a stockout. This value depends on the Desired Service Level. This value will be calculated in Assignment 1B. For now, leave these cells blank.
- Average Inventory plus Safety Stock—Cells C29 through J29 hold the average inventory values plus safety stock. For cell C29, this value is the sum of the Average Inventory in cell C24 and the Safety Stock in cell C28. Copy the completed formula to cells D29 through J29.
- Average Number of Orders per period—For cell C30, this value is the Period Demand in cell C10 divided by the Optimized Order Quantity in cell C23. This formula represents the number of times an item must be reordered during the period. Copy the formula from cell C30 to cells D30 through J30.
- Total Supply Available during period—For cell C31, this value is the Optimized Order Quantity in cell C23 multiplied by the Average Number of Orders per period in cell C30, and then added to the Safety Stock value in cell C28. This value represents the total amount of an item that is available to satisfy its Period Demand; the value will be a limiting constraint in the Solver model. Copy the formula from cell C31 to cells D31 through J31.
- Maximum Cubic Foot Storage Required—For cell C32, this value is the Storage Space Required in cell C18 multiplied by the sum of the Optimized Order Quantity in cell C23 and the Safety Stock in cell C28. This value represents the total amount of storage space required when an item's stock is at its maximum—in other words, just after a newly received reorder quantity plus the safety stock. In reality, the maximum storage required for an item could be greater than the amount produced by this formula if demand is low. Advanced planning software typically adds an allowance factor for this possibility. However, this formula provides a reasonable approximation because of statistical averaging. That is, one item might be overstocked while another item's inventory is running low, which frees up storage space for the overstocked item. Copy the formula from cell C32 to cells D32 through J32. Enter a formula to sum cells C32 through J32 into cell K32, which you will use to define the storage space constraint in Solver.
- Ordering Cost per period—For cell C33, this value equals the Ordering Cost (per order) in cell C14 multiplied by the Average Number of Orders per period in cell C30. Copy the formula from cell C33 to cells D33 through J33. Enter a formula to sum cells C33 through J33 into cell K33, which represents the total Ordering Cost per period.
- Holding Cost per period—For cell C34, this value equals the Holding Cost in cell C13 multiplied by the Average Inventory, plus Safety Stock in cell C29. Copy the formula from cell C34 to cells D34 through J34. Enter a formula to sum cells C34 through J34 into cell K34, which represents the total Holding Cost per period.
- Expected Stockout Penalty Cost per period—Cells C35 through J35 hold penalty costs, with the total penalty cost reported in cell K35. This value is the expected stockout penalty cost that Appliance World incurs for stockouts of each item. You will use these values in Assignment 1B, but leave the cells blank for now.
- Inventory Operating Cost per period—For cell C36, this value is the sum of the Ordering Cost per period in cell C33, the Holding Cost per period in cell C34, and the Expected Stockout Penalty Cost per period in cell C35. Copy the formula from cell C36 into cells D36 through J36.

Enter a formula to sum cells C36 through J36 into cell K36, which is the total Inventory Operating Cost per period for the warehouse. This cell is also the optimization cell for the case. Fill the cell with the appropriate color from the Constants section.

- Average Inventory Value—For cell C37, this value is the Purchase Price per unit in cell C12 multiplied by the Average Inventory plus Safety Stock in cell C29. This value represents the average amount of Appliance World's assets that are tied up in inventory. Copy the formula from cell C37 to cells D37 through J37. Enter a formula to sum cells C37 through J37 into cell K37, which is the total Inventory Value of the warehouse. You will use this value when defining a constraint for the average inventory budget in Solver.

Complete the section by placing borders and shading in the appropriate places, as shown in Figure 9-2. If you built the Calculations and Results section correctly, the completed spreadsheet should look like Figure 9-3.

		Range/ Oven	Compact Refrigerator	Standard Refrigerator	Washing Machine	Clothes Dryer	Dishwasher	Microwave Oven	Chest Freezer	Totals
20	Calculations and Results									
22	Economic Order Quantity (EOQ)	300	1112	671	689	715	422	866	274	
23	Optimized Order Quantity	300	1112	671	689	715	422	866	274	
24	Average Inventory (before Safety Stock)	150	556	335	344	357	211	433	137	
25	Average Daily Demand	80.36	196.43	160.71	219.64	219.64	114.29	223.21	53.57	
26	Z-Value for Safety Stock									
27	Std Deviation of Demand during Lead Time									
28	Safety Stock									
29	Average Inventory plus Safety Stock	150	556	335	344	357	211	433	137	
30	Average Number of Orders per period	7.50	4.94	6.71	8.93	8.60	7.59	7.22	5.48	
31	Total Supply Available during period	2250	5500	4500	6150	6150	3200	6250	1500	
32	Maximum Cubic Foot Storage Required	8,400	13,349	36,895	15,153	15,725	6,325	4,330	9,585	109,763
33	Ordering Cost per period	$1,500	$2,225	$3,690	$2,411	$2,323	$1,897	$1,299	$1,095	$16,440
34	Holding Cost per period	$1,500	$2,225	$3,690	$2,411	$2,323	$1,897	$1,299	$1,095	$16,440
35	Expected Stockout Penalty Cost per period									$0
36	Inventory Operating Cost per period	$3,000	$4,450	$7,379	$4,822	$4,646	$3,795	$2,598	$2,191	$32,880
37	Average Inventory Value	$74,250	$111,243	$184,476	$120,538	$116,153	$89,598	$54,127	$54,772	$805,156

Source: Used with permission from Microsoft Corporation

FIGURE 9-3 Completed Calculations and Results section

The EOQ Solution

The spreadsheet you have created is a model of the EOQ inventory policy employed by the Louisville warehouse. By definition, EOQ is the order quantity at which total cost is minimized—in other words, the quantity at which the holding cost equals the ordering cost. You can examine rows 33 and 34 to confirm that the holding cost equals the ordering cost for each item in inventory. However, as stated earlier, EOQ makes several assumptions about inventory behavior that are not necessarily valid in actual operating situations. One such assumption is that the rate of demand is known and spread evenly throughout the period. EOQ also fails to consider realistic constraints imposed by business conditions, such as space available to store inventory and budgetary maximums on the amount of money a firm is willing to tie up in inventory. For Appliance World, you must consider the constraints of available storage space and inventory budget. If you look at cell K32 of the spreadsheet, you will notice that the total storage space exceeds the Warehouse Capacity in cell C5 by more than 9,000 cubic feet (see Figure 9-4). This constraint requires you to adjust order quantities so that the inventory fits into the available storage capacity. For example, the company can place more orders for smaller quantities of certain items.

Next, look at the total Average Inventory Value in cell K37. In the EOQ model, this value exceeds the Average Inventory Budget for the warehouse (cell C6) by more than $5,000, as shown in Figure 9-4. Because the budget constraint for inventory value is an accounting target, it can probably be exceeded by a small amount, but assume that the constraint must be met for the purposes of this analysis.

	Range/ Oven	Compact Refrigerator	Standard Refrigerator	Washing Machine	Clothes Dryer	Dishwasher	Microwave Oven	Chest Freezer	
Appliance World Warehouse--Inventory Management									
Constants									
Warehouse Capacity (cubic feet)	100,000				Legend:		Changing Cells		
Average Inventory Budget	$800,000						Optimization Cell		
Standard Accounting Period (days)	28						Not Used		
	Range/ Oven	Compact Refrigerator	Standard Refrigerator	Washing Machine	Clothes Dryer	Dishwasher	Microwave Oven	Chest Freezer	
Period Demand (units)	2250	5500	4500	6150	6150	3200	6250	1500	
Standard Deviation of Daily Demand	10.2	14.4	13.5	9.2	9.3	16.8	18.3	9.7	
Purchase Price per Unit	$495	$200	$550	$350	$325	$425	$125	$400	
Holding Cost (per unit, period)	$10	$4	$11	$7	$7	$9	$3	$8	
Ordering Cost (per order)	$200	$450	$550	$270	$270	$250	$180	$200	
Order Lead Time, Days	10	5	10	10	10	15	5	15	
Desired Service Level	95%	95%	99%	99%	99%	95%	95%	90%	
Stockout Penalty Cost (per stockout)	$5,000	$2,500	$6,000	$6,000	$6,000	$4,000	$2,500	$3,000	
Storage Space Required (cubic feet per unit)	28	12	55	22	22	15	5	35	
Calculations and Results									
	Range/ Oven	Compact Refrigerator	Standard Refrigerator	Washing Machine	Clothes Dryer	Dishwasher	Microwave Oven	Chest Freezer	Totals
Economic Order Quantity (EOQ)	300	1112	671	689	715	422	866	274	
Optimized Order Quantity	300	1112	671	689	715	422	866	274	
Average Inventory (before Safety Stock)	150	556	335	344	357	211	433	137	
Average Daily Demand	80.36	196.43	160.71	219.64	219.64	114.29	223.21	53.57	
Z-Value for Safety Stock									
Std Deviation of Demand during Lead Time									
Safety Stock									
Average Inventory plus Safety Stock	150	556	335	344	357	211	433	137	
Average Number of Orders per period	7.50	4.94	6.71	8.93	8.60	7.59	7.22	5.48	
Total Supply Available during period	2250	5500	4500	6150	6150	3200	6250	1500	
Maximum Cubic Foot Storage Required	8,400	13,349	36,895	15,153	15,725	6,325	4,330	9,585	109,763
Ordering Cost per period	$1,500	$2,225	$3,690	$2,411	$2,323	$1,897	$1,299	$1,095	$16,440
Holding Cost per period	$1,500	$2,225	$3,690	$2,411	$2,323	$1,897	$1,299	$1,095	$16,440
Expected Stockout Penalty Cost per period									$0
Inventory Operating Cost per period	$3,000	$4,450	$7,379	$4,822	$4,646	$3,795	$2,598	$2,191	$32,880
Average Inventory Value	$74,250	$111,243	$184,476	$120,538	$116,153	$89,598	$54,127	$54,772	$805,156

Source: Used with permission from Microsoft Corporation

FIGURE 9-4   Storage and budget constraints violated

Rename your current worksheet as **Warehouse 1 EOQ** and then make a copy of it. Right-click the worksheet Name tab, and then click Move or Copy from the menu that appears (see Figure 9-5). In the next window, click the Create a copy check box (see Figure 9-6), and then click OK. A copy of the worksheet named Warehouse 1 EOQ (2) appears in a new tab. Rename the worksheet **Warehouse 1 Manual**.

28	Safety Stock	Insert...
29	Average Inventory plus Sa	Delete
30	Average Number of Order:	
31	Total Supply Available du	Rename
32	Maximum Cubic Foot Stor	Move or Copy...
33	Ordering Cost per period	View Code
34	Holding Cost per period	Protect Sheet...
35	Expected Stockout Penalt	Tab Color
36	Inventory Operating Cost	
37	Average Inventory Value	Hide
38		Unhide...
39		Select All Sheets

Skeleton   Warehouse 1 EOQ   Warehouse 1 M

Source: Used with permission from Microsoft Corporation

FIGURE 9-5   Worksheet Move or Copy option

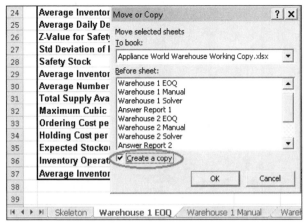

Source: Used with permission from Microsoft Corporation

FIGURE 9-6   The Create a copy check box in the Move or Copy window

Attempting a Manual Solution

Before setting up Solver, try to create a solution that meets the storage and budget constraints in cells C5 and C6 by manually assigning order quantities for the eight appliances in cells C23 through J23. You must reduce the order quantities for some items so that the maximum storage required in cell K32 does not exceed Warehouse Capacity in cell C5 and so the total Average Inventory Value in cell K37 does not exceed the Average Inventory Budget in cell C6.

A good starting point is to examine the storage space required for the different items and reduce order quantities for the larger appliances first. When you find a solution that satisfies the constraints, look at the total Inventory Operating Cost per period in cell K36 and compare it with the value in the EOQ model worksheet. The cost probably did not increase much—the manual solution is probably close to the value that Appliance World managers determined through trial and error when they discovered a shortage of warehouse capacity. However, is it the most cost-effective solution? Solver can probably do a better job.

Setting Up and Running Solver

Make a copy of the manual solution worksheet, and then rename it **Warehouse 1 Solver**. Before you set up Solver, think about the values you need to define:

- The optimization cell is the total Inventory Operating Cost per period in cell K36. You enter this cell in the Set Objective text box of the Solver Parameters window. You want to minimize this value, so click the Min button.
- The Changing Variable Cells are Optimized Order Quantity cells C23 through J23.
- Define the following constraints:
  - The changing cells must be nonnegative integers.
  - The maximum storage required cannot exceed the Warehouse Capacity.
  - The Average Inventory Value cannot exceed the Average Inventory Budget.
  - The Total Supply Available during period must at least equal the Period Demand for each item.

To set up Solver, click the Data tab on the Ribbon, and then click Solver in the Analysis Group at the far-right side of the screen. Enter the preceding cells and constraints in the required fields, and then click Solve. An error message appears, as shown in Figure 9-7.

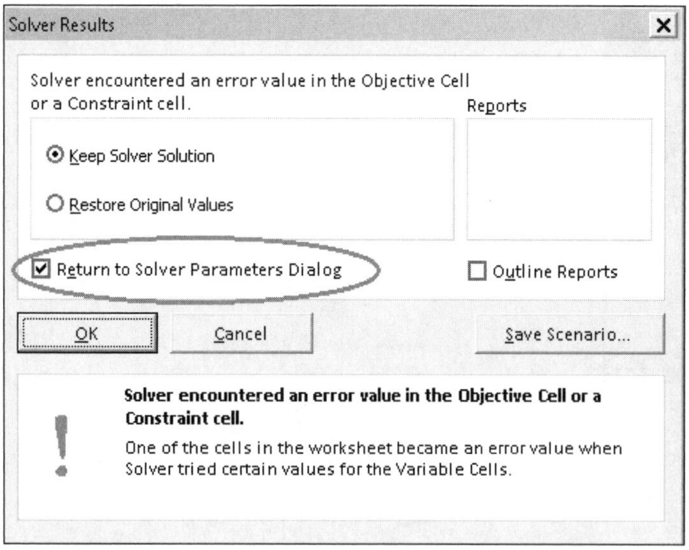

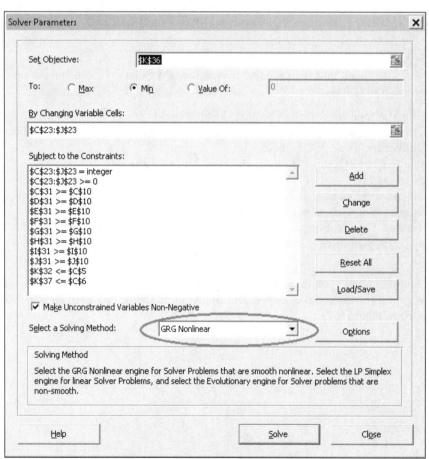

Source: Used with permission from Microsoft Corporation
FIGURE 9-7   Solver error message

The error message is somewhat misleading because there are no errors in the worksheet. Solver tried to use the default Simplex method to solve the problem, but the Simplex method requires linearity in the objective function and constraint equations. The optimization is smooth but nonlinear because the model is multiplying variables in formulas that feed the supply constraints as well as the cost calculations totaled by the optimization cell. However, the error message creates the impression that mistakes exist in one or more worksheet formulas. Fortunately, Solver can optimize nonlinear problems as well. Click the Return to Solver Parameters Dialog check box, and then click OK to return to the Solver Parameters window. Select GRG Nonlinear from the Select a Solving Method check box, and then click Solve (see Figure 9-8).

Source: Used with permission from Microsoft Corporation
FIGURE 9-8   Selecting the GRG Nonlinear solving method

Solver should need two seconds or less to arrive at an optimal solution that meets all the constraints. The order quantities that Solver assigns might be considerably different from the ones you assigned manually, but if you did a good job of assigning order quantities, Solver will probably save only a couple hundred dollars in total inventory operating costs. Click Answer Report to highlight it in the Solver Results window, and then click OK to change the spreadsheet values to the solution.

Assignment 1B: Adding Safety Stock and Penalty Costs to the DSS Model

Make a copy of the Warehouse 1 EOQ worksheet, and then rename it **Warehouse 2 EOQ**. This model will include additional calculations to determine the safety stock levels for each item that are needed to meet Marketing's desired service levels.

As you learned earlier, Appliance World currently carries no safety stocks. Depending on variations in demand and assuming that demand is distributed normally, carrying no safety stock means that Appliance World has a 50 percent chance in every reorder cycle of "stocking out" on an item before it is replenished. Because demand is normally distributed, you can use a special Excel function to determine how much safety stock Appliance World needs to carry for each of the eight appliances at desired service levels. Figure 9-9 shows the four rows that have yet to be completed in the model.

	A	B	C	D	E	F	G	H	I	J	K
19											
20		Calculations and Results									
21			Range/ Oven	Compact Refrigerator	Standard Refrigerator	Washing Machine	Clothes Dryer	Dishwasher	Microwave Oven	Chest Freezer	Totals
22		Economic Order Quantity (EOQ)	300	1112	671	689	715	422	866	274	
23		Optimized Order Quantity	300	1112	671	689	715	422	866	274	
24		Average Inventory (before Safety Stock)	150	556	335	344	357	211	433	137	
25		Average Daily Demand	80.36	196.43	160.71	219.64	219.64	114.29	223.21	53.57	
26		Z-Value for Safety Stock									
27		Std Deviation of Demand during Lead Time									
28		Safety Stock									
29		Average Inventory plus Safety Stock	150	556	335	344	357	211	433	137	
30		Average Number of Orders per period	7.50	4.94	6.71	8.93	8.60	7.59	7.22	5.48	
31		Total Supply Available during period	2250	5500	4500	6150	6150	3200	6250	1500	
32		Maximum Cubic Foot Storage Required	8,400	13,349	36,895	15,153	15,725	6,325	4,330	9,585	109,763
33		Ordering Cost per period	$1,500	$2,225	$3,690	$2,411	$2,323	$1,897	$1,299	$1,095	$16,440
34		Holding Cost per period	$1,500	$2,225	$3,690	$2,411	$2,323	$1,897	$1,299	$1,095	$16,440
35		Expected Stockout Penalty Cost per period									$0
36		Inventory Operating Cost per period	$3,000	$4,450	$7,379	$4,822	$4,646	$3,795	$2,598	$2,191	$32,880
37		Average Inventory Value	$74,250	$111,243	$184,476	$120,538	$116,153	$89,598	$54,127	$54,772	$805,156

Source: Used with permission from Microsoft Corporation

FIGURE 9-9   Calculations and Results section before adding safety stock calculations

To understand the calculations for safety stock, you need a basic understanding of the normal probability distribution and how it works. For example, consider an item with a 95 percent desired service level, such as the range/oven. At a 50 percent service level (or no safety stock), you are basically flipping a coin as to whether the item will stock out before the reorder arrives. At a 95 percent service level, you are adding enough stock to cover all but the last 5 percent of probability under the curve, as shown in Figure 9-10.

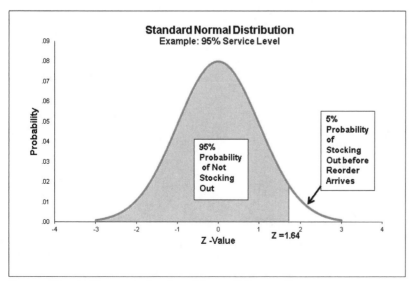

Source: Used with permission from Microsoft Corporation

FIGURE 9-10   Standard normal distribution and Z-value for 95 percent service level

A standard normal distribution is a special form of the normal distribution that has a mean of zero and a standard deviation of 1. It allows you to use an Excel function to figure out how much variation you need to cover with safety stock to improve the probability to the desired service level. Explanations of the remaining incomplete rows in the worksheet follow:

- Z-Value for Safety Stock—Excel has a special function called NORMSINV that can provide a Z-value for any probability. In cell C26, enter a formula of =NORMSINV(C16); a value of 1.64 will appear in the cell. This is the Z-value for a 95 percent desired service level. Copy the formula from cell C26 to cells D26 through J26.
- Std Deviation of Demand during Lead Time—For cell C27, the standard deviation of demand during the lead time equals the Standard Deviation of Daily Demand in cell C11 multiplied by the *square root* of the Order Lead Time in days (cell C15). The reason for using the square root of the order lead time is mathematically complex, but a simpler explanation is that the standard deviation of demand over a period of days increases only by the square root of the number of days because of a greater opportunity for statistical averaging. In other words, days with higher demand are averaged out by days with lower demand. The Excel function for square roots is SQRT. Copy the completed formula from cell C27 to cells D27 through J27.
- Safety Stock—For cell C28, Safety Stock equals the Z-Value for Safety Stock in cell C26 multiplied by Std Deviation of Demand during Lead Time in cell C27. Copy the completed formula from cell C28 to cells D28 through J28. The resulting safety stocks should give Appliance World the desired service levels to avoid stockouts.
- Expected Stockout Penalty Cost per period—For cell C35, the Expected Stockout Penalty Cost equals the probability of stocking out, which is 1 minus the Desired Service Level (1 − cell C16), multiplied by the Stockout Penalty Cost per stockout (cell C17), multiplied by the Average Number of Orders per period (cell C30). Copy the formula from cell C35 to cells D35 through J35, and then sum cells C35 through J35 in cell K35.

Look at your model—you may be surprised to learn how much Stockout Penalty Cost contributes to the overall inventory operating cost. If you want to see how much money Appliance World might have lost over time, enter 50% for service level into each of cells C16 through J16. The resulting Stockout Penalty Cost should be staggering—more than $150,000 per month! You will also notice that a 50 percent service level turns the Z-values and Safety Stock levels to zero in rows 26 and 28.

If you entered the formulas correctly in the spreadsheet, the values in Warehouse 2 EOQ should look like Figure 9-11.

	Range/ Oven	Compact Refrigerator	Standard Refrigerator	Washing Machine	Clothes Dryer	Dishwasher	Microwave Oven	Chest Freezer	

Appliance World Warehouse--Inventory Management

Constants

			Legend:			Changing Cells			
Warehouse Capacity (cubic feet)	100,000					Optimization Cell			
Average Inventory Budget	$800,000					Not Used			
Standard Accounting Period (days)	28								

	Range/ Oven	Compact Refrigerator	Standard Refrigerator	Washing Machine	Clothes Dryer	Dishwasher	Microwave Oven	Chest Freezer
Period Demand (units)	2250	5500	4500	6150	6150	3200	6250	1500
Standard Deviation of Daily Demand	10.2	14.4	13.5	9.2	9.3	16.8	18.3	9.7
Purchase Price per Unit	$495	$200	$550	$350	$325	$425	$125	$400
Holding Cost (per unit, period)	$10	$4	$11	$7	$7	$9	$3	$8
Ordering Cost (per order)	$200	$450	$550	$270	$270	$250	$180	$200
Order Lead Time, Days	10	5	10	10	10	15	5	15
Desired Service Level	95%	95%	99%	99%	99%	95%	95%	90%
Stockout Penalty Cost (per stockout)	$5,000	$2,500	$6,000	$6,000	$6,000	$4,000	$2,500	$3,000
Storage Space Required (cubic feet per unit)	28	12	55	22	22	15	5	35

Calculations and Results

	Range/ Oven	Compact Refrigerator	Standard Refrigerator	Washing Machine	Clothes Dryer	Dishwasher	Microwave Oven	Chest Freezer	Totals
Economic Order Quantity (EOQ)	300	1112	671	689	715	422	866	274	
Optimized Order Quantity	300	1112	671	689	715	422	866	274	
Average Inventory (before Safety Stock)	150	556	335	344	357	211	433	137	
Average Daily Demand	80.36	196.43	160.71	219.64	219.64	114.29	223.21	53.57	
Z-Value for Safety Stock	1.64	1.64	2.33	2.33	2.33	1.64	1.64	1.28	
Std Deviation of Demand during Lead Time	32.26	32.20	42.69	29.09	29.41	65.07	40.92	37.57	
Safety Stock	53	53	99	68	68	107	67	48	
Average Inventory plus Safety Stock	203	609	435	412	426	318	500	185	
Average Number of Orders per period	7.50	4.94	6.71	8.93	8.60	7.59	7.22	5.48	
Total Supply Available during period	2303	5553	4599	6218	6218	3307	6317	1548	
Maximum Cubic Foot Storage Required	9,886	13,985	42,357	16,642	17,231	7,930	4,667	11,270	123,967
Ordering Cost per period	$1,500	$2,225	$3,690	$2,411	$2,323	$1,897	$1,299	$1,095	$16,440
Holding Cost per period	$2,031	$2,437	$4,782	$2,885	$2,768	$2,861	$1,501	$1,481	$20,744
Expected Stockout Penalty Cost per period	$1,875	$618	$402	$536	$516	$1,518	$902	$1,643	$8,011
Inventory Operating Cost per period	$5,406	$5,280	$8,874	$5,831	$5,607	$6,276	$3,702	$4,219	$45,194
Average Inventory Value	$100,512	$121,836	$239,098	$144,226	$138,388	$135,083	$62,540	$74,030	$1,015,714

Source: Used with permission from Microsoft Corporation

FIGURE 9-11   Warehouse 2 EOQ spreadsheet with Safety Stock calculations complete

Make a copy of the worksheet, rename it **Warehouse 2 Manual**, and then attempt to manually assign order quantities for the eight items again. Observe the warehouse space and inventory budget constraints. Notice that adding safety stocks causes you to run out of warehouse space quickly, which requires more frequent reorders in smaller quantities for most of your appliances. Make a copy of the Warehouse 2 Manual worksheet, and rename it **Warehouse 2 Solver**. Set up and run Solver to find a more optimal solution, and create Answer Report 2.

Assignment 1C: Cost Savings Analysis for Additional Leased Storage Space

The DSS model has shown the management team at Appliance World that warehouse capacity is a serious constraint on inventory operating cost and customer service. Fortunately, Appliance World has been leasing its current warehouse space, and the owners now have an additional 20,000 cubic feet of space available for storage in the same facility. The owners want to charge Appliance World $1,000 more per month for the additional storage space. Harry feels that the extra space is a good investment if the management team at Appliance World allows him to increase his average inventory budget from $800,000 to $900,000. Susan from the Marketing department supports the proposal because the increased inventory budget will help cover the value of the new safety stock.

Make a copy of the Warehouse 2 Solver worksheet, and rename it **Warehouse 3**. Change the Warehouse Capacity in cell C5 to 120,000 cubic feet and the Average Inventory Budget in cell C6 to $900,000. Add a section titled Lease Cost Savings Analysis at the bottom of the worksheet, as shown in Figure 9-12. Next, copy the Warehouse 3 worksheet to a new worksheet and rename it **Warehouse 3 Solver**. Use this worksheet to enter the formulas for the lease cost savings analysis.

Source: Used with permission from Microsoft Corporation

FIGURE 9-12 Warehouse 3 spreadsheet with lease cost savings analysis added

The new section includes the following calculations:

- Monthly Operating Cost Savings—The value in cell K40 is the Inventory Operating Cost per period from cell K36 in the Warehouse 3 worksheet, minus the Inventory Operating Cost per period from cell K36 in the worksheet you just created. The result will be zero until you run Solver. If you do not know how to link a value from one worksheet into a formula in another worksheet within the same Excel workbook, ask your instructor for help.
- Monthly Lease–20,000 Cu. Ft.—Enter −$1,000 into cell K41.
- Net Operating Cost Savings/Month—Cell K42 holds the sum of cells K40 and K41. This number is −$1,000 before you run Solver for the new constraint values. If the lease proposal is profitable, this number will be positive after you run Solver.

Run Solver one more time—you should not have to change any setup parameters. Next, create Answer Report 3. Should Appliance World lease the additional warehouse space and increase its inventory budget?

ASSIGNMENT 2: USING THE WORKBOOK FOR DECISION SUPPORT

You have built a series of worksheets to create an optimal inventory management model for Appliance World, to incorporate safety stock into your inventory, and finally to examine the economic feasibility of added storage capacity. You now complete the case by using your solutions and Answer Reports to gather the data needed for Appliance World to make decisions about its inventory policy. You also document your recommendations in a memorandum.

Assignment 2A: Using Your Workbook to Gather Data

Print each worksheet and Answer Report. Print the worksheets in landscape orientation so each fits on one page. Answer Reports are usually impossible to fit on one page, so your instructor may require you to print at least the first page because it contains the original and optimized values for the inventory operating cost and optimized order quantities. If your instructor requires you to print the entire report, resize the columns so that the report fits in portrait orientation, and then print it horizontally.

Assignment 2B: Documenting Your Recommendations in a Memo

Use Microsoft Word to write a memo to the management team at Appliance World. Summarize the results of your analysis, and include your recommendations for proposed changes to storage capacity and inventory budget. Your memo should conform to the following requirements:

- Set up your memo as described in Tutorial E.
- In the first paragraph, briefly describe the situation and state the purpose of your analysis.
- Next, summarize the results of your analysis and state your recommendation for increasing the storage capacity and inventory budget. Your recommendation should be based on your cost savings analysis.
- Support your recommendations with appropriate screen shots from your worksheets. You can use the Microsoft Paint utility to capture and edit screen shots of your worksheets.

The storage capacity and inventory budget constraints deserve further consideration. What is the ideal amount of storage space to optimize the inventory operating cost? Theoretically, it would be the amount needed to achieve sufficient storage for the economic order quantities of each item, plus their safety stocks at a designated service level. This amount would have to be weighed against the cost of additional storage as well as the amount of cash tied up in inventory assets. You could also recommend that Appliance World examine the issue of inventory turnover in future studies; turnover is a measure of how quickly inventory moves through a distribution system. Your instructor might provide guidance on how to address these issues in your report.

ASSIGNMENT 3: GIVING AN ORAL PRESENTATION

Your instructor may request that you summarize your analysis and recommendations in an oral presentation. If so, prepare a presentation for Harry Murvin, Susan Barnes, and other managers that lasts 10 minutes or less. When preparing your presentation, use Microsoft PowerPoint slides or handouts that you think are appropriate. Tutorial F explains how to prepare and give an effective oral presentation.

DELIVERABLES

Prepare the following deliverables for your instructor:

1. A printout of the memo
2. Printouts of all your worksheets and Answer Reports
3. Your Word document, Excel workbook, and PowerPoint presentation on electronic media or sent to your course site as directed by your instructor

Staple the printouts together with the memo on top. If you used more than one Excel workbook file for your case, write your instructor a note that describes the different files.

PART 4

DECISION SUPPORT CASE
USING BASIC EXCEL FUNCTIONALITY

THE SOCIAL SAFETY NET ANALYSIS

Decision Support Using Excel

PREVIEW

A large country operates a retirement program called the Social Safety Net. Via payroll taxes, citizens pay into the fund throughout their working years. When they retire, citizens receive a yearly pension from the fund. A high percentage of the workforce will retire in the next 20 years, and the fund's administrators want to know if the fund has enough money to cover required payments. In this case, you will use Excel to answer that question.

PREPARATION

- Review spreadsheet concepts discussed in class and in your textbook.
- Complete any exercises that your instructor assigns.
- Complete any parts of Tutorials C and D that your instructor assigns, or refer to them as necessary.
- Review the Microsoft Windows file-saving procedures in Tutorial C.
- Refer to Tutorials E and F as necessary.

BACKGROUND

During the Great Depression, large segments of the population lived in poverty, particularly older people, who had few job prospects and little savings. To reduce this hardship, the country's legislature established the Social Safety Net fund. Under the new law, working people agreed to have their wages and salaries taxed to establish the fund. From this fund, retirees were paid an amount that was sufficient to stave off poverty.

Thus, the country's younger people agreed to take care of elderly citizens. The fund was designed to survive in perpetuity, so that sacrifices made by young people would be recouped when they retired.

The country grew and prospered greatly in the ensuing decades. For many years, taxes paid into the fund exceeded payments to retirees, and the fund balance grew to exceed $2 trillion. However, in the past decade, the country's economy has changed and the fund has been affected:

- The country has not fully recovered from the severe recession of 2008 and 2009, although the economy has resumed growing at a low rate. Newly created jobs often offer low pay or are part-time. As wages and salaries have stagnated, so have payroll taxes that go into the fund.
- The payroll tax has remained at 12.4 percent for many years. Of this tax, 6.2 percent is paid by the worker and 6.2 percent is paid by the employer on the first $110,000 of pretax income. The current political climate is anti-tax, so there is no chance of raising the tax rate to increase the size of the fund.
- The birth rate after World War II was high. The post-war generation, also known as the "baby boomers," is very large and has begun to retire. The size of that generation, coupled with declining birth rates in recent decades, means that fewer workers are available to support retirees.

As a result of the preceding factors, it is now possible for fund payments to exceed taxes paid into it. The problem is aggravated by a few other financial factors:

- The trust fund is not actually held in cash. The government's Treasury Department uses money from the fund to pay for national defense, national parks, and other aspects of the country's business. The Treasury gives the fund special, nonnegotiable bonds. If the fund needs cash, the Treasury redeems the bonds. Naturally, the Treasury pays interest on bonds to the fund. The interest rate is substantially higher than the market rate. For example, the Treasury currently borrows at less than 2 percent on the open market, but it pays the fund more than 4 percent on its bonds. In effect, the Treasury is subsidizing the country's retirees with the augmented interest payments each year. In fact, the Treasury takes the same approach with all the trust funds that it administers, including the retired military pension fund. In tight economic times, the question arises: Can the country afford to continue this subsidy?

- Benefit payments are indexed to the increase in the cost of living each year. This is called the COLA, or cost of living adjustment. For example, if inflation is 4 percent in a year, retirees are protected because their benefit payment is increased by 4 percent. The COLA is a generous idea, but it leads to a couple of problems. First, compounded interest, even at a low rate, can add up over time, so benefit payments can increase greatly. Second, the government has often erred on the high side when setting the COLA rate. Thus, if the inflation rate is 3 percent, for example, the safety net COLA might be set at 3.5 percent. This practice satisfies retirees, but it also tends to inflate fund payments.

- Taxes paid into the fund are based on the first $110,000 of pretax earnings. Thus, if a household's pretax income is $110,000, it pays Social Safety Net taxes of $13,640 ($110,000 multiplied by the payroll tax of 12.4 percent). If a household's pretax income is $110 *million*, the taxes are the same: $13,640. This policy strikes some observers as unfair, especially in recent years, when the distribution of income and wealth in the country has become more skewed. The top 20 percent of households, or the top quintile, earns over 50 percent of the country's pretax income. Some people argue that the pretax earnings threshold should be raised to increase payroll tax receipts.

Administrators of the Social Safety Net have just finished a multiyear projection of economic and demographic factors that affect the fund. Administrators are most interested in the fund's ability to pay benefits through 2030. If the fund can remain solvent until then, administrators think the fund probably can survive because the baby boomers will be retired and their effects will be mostly absorbed. However, if the fund cannot remain solvent through 2030, major changes will be needed to save it. The administrators have asked you to create a spreadsheet model of the fund's status as a way to guide decision making.

ASSIGNMENT 1: CREATING A SPREADSHEET FOR DECISION SUPPORT

In this assignment, you produce a spreadsheet that models the problem. Then, in Assignment 2, you write a memorandum that explains your findings. In Assignment 3, you may be asked to prepare an oral presentation of your analysis.

A spreadsheet has been started and is available for you to use. If you want to use the spreadsheet skeleton, locate Case 10 in your data files and then select **SSNEstimator.xlsx**. Your worksheet should contain the following sections:

- Constants
- Inputs
- Summary of Key Results
- Calculations

A discussion of each section follows.

Constants Section

Your spreadsheet should include the constants shown in Figure 10-1. An explanation of the line items follows the figure. Your model runs through 2030, but values through 2018 only are shown in Figure 10-1.

	A	B	C	D	E	F	G
1	SOCIAL SAFETY NET ESTIMATOR						
2							
3	CONSTANTS	2013	2014	2015	2016	2017	2018
4	Expected Payroll tax receipts	NA	$ 665,900,000,000	$ 706,000,000,000	$ 749,500,000,000	$ 796,400,000,000	$ 844,400,000,000
5	Expected total Benefit payments, no COLA	NA	$ 712,941,176,471	$ 758,627,450,980	$ 807,156,862,745	$ 860,294,117,647	$ 919,117,647,059
6	Number of households in top quintile	NA	25,250,000	25,502,500	25,757,525	26,015,100	26,275,251
7	Average pretax household income, top quintile	NA	$ 324,450	$ 334,184	$ 344,209	$ 354,535	$ 365,171
8	Number of households in fourth quintile	NA	$ 24,240,000	$ 24,482,400	$ 24,727,224	$ 24,974,496	25,224,241
9	Average pretax household income, fourth quintile	NA	$ 113,300	$ 116,699	$ 120,200	$ 123,806	$ 127,520
10	Interest rate, trust fund	NA	4%	4%	4%	4%	4%
11	Expected income tax rate on benefit payments	NA	5%	5%	5%	5%	5%

Source: Used with permission from Microsoft Corporation

FIGURE 10-1   Constants section

- Expected Payroll tax receipts—Administrators have projected the number of workers contributing to the fund, the number of retirees drawing from the fund, increases in gross domestic product (GDP), and inflation rates. Based on these projections, payroll taxes paid into the fund have been estimated through 2030. Note that the GDP is projected to increase at an average of 3.5 percent, which is optimistic.
- Expected total Benefit payments, no COLA—This amount is the total payments to retirees each year before COLA is added.
- Number of households in top quintile—This row holds the number of households in the top 20 percent of pretax earnings.
- Average pretax household income, top quintile—This amount is the average income for a household in the top quintile.
- Number of households in fourth quintile—This row holds the number of households in the next highest level of pretax earnings.
- Average pretax household income, fourth quintile—This amount is the average income for a household in the fourth quintile.
- Interest rate, trust fund—The Treasury is projected to pay 4 percent on trust fund bonds in the period under review.
- Expected income tax rate on benefit payments—Retirees do pay taxes on benefits payments, although not much on average. Historically, the rate has been 5 percent. The fund is credited for income taxes collected.

Inputs Section

Your spreadsheet should include the inputs shown in Figure 10-2. An explanation of the line items follows the figure.

	A	B	C	D	E	F
13	INPUTS					
14	COLA percentage in all years (.XX)		GDP Deflator (.XX)		Interest Rate Deflator (.XX)	
15	Additional payroll tax base, all years ($XX)					

Source: Used with permission from Microsoft Corporation

FIGURE 10-2   Inputs section

- COLA percentage in all years (.XX)—Enter a decimal number for the COLA each year. For example, if 4 percent is expected, enter .04. Note that 3.5 percent is anticipated by fund administrators.
- GDP Deflator (.XX)—As noted previously, the payroll tax receipts projection is for a GDP growth of 3.5 percent each year, which might be too high. If you want to see results with a GDP growth of 2 percent, enter .02 in this cell. Your instructor may specify conditional formatting so that values greater than .035 are flagged.
- Interest Rate Deflator (.XX)—As noted previously, interest earnings are projected at 4 percent, which might be too high. If you want to see results with a 2 percent rate, enter .02 in this cell. Your instructor may specify conditional formatting so that values greater than .04 are flagged.

- Additional payroll tax base, all years ($XX)—The tax base, or pretax income threshold, is $110,000. For example, if you want to see results with a base of $115,000, you would enter 5000 in this cell.

The cells should be formatted appropriately for numbers, currency, or percentages. You should insert a comment in each of the cells and use conditional formatting in one or more cells. The existence of a comment is indicated by a diamond in the upper-right corner of a cell; place the mouse pointer over the diamond to see the comment. To enter a comment, right-click a cell and choose Insert Comment from the menu. Conditional formatting is available from the Styles group on the Home tab; click the Conditional Formatting button, and then click Highlight Cells Rules from the drop-down menu.

Summary of Key Results Section

Your worksheet should include the key results shown in Figure 10-3. The values are echoed from other parts of your spreadsheet. An explanation of the line items follows the figure. Your model extends through 2030, but only selected years are shown in this section.

	A	B	C	D
17	**SUMMARY OF KEY RESULTS**	**2017**	**2022**	**2030**
18	Primary Surplus (Deficit)			
19	Total Surplus (Deficit)			
20	Trust Fund Assets			
21	Expected Payments exceed Assets?			

Source: Used with permission from Microsoft Corporation

FIGURE 10-3 Summary of Key Results section

- Primary Surplus (Deficit)—This value is the sum of payroll tax receipts and income taxes on benefits, minus benefits paid in the year.
- Total Surplus (Deficit)—This value is the sum of payroll tax receipts, income taxes on benefits, and interest earned, minus benefits paid in the year.
- Trust Fund Assets—This value is the fund balance at the end of the year.
- Expected Payments exceed Assets?—This value (Yes or No) indicates whether the expected benefit payments in the year exceed the fund assets at the start of the year.

Calculations Section

The first few years of data in the Calculations section are shown in Figure 10-4. Your model runs through 2030, but the figure shows values only through 2018. Values are calculated by formula, not hard-coded. Use cell addresses when referring to constants in formulas, unless otherwise directed. Use absolute addressing properly. An explanation of the line items follows the figure.

	A	B	C	D	E	F	G
23	**CALCULATIONS**	**2013**	**2014**	**2015**	**2016**	**2017**	**2018**
24	Expected total Benefit payments, with COLA	NA					
25	Overstatement of payroll tax receipts -- GDP deflator	NA					
26	Additional payroll tax receipts -- 4th quintile	NA					
27	Additional payroll tax receipts -- top quintile	NA					
28	Total Payroll tax receipts	NA					
29	Income Taxes received on benefits	NA					
30	Payroll tax receipts plus income taxes on benefits	NA					
31	Interest earned on Trust Fund Assets	NA					
32	Total Safety Net Income	NA					
33	Trust Fund balance at year-end	$ 2,687,400,000,000					
34							
35	Primary Surplus (Deficit) in Year	NA					
36	Total Surplus (Deficit) in Year	NA					
37	Expected Payments exceed Assets?	NA					

Source: Used with permission from Microsoft Corporation

FIGURE 10-4 Calculations section

- Expected total Benefit payments, with COLA—This amount is the expected total benefit payments from the Constants section increased by the COLA percentage from the Inputs section.
- Overstatement of payroll tax receipts – GDP deflator—This amount is a function of the GDP Deflator percentage from the Inputs section and expected payroll tax receipts from the Constants section.
- Additional payroll tax receipts – 4th quintile—This amount is a function of the additional payroll tax base from the Inputs section, the tax rate of 0.124, and the number of households in the

fourth quintile, which is a value from the Constants section. Because the tax rate is always 0.124, it can be hard-coded. For example, if the base is increased by $1,000 and 10 million households are paying, the added receipts would be $1,000 * 10,000,000 * 0.124.

- Additional payroll tax receipts – top quintile—This amount is a function of the additional payroll tax base from the Inputs section, the tax rate of 0.124, and the number of households in the top quintile, which is a value from the Constants section. Because the tax rate is always 0.124, it can be hard-coded.
- Total Payroll tax receipts—This value is the expected payroll tax receipts from the Constants section, plus additional receipts from the fourth and top quintiles, minus overstated receipts. Except for the expected payroll tax receipts, all the values for this calculation come from other cells of the Calculations section.
- Income Taxes received on benefits—This amount is a function of benefits paid to retirees (from the Calculations section) and the tax rate from the Constants section.
- Payroll tax receipts plus income taxes on benefits—This value is the sum of total payroll taxes received and income taxes received.
- Interest earned on Trust Fund Assets—This amount is a function of trust fund assets at the beginning of the year and the interest rate from the Constants section.
- Total Safety Net Income—This value is the sum of payroll tax receipts, taxes on benefits, and interest earned.
- Total Fund balance at year-end—This value is the fund balance at the beginning of the year, plus total safety net income, minus expected total benefit payments with COLA.
- Primary Surplus (Deficit) in Year—This amount equals payroll tax receipts plus income taxes on benefits, minus total benefit payments with COLA. Both values in the equation come from other cells of the Calculations section. This value contrasts cash received by the Treasury with cash to be paid out.
- Total Surplus (Deficit) in Year—This amount equals total safety net income minus total benefit payments with COLA. This value contrasts cash received plus interest received with cash to be paid out. The cash plus interest received would be in the form of additional bonds.
- Expected Payments exceed Assets?—If the year's expected benefit payments with COLA exceed trust fund assets at the start of the year, enter YES in this cell. Otherwise, enter NO. This value is an indicator that trust fund assets are nearly exhausted and the fund is operating from year to year.

ASSIGNMENT 2: USING THE SPREADSHEET FOR DECISION SUPPORT

You will now complete the case by (1) using the spreadsheet model to gather data needed to answer questions about the fund, (2) documenting your findings in a memo, and (3) giving an oral presentation if your instructor requires it.

Assignment 2A: Using the Spreadsheet to Gather Data

You have built the spreadsheet to create "what-if" scenarios with the model's input values. The inputs represent the logic of a question, and the outputs provide information needed to answer the question. The scenarios are based on the following seven questions from fund administrators.

Question 1: What is the fund's status in the base case? In other words, assume that things work out according to the administrator's assumptions: The COLA is 3.5 percent, and there is no added tax base, no GDP deflator, and no interest rate deflator. The inputs are shown in Figure 10-5.

COLA percentage in all years	.035
Additional payroll tax base	0
GDP deflator	0
Interest rate deflator	0

Source: © Cengage Learning 2014

FIGURE 10-5 Input data for Question 1

Enter the inputs and then observe the outputs in the Summary of Key Results area. Next, you should record the results in a summary area by copying the data from the Summary of Key Results cells and pasting it into the summary area. Highlight the area to be copied, and click the Copy button in the Clipboard group of the Home tab. Next, click the Paste button in the Clipboard group, select Paste Special from the menu, and then select Values in the next window. You could use a second worksheet for this purpose, as shown in Figure 10-6.

▲	A	B	C	D
1				
2	**Base Case (COLA=3.5%; Added base=$0; GDP deflator=0%; Interest rate deflator=0%)**			
3		**2017**	**2022**	**2030**
4	Primary Surplus (Deficit)			
5	Total Surplus (Deficit)			
6	Trust Fund Assets			
7	Expected Payments exceed Assets?			

Source: Used with permission from Microsoft Corporation

FIGURE 10-6 Results for Question 1 recorded in summary area

Question 2: What is the fund's status if the legislature decides to eliminate the interest rate subsidy by reducing the rate two percentage points? In other words, how important is interest income to the fund? The inputs are shown in Figure 10-7.

COLA percentage in all years	.035
Additional payroll tax base	0
GDP deflator	0
Interest rate deflator	.02

Source: © Cengage Learning 2014

FIGURE 10-7 Input data for Question 2

Enter the inputs and then observe the outputs in the Summary of Key Results area. Next, you should record the results in a summary area.

Question 3: What is the fund's status if the GDP does not grow as assumed? This scenario would result in lower payroll tax receipts. The inputs are shown in Figure 10-8.

COLA percentage in all years	.035
Additional payroll tax base	0
GDP deflator	.02
Interest rate deflator	0

Source: © Cengage Learning 2014

FIGURE 10-8 Input data for Question 3

You should record the results in a summary area.

Question 4: What is the fund's status if the COLA is eliminated? The legislature might need to eliminate the COLA as an extreme measure if the economy is suffering, which would result in lower benefit payments. The inputs are shown in Figure 10-9.

COLA percentage in all years	0
Additional payroll tax base	0
GDP deflator	0
Interest rate deflator	0

Source: © Cengage Learning 2014

FIGURE 10-9 Input data for Question 4

You should record the results in a summary area.

Question 5: What is the fund's status if the tax base (pretax income threshold) is increased? The legislature might be able to increase receipts using this "back door" method, as opposed to legislating an increase in the tax *rate*. The inputs are shown in Figure 10-10.

COLA percentage in all years	.035
Additional payroll tax base	10000
GDP deflator	0
Interest rate deflator	0

Source: © Cengage Learning 2014

FIGURE 10-10   Input data for Question 5

You should record the results in a summary area.

Question 6: Some members of the legislature might want to make changes to the fund in accordance with their own fiscal philosophy: Pay no COLA, do not increase the tax base, use a realistic GDP estimate, and do not subsidize interest. What is the fund's status if these changes are made? The inputs are shown in Figure 10-11.

COLA percentage in all years	0
Additional payroll tax base	0
GDP deflator	.02
Interest rate deflator	.02

Source: © Cengage Learning 2014

FIGURE 10-11   Input data for Question 6

You should record the results in a summary area.

Question 7: What would be a presumed worst-case scenario for the fund's health? In other words, what would happen if the legislature pays the COLA, does not increase the tax base, assumes little or no GDP increase, and does not subsidize interest? The inputs are shown in Figure 10-12.

COLA percentage in all years	.035
Additional payroll tax base	0
GDP deflator	.03
Interest rate deflator	.02

Source: © Cengage Learning 2014

FIGURE 10-12   Input data for Question 7

You should record the results in a summary area.

When you finish gathering data for the preceding seven questions, print the model's worksheet with any set of inputs. Then save the spreadsheet for the final time by selecting Save from the File menu.

Assignment 2B: Documenting your Findings and Recommendation in a Memorandum

Document your findings in a memo that answers the seven questions from the preceding section. The memo should also summarize your general conclusions about the health of the fund. Use the following guidelines to prepare your memo in Microsoft Word:

- Your memo should have proper headings such as Date, To, From, and Subject. You can address the memo to the administrators of the Social Safety Net fund. You should set up the memo as discussed in Tutorial E.
- Briefly outline the situation. However, you need not provide much background—you can assume that readers are familiar with the situation.
- List and then answer the seven questions in the body of the memo.
- What general conclusions can you draw about the status of the fund? What factors seem most important for financial health? Administrators would be concerned if the fund balance dipped

below $1 trillion or if expected payments exceeded fund assets. What situations would cause these events?

- State your general conclusion about the health of the Social Safety Net fund. Does the program require major changes, or will it survive in its current state?
- Include tables and charts to support your claims, as your instructor specifies. Tutorial E explains how to create a table in Microsoft Word.

ASSIGNMENT 3: GIVING AN ORAL PRESENTATION

Your instructor may also request that you explain your analysis and results in an oral presentation. If so, assume that the Social Safety Net administrators are impressed by your analysis and want you to give a presentation to explain the results. Prepare to talk to the group for 10 minutes or less. Use visual aids or handouts that you think are appropriate. Tutorial F explains how to prepare and give an oral presentation.

DELIVERABLES

Assemble the following deliverables for your instructor:

1. Printout of your memo
2. Spreadsheet printouts
3. Flash drive or CD that includes your Word file and Excel file

Staple the printouts together with the memo on top. If you have more than one .xlsx file on your flash drive or CD, write your instructor a note that identifies your spreadsheet model's .xlsx file.

PART 5

INTEGRATION CASES USING ACCESS AND EXCEL

CASE **11**

THE GOLF PRO EFFECTIVENESS ANALYSIS

Decision Support with Microsoft Access and Excel

PREVIEW

Four professional golfers gave lessons to members of a country club. Is it possible to determine how well the golfers taught their students? You can get an answer by computing the students' handicaps before and after the lessons to see which students benefitted from instruction and which did not. In this case, you will use Access and Excel to perform the analysis.

PREPARATION

- Review database and spreadsheet concepts discussed in class and in your textbook.
- Complete any exercises that your instructor assigns.
- Complete any parts of Tutorials B, C, and D that your instructor assigns, or refer to them as necessary.
- Review file-saving procedures for Windows programs, as discussed in Tutorial C.
- Refer to Tutorials E or F as necessary.

BACKGROUND

A country club might consider hiring a golf professional for several reasons:

- A golf swing is a difficult athletic move.
- Golfers want to improve.
- Self-diagnosis rarely works to cure swing problems.
- The advice of a teaching professional can help a golfer cure swing problems.

Based on these reasons, you might think that improvement in golf would be straightforward: A golfer who is having problems with his game should simply consult a pro for lessons. Unfortunately, not all golf pros are created equal. Some pros are good teachers, and others are not. Some professionals talk a good game but cannot help lesser golfers improve.

Your local country club has nearly 1000 members, and its hard-working pro is retiring soon. Club management has decided to hire two professionals to replace him. Many local teaching pros applied, and the list of candidates has been reduced to four. Each candidate plays the game well, can talk about golf in an interesting way, and seems to establish rapport quickly.

To select the two pros for the job, club management decides to have the four candidates audition by assigning them to teach members who need help. Each amateur golfer's handicap will be computed two months before the end of the golfing season, and then each amateur will receive lessons from one of the professionals. Each amateur's handicap will be calculated again during the next golf season, and the two pros whose student players improve the most will be hired.

Before you learn more about the country club's job competition, you need to understand some basic principles of golf. If you already play golf, you can skip the following information:

- A golf course is divided into 18 segments called holes. Each hole begins at a tee box, proceeds along an expanse of cut grass called the fairway, and ends at a small area called a green.
- Golfers strike the ball with various clubs to advance from tee to green. The green is a roughly circular, closely mown area that contains a hole. The object is to get the ball in the hole, usually

by means of a putt—a short stroke with a flat club called a putter that is used exclusively on the green. A golfer's score on a hole is the number of strokes required to get the ball in the hole.

- Each hole has a par, which is an established number of strokes needed to complete the hole. Par can be 3, 4, or 5 for a hole, depending on the distance between tee and green. On a par 3 hole, the golfer is expected to hit the ball from the tee box to the green, and then putt the ball twice to complete the hole in par. On the longer par 4 and par 5 holes, the golfer needs additional shots to reach the green.
- The better a golfer plays, the fewer strokes he or she needs to complete a hole. Most amateurs struggle to make pars.
- Golf courses have a variety of par 3, par 4, and par 5 holes. Typically, the total par for a round of 18 holes is 70, 71, or 72.
- A player's score for a round is the total number of strokes needed to complete the 18 holes. Golfers typically characterize their score in relationship to par for the course. For example, if par is 72 and a golfer completes the course in 85 strokes, the golfer played the round in 13 strokes over par.
- A handicap is a customized number that indicates how well a particular golfer is expected to play. A skillful golfer might have a handicap at or near zero, meaning that the golfer would be expected to shoot at or near par for 18 holes. A poor player would have a high handicap. For example, a golfer with a 35 handicap would be expected to shoot at least 35 over par for 18 holes.

You might think that calculating a golfer's handicap would be straightforward: Subtract par from the golfer's score for each round played, sum the differences, and then divide the total by the number of rounds played. However, this approach has three problems:

- First, golfers do not keep a record of every round they have ever played. More important, a golfer's ability can change over time, so it's necessary to base the handicap on recent scores. Most handicapping systems use the most recent 20 to 30 rounds played to compute a golfer's handicap.
- The next problem is presumably one of vanity. Many players think their good rounds are more representative of their abilities than their poor rounds. Thus, most handicapping systems exclude a certain number of a golfer's higher scores in recent rounds.
- A handicap should be predictive of a golfer's score on any course, not just the golfer's home course. However, courses have varying levels of difficulty. For example, a golfer who has a handicap of 5 based on playing an easier course would probably not shoot a similar score on a more difficult course. Handicapping systems deal with this problem by adjusting their calculations to compare a course's perceived difficulty with that of a "standard" course. The difficulty rating for a course is called its slope. Periodically, each course is studied to determine its slope. The more difficult the course is perceived to be, the higher the slope rating. A course with average difficulty has a slope rating of 113. The slope rating at your country club is 120, which means that your course is considered more difficult than the average golf course.

Now that you have learned a few principles of golf, you can return to the example of your country club. The club has the following method for computing a golfer's handicap:

- Identify the most recent 20 scores.
- Select the 10 lowest scores from the 20 rounds.
- For each of the 10 lowest scores, determine the "handicap differential": the score minus par for the course.
- Multiply each of the 10 handicap differentials by .96.
- Compute the average of the 10 reduced handicap differentials.
- Multiply the average by the ratio of 113 to the course's slope rating, 120. The result is the golfer's handicap based on play at your club.

For example, assume that your friend has played the following low 10 rounds in his most recent 20 outings: 80, 86, 80, 86, 80, 86, 80, 86, 80, and 86. Because par at your club course is 70, the handicap differentials are 80–70, 86–70, 80–70, 86–70, 80–70, 86–70, 80–70, 86–70, 80–70, and 86–70. The sum of the reduced handicap differentials is 124.8, and the average is 12.48. Therefore, your friend's handicap is (12.48) * (113/120) = 11.75. Multiplying by the ratio 113/120 reduces the handicap to account for the fact that your course is harder than average.

Your club's management believes that golfers will benefit from lessons given by a skilled teaching pro; the proof of effective instruction will be lower scores by club golfers after the lessons. If a golfer takes lessons from a poor teaching pro, scores should not improve. The club's management has decided to use the following procedure to determine a pro's effectiveness:

1. Compute the golfer's handicap before instruction.
2. Assign a pro to the golfer for a set of lessons.
3. Have the golfer play 20 rounds. Compute a new handicap for the golfer based only on rounds played after the lessons.
4. Subtract the post-instruction handicap from the pre-instruction handicap. If the result is positive, the golfer benefitted from the pro's instruction. If the result is negative, the golfer did not benefit.

Thirty golfers have volunteered for lessons to help determine which two pros the club will hire. Their scores in the year before lessons begin are recorded in an Excel spreadsheet called **FirstYearScores.xlsx**. If you want to use this file to work through the case, select Case 11 from your data files and open the spreadsheet. The data for player 1 looks like that in Figure 11-1.

	A	B	C
1	**Player Number**	**Score Date**	**Score**
2	1	23-May-12	80
3	1	30-May-12	78
4	1	6-Jun-12	78
5	1	19-Jun-12	80
6	1	27-Jun-12	77
7	1	6-Jul-12	83
8	1	9-Jul-12	82
9	1	18-Jul-12	83
10	1	22-Jul-12	80
11	1	26-Jul-12	81
12	1	31-Jul-12	77
13	1	8-Aug-12	80
14	1	20-Aug-12	81
15	1	22-Aug-12	78
16	1	29-Aug-12	79
17	1	11-Sep-12	82
18	1	14-Sep-12	78
19	1	15-Sep-12	80
20	1	22-Sep-12	81
21	1	5-Oct-12	79

Source: Used with permission from Microsoft Corporation

FIGURE 11-1 Pre-instruction scores for player 1 in FirstYearScores spreadsheet file

For example, on May 23, 2012, player 1 shot a score of 80 at your course. One week later, he shot a 78. The spreadsheet contains a worksheet of 20 recent scores for each of the 30 players.

A corresponding Excel spreadsheet called **SecondYearScores.xlsx** contains scores for the 30 players after they had lessons from the four pros. This file is also available in your Case 11 data files. The organization of the two spreadsheets is identical: Each contains separate worksheets for 30 players, with 20 scores in each worksheet.

The FirstYearScores spreadsheet also includes a worksheet that assigns a number to each player and shows which pro is assigned to each player. The top part of the worksheet looks like Figure 11-2.

	A	B	C
1	**Player Number**	**Player Name**	**Pro**
2	1	Brody	Palmer
3	2	Smith	Snead
4	3	Xavier	Nicklaus
5	4	Mathis	Palmer
6	5	Mantle	Hogan
7	6	Maris	Snead

Source: Used with permission from Microsoft Corporation

FIGURE 11-2   Players and their assigned professionals

For example, player 1 is named Brody. Brody's professional is named Palmer.

Your analysis begins in Excel. To help you determine handicaps, you will use an Excel macro to sort the data and eliminate the highest 10 scores for each golfer. Then you will use Microsoft Access to finish your analysis.

ASSIGNMENT 1: SORTING SCORE DATA AND ELIMINATING THE 10 WORST SCORES

In this assignment, you create a macro that sorts a player's 20 scores from lowest to highest and then eliminates the highest 10 scores. Figure 11-3 shows player 1's worksheet in FirstYearScores.xlsx after you have run the macro.

	A	B	C
1	**Player Number**	**Score Date**	**Score**
2	1	27-Jun-12	77
3	1	31-Jul-12	77
4	1	30-May-12	78
5	1	6-Jun-12	78
6	1	22-Aug-12	78
7	1	14-Sep-12	78
8	1	29-Aug-12	79
9	1	5-Oct-12	79
10	1	23-May-12	80
11	1	19-Jun-12	80

Source: Used with permission from Microsoft Corporation

FIGURE 11-3   Player 1 data after sorting scores and deleting top 10 scores

As you can see, your macro will change the contents of the Excel files. You want to be able to recover the original data in case you make a mistake. Thus, your first step is to create a backup folder. Next, save copies of the FirstYearScores.xlsx and SecondYearScores.xlsx files in this folder.

You will perform the same steps of sorting scores and eliminating the 10 highest scores in each of the 30 worksheets. First, you sort 20 records on the Score field, from lowest to highest. Then you clear the contents of the bottom 10 records. Performing these steps manually in all 30 worksheets would be laborious and lead to errors.

Excel macros let you automate such repetitive tasks. You create the macro in the macro recorder, which captures the Excel steps in sequence as a stored program. After you have recorded the macro, you can run it in each worksheet. Record the macro now.

If necessary, see Tutorial E for instructions on recording macros. The discussion there is closely aligned with the requirements of this case.

When you save a macro in a spreadsheet, Excel saves and closes the .xlsx file and creates a new macro-enabled .xlsm file. Thus, your FirstYearScores.xlsx file is replaced by a corresponding **FirstYearScores.xlsm** file.

Apply your macro to the 30 player worksheets in FirstYearScores.xlsm.

In Assignment 2, you import this data into Access. Rather than import from 30 worksheets, you want to import all the data from just one worksheet. Therefore, you need to create a new worksheet and give it an appropriate name, such as PlayerSummary. Manually copy the player 1 column headers into row 1 of the new worksheet. Then, copy the 10 player 1 records into the new worksheet. Perform the same steps manually for each player; in other words, do not use a macro. When you finish, you should have one header row and 300 score records. Check your work before you continue.

Save and close the FirstYearScores.xlsm file.

Next, record the same macro in the **SecondYearScores.xlsm** file. Apply the macro to the 30 player worksheets in the spreadsheet. Create a new worksheet, manually copy the score records into this summary sheet, and verify that you have 300 rows of player scores when you finish.

Save and close the SecondYearScores.xlsm file.

ASSIGNMENT 2: COMPUTING HANDICAPS AND RANKING TEACHING PROFESSIONALS

You can now compute handicaps for each player and determine which teaching professionals were most effective. An Access file called **GolfPros.accdb** has been provided to assist you. It is available in your Case 11 data files.

The file has three tables. The Players table shows which pro is assigned to each player. The FirstYearScores table holds year 1 player scores before they received instructions from the pros. The SecondYearScores table holds year 2 player scores after golfers received professional instruction.

The design of the Players table is shown in Figure 11-4.

Players	
Field Name	Data Type
Player Number	Text
Player Name	Text
Pro	Text

Source: Used with permission from Microsoft Corporation

FIGURE 11-4   Design of Players table

Notice that a primary key is not designated, which is acceptable for this case. The design of the FirstYearScores table is shown in Figure 11-5.

FirstYearScores	
Field Name	Data Type
Player Number	Text
Score Date	Date/Time
Score	Number

Source: Used with permission from Microsoft Corporation

FIGURE 11-5   Design of FirstYearScores table

Again, a primary key is not designated, which is acceptable for this case. The SecondYearScores table has the same design.

Next, you need to import data from Excel into your Access tables as follows:

- Import the data from the Players worksheet in the FirstYearScores.xlsm file to the Players table in the GolfPros.accdb file.
- Import the data from the summary worksheet in the FirstYearScores.xlsm file to the FirstYearScores table in the GolfPros.accdb file. This import moves 300 records at once.
- Import the data from the summary worksheet in the SecondYearScores.xlsm file to the SecondYearScores table in the GolfPros.accdb file. This import also moves 300 records at once.

If you need help with the preceding steps, Tutorial B explains how to import data from Excel into Access. Next, you design and run several queries.

Queries for Year 1 and Year 2 Handicap Differentials

Create a query to compute handicap differentials for each golfer in year 1. As you learned earlier, the formula is: ((score – 70) * (113/120)). Use the built-in Round function in the calculated field to show one decimal in the following formula:

$$\text{Round}((\text{golfer's score} - 70) * (113/120)), 1)$$

Format the output field as a Standard number with one decimal. Your output should look like that in Figure 11-6.

HDYr1	
Player Number	HD1
1	9.4
1	6.6
1	6.6
1	7.5
1	7.5
1	7.5
1	7.5
1	8.5
1	8.5
1	9.4
10	11.3
10	14.1

Source: Used with permission from Microsoft Corporation

FIGURE 11-6   Year 1 handicap differentials

The query creates 300 output records, although only a few are shown in Figure 11-6. Save the query as HDYr1.

You should also create a query for year 2 handicap differentials named HDYr2.

Queries for Handicaps in Year 1 and Year 2

Create a query to compute year 1 handicaps. These handicaps are based on rounds played before professional instruction. Recall from Tutorial B that you can base a query on output from another query. Here, you can base your query on the HDYr1 query from the preceding section. As you learned earlier, the club's procedure is to multiply handicap differentials by .96 and then take the average for each player. Format the output field as a Standard number with one decimal. Your output should look like that in Figure 11-7.

HandicapsYr1	
Player Number	Handicap1
1	7.6
10	11.1
11	8.0

Source: Used with permission from Microsoft Corporation

FIGURE 11-7   Pre-instruction handicaps

The query creates 30 output records, although only a few are shown in Figure 11-7. Save the query as HandicapsYr1.

You should also create a query for year 2 handicaps. These handicaps are based on rounds played after professional instruction. Name the query HandicapsYr2.

Query for Year 1 (Pre-Instruction) and Year 2 (Post-Instruction) Results

Create a query to show year 1 results from all players along with the players' names. Recall from Tutorial B that you can base a query on a combination of table records and output from another query. Here, you can base your query on the HandicapsYr1 query and the Players table. Format the numerical output as a Standard number with one decimal. Your output should look like that in Figure 11-8.

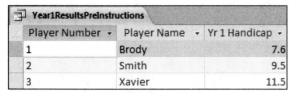

Source: Used with permission from Microsoft Corporation

FIGURE 11-8   Year 1 results query output

The query creates 30 output records, although only a few are shown in Figure 11-8. Save the query as Year1ResultsPreInstructions.

You should also create a query for year 2 results. These results are from rounds played after professional instruction. Save the query as Year2ResultsPostInstruction.

Query for Year 1 to Year 2 Improvement

Create a query to show the handicap change for each golfer from year 1 to year 2. You can base this query on your queries of year 1 results and year 2 results from the previous section. Format the numerical output as a Standard number with one decimal. Your output should look like that in Figure 11-9.

Player Number	Player Name	Hcap Change	Pro
1	Brody	0.3	Palmer
10	Unitis	1.5	Snead
11	Luckman	-0.6	Nicklaus

Source: Used with permission from Microsoft Corporation

FIGURE 11-9   Change in handicaps

The query creates 30 output records, although only a few are shown in Figure 11-9. Save the query as Yr1 to Yr2 Improvement.

Query for Average Handicap Improvement by Teaching Professional

Create a query to show the average handicap improvement as a result of instruction from each of the four teaching pros. You can base this query on the output of the Yr1 to Yr2 Improvement query. Format the numerical output as a Standard number with one decimal. Your output should consist of four records, as shown by the illustrative data in Figure 11-10.

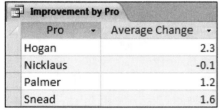

Pro	Average Change
Hogan	2.3
Nicklaus	-0.1
Palmer	1.2
Snead	1.6

Source: Used with permission from Microsoft Corporation

FIGURE 11-10   Handicap improvement by professional

Save the query as Improvement by Pro.

ASSIGNMENT 3: DOCUMENTING FINDINGS IN A MEMORANDUM

In this assignment, you write a memo in Microsoft Word that documents your findings and recommendations. You should briefly describe the country club's hiring decision and the procedure used to select the two new pros. Your memo should then state the results of the analysis and recommend which two pros should get the two positions.

In your memo, observe the following requirements:

- Your memo should have proper headings such as Date, To, From, and Subject. You can address the memo to the managers of the country club. Set up the memo as discussed in Tutorial E.
- Briefly outline the situation. However, you need not provide much background—you can assume that readers are familiar with the situation.
- State your findings and recommendations in the body of the memo.
- Support your work by showing your recommendations in a table. Tutorial E describes how to create a table in Microsoft Word.

Your instructor might also request that you summarize your analysis and results to club management in an oral presentation. Prepare to talk to the group for 10 minutes or less. Use visual aids or handouts that you think are appropriate. See Tutorial F for tips on preparing and giving an oral presentation.

DELIVERABLES

Assemble the following deliverables for your instructor:

1. Printout of your memo
2. Printout of your Excel macro; in the macro editor, select File and then select Print
3. Spreadsheet and/or database printouts, as required by your instructor
4. Electronic media such as flash drive or CD, which should contain your Word file, Access file, and Excel files

Staple the printouts together with the memo on top. If you have .xlsx, .xlsm, or .accdb files from other assignments stored on your electronic media, write your instructor a note that identifies the files for this assignment.

THE STATE LOTTERY ANALYSIS

Decision Support with Microsoft Access and Excel

PREVIEW

The winning numbers in your state's pick-five lottery seem nonrandom to a well-known investment advisor. Is there a problem with the way winning numbers are selected? In this case, you will use Access and Excel to analyze the soundness of your state's lottery system.

PREPARATION

- Review database and spreadsheet concepts discussed in class and in your textbook.
- Complete any exercises that your instructor assigns.
- Complete any parts of Tutorials B, C, and D that your instructor assigns, or refer to them as necessary.
- Review file-saving procedures for Windows programs, as discussed in Tutorial C.
- Refer to Tutorials E and F as necessary.

BACKGROUND

Your state has run a pick-five lottery for several years. A player pays one dollar for a ticket and picks five numbers between 1 and 56. The lottery system determines winners by generating five numbers between 1 and 56 and displaying them on ping-pong balls, which are rolled out one at a time during a televised drawing for dramatic effect. If all five numbers match a player's numbers, the player wins a tremendous amount of money. Players who match fewer numbers win less money. Of course, most players match no numbers and gain nothing, except for some entertainment.

Many states have a lottery, and most people think it's good fun. Lottery proceeds are used by state governments to fund programs in education, health care, and other areas.

Most people also think that state lotteries are honest in the sense that winning numbers are randomly chosen. In a random selection, the number 2 is as likely to be chosen as the number 56, for example. People would be disturbed to learn that some numbers are more likely to be chosen than others, and they would probably avoid a lottery that was shown to select winning numbers in a nonrandom manner. So, state lottery officials were alarmed when Dr. Bern E. Ooley, a well-known financial advisor, wrote a recent newspaper article making just such a claim.

Bern Ooley has long enjoyed a solid reputation as an advocate for the common man. He writes a weekly financial advice column in the state's most widely read daily newspaper, and he has a Saturday morning advice show on a popular AM radio station that reaches all corners of your state. Bern has often been correct about identifying turning points in the investment markets and claims that he cannot recall the times he was wrong. "Always interesting and never in doubt" is his motto.

Your state's lottery has had drawings twice a week for almost seven years. Ooley researched each drawing and was troubled by the distribution of numbers in the 706 winning combinations. The distribution is shown graphically in Figure 12-1.

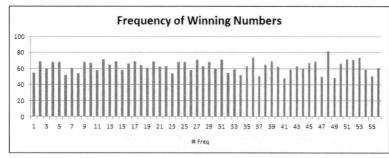

Source: Used with permission from Microsoft Corporation

FIGURE 12-1 Bar graph showing distribution of numbers in winning combinations

For example, the number 48 appeared 82 times in the 706 five-number combinations, but the number 55 appeared only 51 times.

The distribution is shown in tabular form in Figure 12-2.

Num	Freq	Num	Freq	Num	Freq	Num	Freq
1	55	15	58	29	68	43	63
2	69	16	66	30	60	44	60
3	60	17	69	31	71	45	67
4	68	18	64	32	55	46	69
5	68	19	61	33	59	47	50
6	52	20	69	34	52	48	82
7	61	21	62	35	63	49	49
8	54	22	63	36	74	50	66
9	68	23	54	37	51	51	72
10	67	24	68	38	65	52	71
11	58	25	68	39	69	53	74
12	72	26	58	40	62	54	59
13	65	27	71	41	48	55	51
14	69	28	63	42	59	56	61

Source: © Cengage Learning 2014

FIGURE 12-2 Table showing distribution of numbers in winning combinations

Figure 12-3 shows the five numbers that appeared most frequently in the 706 winning combinations.

	Number	Frequency
Most frequent	48	82
2nd most frequent	36	74
3rd most frequent	53	74
4th most frequent	12	72
5th most frequent	51	72

Source: © Cengage Learning 2014

FIGURE 12-3 Table showing five numbers most likely to appear in winning combinations

How can it be, Ooley asks, that these five of the 56 numbers appear much more frequently than other numbers? Such an anomaly might be expected if the lottery had been running only a few weeks, he wrote in his article. "You'd expect lumpiness in the distribution early on," Ooley said, but the lottery has now drawn a total of 3530 numbers in the winning combinations after seven years. The distribution should be evening out after thousands of numbers, but Ooley says it is not. Bern's advice to his listeners and readers is to find another lottery if you must gamble. If you decide to play the state lottery, include some or all of the numbers 48, 36, 53, 12, and 51 in your picks, because these numbers are obviously favored by the lottery's number-selection method.

The problem for state lottery officials is that many people seem to believe Ooley's advice. These officials are convinced that the lottery's methods are sound, and they need some way to reassure the public. You have been called in to help.

You inquire about the lottery's method for selecting numbers, and you learn that the state has a contract with a company that generates random numbers as needed. Five random numbers between 1 and 56 are selected and then transmitted over secure telecommunications channels to the state lottery office. The same company services many state lotteries. The company's random number algorithm is well known and is used by many major software vendors. The company's implementation of the algorithm is certified periodically by independent statisticians. Your state's lottery officials do not understand what could go wrong within that system.

In theory, nothing should go wrong, and professional statisticians would be comfortable with the lottery. But most of the state's citizens are not statisticians, and at first glance, Ooley's criticisms seem credible. Why are some numbers picked more frequently than others? Wouldn't 3530 picks in more than 700 lottery drawings be enough to even out the distribution? You decide to conduct the following test:

1. Obtain recent winning lottery numbers from other states.
2. Examine whether the selection of these numbers is distributed like your state's lottery numbers.
3. If the numbers are similarly distributed, inform the public to reassure them. However, if the numbers are not distributed similarly, further investigation into your state's number-selection method is required.

Lottery officials have provided winning number combinations from pick-five lotteries in three other states. In each case, the most recent 706 winning numbers are provided. You have not been told the names of the states; you know only that the winning numbers come from Big State, Medium State, and Small State. An Access database named LotteryAnalysis.accdb contains the data. To use this database, select Case 12 in your data files, and then select **LotteryAnalysis.accdb**.

Figure 12-4 shows the design of the BigState data in the database file.

Field Name	Data Type
Play Num	Text
Ball 1	Number
Ball 2	Number
Ball 3	Number
Ball 4	Number
Ball 5	Number

Source: Used with permission from Microsoft Corporation
FIGURE 12-4   BigState table design

The values in the Play Num field range from 1 to 706 to represent each lottery drawing. Ball 1 values represent the first number drawn in each lottery; these values range from 1 to 56. The Ball 2, Ball 3, Ball 4, and Ball 5 fields use the same logic as the Ball 1 field.

Figure 12-5 shows a few records in the BigState table.

Play Num	Ball 1	Ball 2	Ball 3	Ball 4	Ball 5
1	43	22	4	9	11
10	18	21	27	19	40
100	3	49	20	13	40
101	3	30	21	22	12

Source: Used with permission from Microsoft Corporation
FIGURE 12-5   BigState table data records

For example, in the first lottery shown, the winning numbers were 43, 22, 4, 9, and 11. In the tenth lottery drawing, the winning numbers were 18, 21, 27, 19, and 40.

ASSIGNMENT 1: USING EXCEL FOR DECISION SUPPORT

In this assignment, you import the data from the three Access tables into Excel worksheets. Then you develop information needed to compare the three state lotteries with your state's lottery.

Importing Table Data

Open a new file in Excel and save it as **LotteryAnalysis.xlsx**.

Import the BigState table records into Excel. Click the Data tab, and then select From Access in the Get External Data group. Specify the Access filename, the BigState table name, and where to place the data in the worksheet (cell A1 is recommended).

The data is imported into Excel as an Excel data table, which is the format you want. If cell A1 is not already selected, click it. In the Table Style Options group of the Table Tools Design tab, select Total Row to add a totals row to the bottom of the table. Rename the worksheet **Big State**. The first few rows of your worksheet should look like Figure 12-6.

	A	B	C	D	E	F
1	Play Num	Ball 1	Ball 2	Ball 3	Ball 4	Ball 5
2	1	43	22	4	9	11
3	2	22	24	51	14	4
4	3	4	56	22	7	24
5	4	4	45	50	39	3
6	5	38	15	45	39	5

Source: Used with permission from Microsoft Corporation
FIGURE 12-6   Rows in the Big State worksheet

Import the Medium State and Small State table records into other worksheets in the same way. Add a totals row to each table. Name the worksheets **Medium State** and **Small State**.

Using Data Tables, Pivot Tables, and Charts to Gather Data

The data tables help you gather some of the information you need. For each of the three states, you use the data tables to develop pivot tables and then use them to tabulate frequencies of winning numbers. You then develop bar graphs to illustrate these frequencies, similar to the bar graph shown in Figure 12-1.

In the Big State worksheet, use the Total row to compute the average of the values shown for each ball. Then, below the Total row, use the =Average() function to compute the average of all values, which should be in the range B2..F707. Your results should look like the illustrative data shown in Figure 12-7.

	A	B	C	D	E	F
703	702	19	44	52	52	18
704	703	28	19	8	54	32
705	704	39	55	45	6	10
706	705	18	36	27	2	1
707	706	23	17	8	45	54
708	Total	28.5949	28.64589	28.43059	29.10057	28.32295
709						
710	Average:	28.61898				

Source: Used with permission from Microsoft Corporation
FIGURE 12-7   Summary data for a state's lottery numbers

There are 56 possible integers. Logically, half of the balls drawn should be numbered between 1 and 28 and the other half should be numbered between 29 and 56. Thus, you would expect the average for any ball's values, and the overall average, to be near 28 or 29.

You need to develop a frequency table for all values between 1 and 56. To build this table, you can develop frequency tables for each ball and then combine the values for the five balls.

To create a frequency table for a ball, use a pivot table. First, create the pivot table based on the entire data table. Insert the pivot table in the existing worksheet; choose cell location H1. To create a pivot table based only on ball 1 data, drag the Ball 1 field label from the pivot table field list to the Row Labels window and to the Sigma Values window. Both windows are in the lower-right corner of the worksheet. Change the Value Field Settings from Sum to Count to get a count of the 56 ball values, which is equivalent to a frequency chart. The results should resemble the illustrative data shown in Figure 12-8.

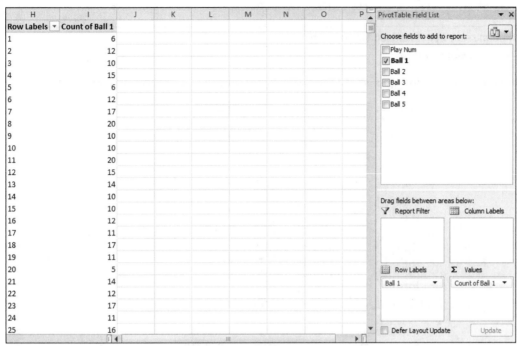

Source: Used with permission from Microsoft Corporation

FIGURE 12-8   Developing a pivot table for ball 1 values

Use the same steps to develop pivot tables for all five balls, as shown in Figure 12-9.

Num	Freq	Num	Freq	Num	Freq	Num	Freq	Num	Freq
1	14	1	17	1	10	1	14	1	10
2	13	2	13	2	8	2	15	2	16
3	13	3	13	3	18	3	13	3	11

Source: Used with permission from Microsoft Corporation

FIGURE 12-9   Developing pivot tables for all ball values

You can then create a summary of the frequencies for all five balls using the =SUM() function, as shown in Figure 12-10.

Num	Freq	Num	Freq	Num	Freq	Num	Freq	Num	Freq	Number	Frequency
1	14	1	17	1	10	1	14	1	10	1	65
2	13	2	13	2	8	2	15	2	16	2	65
3	13	3	13	3	18	3	13	3	11	3	68
4	14	4	13	4	14	4	12	4	13	4	66

Source: Used with permission from Microsoft Corporation

FIGURE 12-10   Developing a frequency table for numbers 1 to 56

Note that Figure 12-10 shows only partial data.

Check your work by summing the Frequency column. For example, a likely cause of error would be forgetting to change from Sum to Count when creating a pivot table. The total should be 3530 (706 lotteries multiplied by five ball values for each lottery). You can also check the table by adding the grand totals for the five pivot tables, which should equal 3530 as well. See the illustrative data in Figure 12-11.

H	I	J	K	L	M	N	O	P	Q	R	S	T	U	V	W	X
53	11	53		12	53		13	53	8	53		10	53		54
54	9	54		13	54		7	54	19	54		14	54		62
55	9	55		19	55		20	55	11	55		11	55		70
56	19	56		12	56		9	56	8	56		19	56		67
Grand Total	**706**	**Grand Total**		**706**	**Grand Total**		**706**	**Grand Total**	**706**	**Grand Total**		**706**			3530
														check:	3530

Source: Used with permission from Microsoft Corporation

FIGURE 12-11 Checking frequency data

Use the frequency table data to begin creating a stacked column chart, as shown in Figure 12-12.

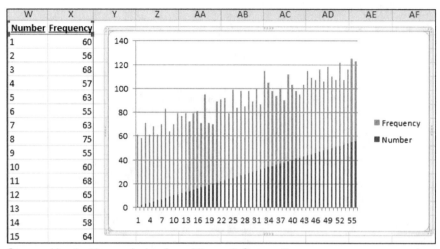

Source: Used with permission from Microsoft Corporation

FIGURE 12-12 Start of column chart

The columns are a combination of the Number and Frequency values, as indicated by the colored legends. You do not want to include the Number values in the column, so click in the Number section of the chart's bars and then press the Delete key. Your chart should resemble the one shown in Figure 12-13, which is based on illustrative data.

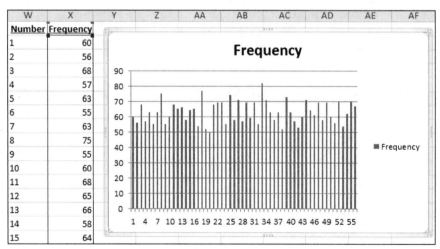

Source: Used with permission from Microsoft Corporation

FIGURE 12-13 Developing the column chart

You can make the chart look more professional by adding axis labels and a better title. You could also consider moving the data legend ("Frequency"), as shown in Figure 12-14.

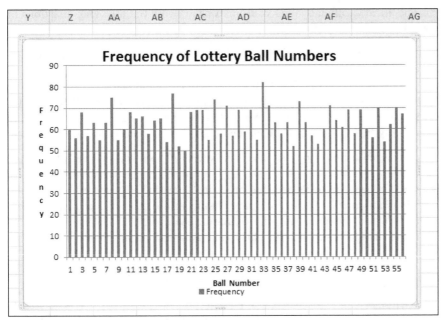

Source: Used with permission from Microsoft Corporation

FIGURE 12-14   Frequency chart for the 56 possible lottery values

Values in the figure are for illustrative purposes only. To add axis labels, use options in the Labels group of the chart's Layout tab. Click the Chart Title button to create a title, and click the Axis Title button to create a horizontal or vertical axis label. You can move a legend by right-clicking it, selecting Format Legend from the menu, and then selecting the position you want in the Legend Options window.

By visual inspection, you can see which five values were chosen most frequently in the lottery. Just examine the frequency values on the Y-axis of the chart. You can also verify these values by placing the cursor at the top of a column: Excel will show the related X and Y values.

You should also develop a list of the five most frequently selected numbers by sorting the frequency data. Select the frequency data, and then copy it to the right of the chart. (Be sure to make the copy using the Paste–Values option.) Then use the Data–Sort option to sort the values on the Frequency field from largest to smallest. You should see results similar to those in Figure 12-15.

AH	AI
Number	**Frequency**
33	82
18	77
8	75
25	74
39	73
27	71
34	71
44	71

Source: Used with permission from Microsoft Corporation

FIGURE 12-15   Five most frequently selected numbers developed by sorting frequency values

As you can see in the figure, the five most frequently selected numbers were 33 (82 times), 18 (77), 8 (75), 25 (74), and 39 (73).

Use the same procedures for the data in the Medium State and Small State worksheets.

Summarizing Data for Three States

Your state's lottery officials hope your analysis shows that the state lottery is working properly, and that Dr. Bern Ooley's objections are nothing to worry about. To dispel those objections, your analysis must show that the number-generating algorithm used by your state's lottery and by other states produces reasonable results in a nonrepeating manner. Specifically, officials hope your analysis answers the following questions in clear-cut ways that impress lay people:

- Does the algorithm produce many unusual ball values? Ball averages should be close to each other and in the range of 27 to 29.
- Does the algorithm favor the five numbers that Ooley says people should select?
- Do the top five values in the other three states follow your state's top five pattern? The five most common frequencies should range from 71 to 85, which is your state's range.
- Does the algorithm favor any ball values? Do values repeat across the state lotteries?

If your analysis cannot dispel doubts, then more sophisticated statistical analysis will be needed to refute Ooley.

Summarize your data in a way that allows you to answer the preceding questions. Create a new worksheet named **Summary** in which you manually gather summary data from the other three worksheets.

In one part of the summary sheet, record the average of the values for each of the five balls in the three states, as shown in the illustrative data in Figure 12-16. This data has been computed elsewhere (see Figure 12-7).

	A	B	C	D	E	F	G	H	I	J	K
11		Small State:				Medium State:				Big State:	
12			Average				Average				Average
13		Ball 1	28.85			Ball 1	27.85			Ball 1	27.96
14		Ball 2	28.47			Ball 2	27.39			Ball 2	27.53
15		Ball 3	29.66			Ball 3	28.06			Ball 3	28.26
16		Ball 4	29.12			Ball 4	28.26			Ball 4	28.56
17		Ball 5	28.50			Ball 5	28.36			Ball 5	27.84
18		All	28.83			All	27.99			All	27.83

Source: Used with permission from Microsoft Corporation

FIGURE 12-16   Average values for five balls in each state

This data will be useful when you address the question of whether the number-generating algorithm produces many unusual values. Are ball averages close to each other and in the range of 27 to 29?

In another part of the summary sheet, record the five balls drawn most frequently as winning numbers, including the frequencies for your state, as shown in the illustrative data in Figure 12-17.

	A	B	C	D	E	F	G	H	I	J	K	L	M	N	O	P
1																
2		Small State:				Medium State:				Big State:				Your State:		
3		Top 5:				Top 5:				Top 5:				Top 5:		
4			Number	Frequency			Number	Frequency			Number	Frequency			Number	Frequency
5			4	81			27	82			52	84			48	82
6			40	80			12	81			4	82			36	74
7			23	77			13	80			21	80			53	74
8			45	76			17	80			16	75			12	72
9			33	74			44	78			34	74			51	72

Source: Used with permission from Microsoft Corporation

FIGURE 12-17   Data for the five balls drawn most frequently as winning numbers

This data will be useful when you answer the following questions about the number-generating algorithm: (1) Does the algorithm favor your state's five most frequently selected winning numbers? (2) Do the other three state lotteries have the same pattern as your state? In other words, do the top five frequencies range from 71 to 85?

In another part of the worksheet, you should copy the top five frequency values for the small, medium, and large states and then sort that data from largest to smallest. This data shows whether any ball values repeated in the three state lotteries. The illustrative data in Figure 12-18 shows how this part of the summary sheet should look.

⊿	R	S
1		
2	**Big, Medium, Small:**	
3	**Top 5:**	
4	**Number**	**Frequency**
5	52	84
6	45	76
7	44	78
8	40	80
9	34	74
10	33	74
11	27	82
12	23	77
13	21	80
14	17	80
15	16	75
16	13	80
17	12	81
18	4	81
19	4	82

Source: Used with permission from Microsoft Corporation

FIGURE 12-18   Sorted top five frequencies for all three states combined

This data helps you determine whether the algorithm favors any ball values. In the data, ball number 4 is chosen frequently in two states. One repeating value would probably not be surprising, but if other ball values repeated, further statistical analysis might be needed to show that the algorithm's number generation is random.

In your summary sheet, manually enter notes that answer the preceding list of questions. You need these notes for later reference in Assignment 2.

ASSIGNMENT 2: DOCUMENTING FINDINGS IN A MEMORANDUM

In this assignment, you write a memo in Microsoft Word that documents your findings. In your memo, observe the following requirements:

- Your memo should have proper headings such as Date, To, From, and Subject. You can address the memo to state lottery officials. Set up the memo as discussed in Tutorial E.
- Briefly outline the situation. However, you need not provide much background—you can assume that readers are generally familiar with your task.
- In the body of the memo, summarize the four questions from the preceding section and how your analysis addresses the questions.
- State the answers to the four questions. Then tell lottery officials whether you think more sophisticated statistical analysis is needed to dispel Ooley's objections.

Your instructor will tell you if you need to import charts or tabular data into your memo from Excel.

DELIVERABLES

Assemble the following deliverables for your instructor:

1. Printout of your memo
2. Spreadsheet printouts, if required by your instructor
3. Electronic media such as flash drive or CD, which should include your Word file, Access file, and Excel file

Staple the printouts together with the memo on top. If you have more than one .xlsx file or .accdb file on your electronic media, write your instructor a note that identifies the files for this assignment.

PART 6

ADVANCED SKILLS
USING EXCEL

GUIDANCE FOR EXCEL CASES

The Microsoft Excel cases in this book require the student to write a memorandum that includes a table. Guidelines for preparing a memo in Microsoft Word and instructions for entering a table in a Word document are provided to begin this tutorial. Also, some of the cases in this book require the use of advanced Excel techniques. Those techniques are explained in this tutorial rather than in the cases themselves:

- Using data tables
- Using pivot tables
- Using built-in functions
- Creating macros

You can refer to Sheet 1 of TutEData.xlsx when reading about data tables. Refer to Sheet 2 when reading about pivot tables. When reading about macros, you can refer to macro1.xlsx.

PREPARING A MEMORANDUM IN WORD

A business memo should include proper headings, such as TO, FROM, DATE, and SUBJECT. If you want to use a Word memo template, follow these steps:

1. In Word, click File.
2. Click New.
3. Click the Memos button in the Office.com or Microsoft Office Online Templates section.
4. Double-click the Contemporary design memo.

The first time you do this, you may need to click Download to install the template.

ENTERING A TABLE INTO A WORD DOCUMENT

Enter a table into a Word document using the following procedure:

1. Click the cursor where you want the table to appear in the document.
2. In the Insert group, click the Table drop-down menu.
3. Click Insert Table.
4. Choose the number of rows and columns.
5. Click OK.

DATA TABLES

An Excel data table is a contiguous range of data that has been designated as a table. Once you make this designation, the table gains certain properties that are useful for data analysis. (Note that in some previous versions of Excel, data tables were called *data lists*.) Suppose you have a list of runners who have completed a race, as shown in Figure E-1.

	A	B	C	D	E	F
1	**RUNNER#**	**LAST**	**FIRST**	**AGE**	**GENDER**	**TIME (MIN)**
2	100	HARRIS	JANE	O	F	70
3	101	HILL	GLENN	Y	M	70
4	102	GARCIA	PEDRO	M	M	85
5	103	HILBERT	DORIS	M	F	90
6	104	DOAKS	SALLY	Y	F	94
7	105	JONES	SUE	Y	F	95
8	106	SMITH	PETE	M	M	100
9	107	DOE	JANE	O	F	100
10	108	BRADY	PETE	O	M	100
11	109	BRADY	JOE	O	M	120
12	110	HEEBER	SALLY	M	F	125
13	111	DOLTZ	HAL	O	M	130
14	112	PEEBLES	AL	Y	M	63

Source: Used with permission from Microsoft Corporation

FIGURE E-1   Data table example

To turn the information into a data table, highlight the data range, including headings, and click the Insert tab. Then click Table in the Tables group. The Create Table window appears, as shown in Figure E-2.

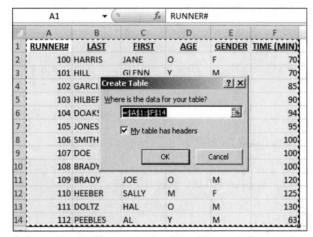

Source: Used with permission from Microsoft Corporation

FIGURE E-2   Create Table window

When you click OK, the data range appears as a table. On the Design tab, click Total Row to add a totals row to the data table. You also can select a light style in the Table Styles list to get rid of the contrasting color in the table's rows. Figure E-3 shows the results.

	A	B	C	D	E	F
1	**RUNNER#** ▾↑	**LAST** ▾	**FIRST** ▾	**AGE** ▾	**GENDER** ▾	**TIME (MIN)** ▾
2	100	HARRIS	JANE	O	F	70
3	101	HILL	GLENN	Y	M	70
4	102	GARCIA	PEDRO	M	M	85
5	103	HILBERT	DORIS	M	F	90
6	104	DOAKS	SALLY	Y	F	94
7	105	JONES	SUE	Y	F	95
8	106	SMITH	PETE	M	M	100
9	107	DOE	JANE	O	F	100
10	108	BRADY	PETE	O	M	100
11	109	BRADY	JOE	O	M	120
12	110	HEEBER	SALLY	M	F	125
13	111	DOLTZ	HAL	O	M	130
14	112	PEEBLES	AL	Y	M	63
15	Total					1242

Source: Used with permission from Microsoft Corporation

FIGURE E-3   Data table example

The headings have acquired drop-down menu tabs, as you can see in Figure E-3.

You can sort the data table records by any field. Perhaps you want to sort by times. If so, click the drop-down menu in the TIME (MIN) heading, select Sort, and then click Smallest to Largest. You get the results shown in Figure E-4.

	A	B	C	D	E	F
1	RUNNER ▼	LAST ▼	FIRST ▼	AGE ▼	GENDE ▼	TIME (MIN ▼↓
2	112	PEEBLES	AL	Y	M	63
3	100	HARRIS	JANE	O	F	70
4	101	HILL	GLENN	Y	M	70
5	102	GARCIA	PEDRO	M	M	85
6	103	HILBERT	DORIS	M	F	90
7	104	DOAKS	SALLY	Y	F	94
8	105	JONES	SUE	Y	F	95
9	106	SMITH	PETE	M	M	100
10	107	DOE	JANE	O	F	100
11	108	BRADY	PETE	O	M	100
12	109	BRADY	JOE	O	M	120
13	110	HEEBER	SALLY	M	F	125
14	111	DOLTZ	HAL	O	M	130
15	Total					1242

Source: Used with permission from Microsoft Corporation

FIGURE E-4   Sorting list by drop-down menu

You can see that Peebles had the best time and Doltz had the worst time. You also can sort from Largest to Smallest.

In addition, you can sort by more than one criterion. Assume that you want to sort first by gender and then by time (within gender). You first sort from Smallest to Largest in Gender. Then you again click the Gender drop-down tab, click Sort By Color, and then click Custom Sort. In the Sort window that appears, click Add Level and choose Time as the next criterion. See Figure E-5.

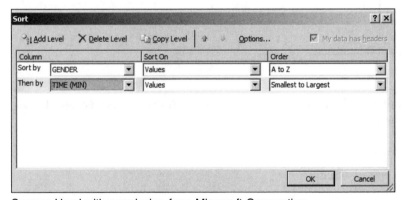

Source: Used with permission from Microsoft Corporation

FIGURE E-5   Sorting on multiple criteria

Click OK to get the results shown in Figure E-6.

	A	B	C	D	E	F
1	RUNNER ▼	LAST ▼	FIRST ▼	AGE ▼	GENDE ▼	TIME (MIN ▼
2	100	HARRIS	JANE	O	F	70
3	103	HILBERT	DORIS	M	F	90
4	104	DOAKS	SALLY	Y	F	94
5	105	JONES	SUE	Y	F	95
6	107	DOE	JANE	O	F	100
7	110	HEEBER	SALLY	M	F	125
8	112	PEEBLES	AL	Y	M	63
9	101	HILL	GLENN	Y	M	70
10	102	GARCIA	PEDRO	M	M	85
11	106	SMITH	PETE	M	M	100
12	108	BRADY	PETE	O	M	100
13	109	BRADY	JOE	O	M	120
14	111	DOLTZ	HAL	O	M	130
15	Total					1242

Source: Used with permission from Microsoft Corporation

FIGURE E-6   Sorting by gender and time (within gender)

You can see that Harris had the best female time and that Peebles had the best male time.

Perhaps you want to see the top n listings for some attribute; for example, you may want to see the top five runners' times. Select the Time column's drop-down menu, and select Number Filters. From the menu that appears, select Top 10. The Top 10 AutoFilter window appears, as shown in Figure E-7.

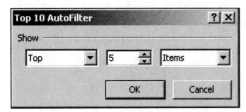

Source: Used with permission from Microsoft Corporation

FIGURE E-7   Top 10 AutoFilter window

This window lets you specify the number of values you want. In Figure E-7, five values were specified. Click OK to get the results shown in Figure E-8.

	A	B	C	D	E	F
1	RUNNER ▼	LAST ▼	FIRST ▼	AGE ▼	GENDE ▼	TIME (MIN ▼
6	107	DOE	JANE	O	F	100
7	110	HEEBER	SALLY	M	F	125
11	106	SMITH	PETE	M	M	100
12	108	BRADY	PETE	O	M	100
13	109	BRADY	JOE	O	M	120
14	111	DOLTZ	HAL	O	M	130
15	Total					675

Source: Used with permission from Microsoft Corporation

FIGURE E-8   Top 5 times

The output contains more than five data records because there are ties at 100 minutes. If you want to see all of the records again, click the Time drop-down menu and click Clear Filter. The full table of data reappears, as shown in Figure E-9.

	A	B	C	D	E	F
1	**RUNNER** ▾	**LAST** ▾	**FIRST** ▾	**AGE** ▾	**GENDE** ▾↑	**TIME (MIN** ▾↑
2	100	HARRIS	JANE	O	F	70
3	103	HILBERT	DORIS	M	F	90
4	104	DOAKS	SALLY	Y	F	94
5	105	JONES	SUE	Y	F	95
6	107	DOE	JANE	O	F	100
7	110	HEEBER	SALLY	M	F	125
8	112	PEEBLES	AL	Y	M	63
9	101	HILL	GLENN	Y	M	70
10	102	GARCIA	PEDRO	M	M	85
11	106	SMITH	PETE	M	M	100
12	108	BRADY	PETE	O	M	100
13	109	BRADY	JOE	O	M	120
14	111	DOLTZ	HAL	O	M	130
15	Total					1242

Source: Used with permission from Microsoft Corporation
FIGURE E-9   Restoring all data to screen

Each of the cells in the Total row has a drop-down menu. The menu choices are statistical operations that you can perform on the totals—for example, you can take a sum, take an average, take a minimum or maximum, count the number of records, and so on. Assume that the Time drop-down menu was selected, as shown in Figure E-10. Note that the Sum operator is highlighted by default.

	A	B	C	D	E	F
1	**RUNNER** ▾	**LAST** ▾	**FIRST** ▾	**AGE** ▾	**GENDE** ▾↑	**TIME (MIN** ▾↑
2	100	HARRIS	JANE	O	F	70
3	103	HILBERT	DORIS	M	F	90
4	104	DOAKS	SALLY	Y	F	94
5	105	JONES	SUE	Y	F	95
6	107	DOE	JANE	O	F	100
7	110	HEEBER	SALLY	M	F	125
8	112	PEEBLES	AL	Y	M	63
9	101	HILL	GLENN	Y	M	70
10	102	GARCIA	PEDRO	M	M	85
11	106	SMITH	PETE	M	M	100
12	108	BRADY	PETE	O	M	100
13	109	BRADY	JOE	O	M	120
14	111	DOLTZ	HAL	O	M	130
15	Total					1242 ▾
16						None
17						Average
18						Count
19						Count Numbers
20						Max
21						Min
22						Sum
						StdDev
						Var
						More Functions...

Source: Used with permission from Microsoft Corporation
FIGURE E-10   Selecting Time drop-down menu in Total row

By changing from Sum to Average, you find that the average time for all runners was 95.5 minutes, as shown in Figure E-11.

	A	B	C	D	E	F
1	RUNNER ▾	LAST ▾	FIRST ▾	AGE ▾	GENDE ▾↑	TIME (MIN ▾↑
2	100 HARRIS		JANE	O	F	70
3	103 HILBERT		DORIS	M	F	90
4	104 DOAKS		SALLY	Y	F	94
5	105 JONES		SUE	Y	F	95
6	107 DOE		JANE	O	F	100
7	110 HEEBER		SALLY	M	F	125
8	112 PEEBLES		AL	Y	M	63
9	101 HILL		GLENN	Y	M	70
10	102 GARCIA		PEDRO	M	M	85
11	106 SMITH		PETE	M	M	100
12	108 BRADY		PETE	O	M	100
13	109 BRADY		JOE	O	M	120
14	111 DOLTZ		HAL	O	M	130
15	Total					95.53846154 ▾

Source: Used with permission from Microsoft Corporation

FIGURE E-11   Average running time shown in Total row

PIVOT TABLES

Suppose you have data for a company's sales transactions by month, by salesperson, and by amount for each product type. You would like to display each salesperson's total sales by type of product sold and by month. You can use a pivot table in Excel to tabulate that summary data. A pivot table is built around one or more dimensions and thus can summarize large amounts of data. Figure E-12 shows total sales cross-tabulated by salesperson and by month.

	A	B	C	D	E
1	**Name**	**Product**	**January**	**February**	**March**
2	Jones	Product1	30,000	35,000	40,000
3	Jones	Product2	33,000	34,000	45,000
4	Jones	Product3	24,000	30,000	42,000
5	Smith	Product1	40,000	38,000	36,000
6	Smith	Product2	41,000	37,000	38,000
7	Smith	Product3	39,000	50,000	33,000
8	Bonds	Product1	25,000	26,000	25,000
9	Bonds	Product2	22,000	25,000	24,000
10	Bonds	Product3	19,000	20,000	19,000
11	Ruth	Product1	44,000	42,000	33,000
12	Ruth	Product2	45,000	40,000	30,000
13	Ruth	Product3	50,000	52,000	35,000

Source: Used with permission from Microsoft Corporation

FIGURE E-12   Excel spreadsheet data

You can create pivot tables and many other kinds of tables with the Excel PivotTable tool. To create a pivot table from the data in Figure E-12, follow these steps:

1. Starting in the spreadsheet in Figure E-12, click a cell in the data range, and then click the Insert tab. In the Tables group, choose PivotTable. You see the screen shown in Figure E-13.

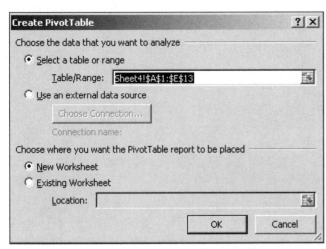

Source: Used with permission from Microsoft Corporation

FIGURE E-13   Creating a pivot table

2.   Make sure New Worksheet is checked under "Choose where you want the PivotTable report to be placed." Click OK. The screen shown in Figure E-14 appears. If it does not, right-click in a cell in the pivot table area, and click PivotTable Options from the menu that appears. Click the Display tab and then check the Classic PivotTable layout.

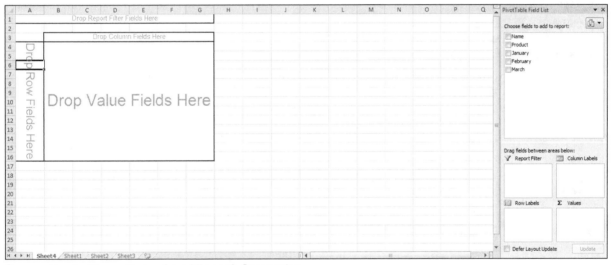

Source: Used with permission from Microsoft Corporation

FIGURE E-14   PivotTable design screen

The data range's column headings are shown in the PivotTable Field List on the right side of the screen. From there, you can click and drag column headings into the Row, Column, and Value areas that appear in the spreadsheet.

3.   If you want to see the total sales by product for each salesperson, drag the Name field to the Drop Column Fields Here area in the spreadsheet. You should see the result shown in Figure E-15.

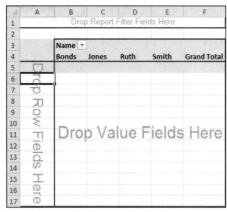

Source: Used with permission from Microsoft Corporation

FIGURE E-15   Column fields

4.   Next, drag the Product field to the Drop Row Fields Here area. You should see the result shown in Figure E-16.

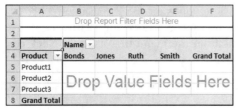

Source: Used with permission from Microsoft Corporation

FIGURE E-16   Row fields

5.   Finally, drag the month fields (January, February, and March) individually to the Drop Value Fields Here area to produce the finalized pivot table. You should see the result shown in Figure E-17.

| Product | Values | | Name | | | | |
		Bonds	Jones	Ruth	Smith	Grand Total
Product1	Sum of January	25000	30000	44000	40000	139000
	Sum of February	26000	35000	42000	38000	141000
	Sum of March	25000	40000	33000	36000	134000
Product2	Sum of January	22000	33000	45000	41000	141000
	Sum of February	25000	34000	40000	37000	136000
	Sum of March	24000	45000	30000	38000	137000
Product3	Sum of January	19000	24000	50000	39000	132000
	Sum of February	20000	30000	52000	50000	152000
	Sum of March	19000	42000	35000	33000	129000
Total Sum of January		66000	87000	139000	120000	412000
Total Sum of February		71000	99000	134000	125000	429000
Total Sum of March		68000	127000	98000	107000	400000

Source: Used with permission from Microsoft Corporation

FIGURE E-17   Data items

By default, Excel adds all of the sales for each salesperson by month for each product. At the bottom of the pivot table, Excel also shows the total sales for each month for all products.

Refer back to Figure E-14 and note the four small panes in the lower-right corner. The Values pane lets you easily change from the default Sum operator to another one (Min, Max, Average, and so on). Click the drop-down arrow, select Value Fields Setting, and then select the desired operator.

BUILT-IN FUNCTIONS

You might need to use some of the following functions when solving the Excel cases elsewhere in this text:

- MIN, MAX, AVERAGE, COUNTIF, ROUND, ROUNDUP, and RANDBETWEEN

The syntax of these functions is discussed in this section. The following examples are based on the runner data shown in Figure E-18.

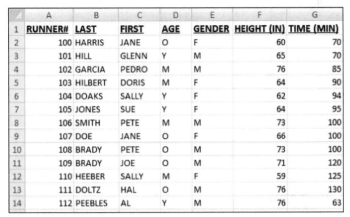

	A	B	C	D	E	F	G
1	RUNNER#	LAST	FIRST	AGE	GENDER	HEIGHT (IN)	TIME (MIN)
2	100	HARRIS	JANE	O	F	60	70
3	101	HILL	GLENN	Y	M	65	70
4	102	GARCIA	PEDRO	M	M	76	85
5	103	HILBERT	DORIS	M	F	64	90
6	104	DOAKS	SALLY	Y	F	62	94
7	105	JONES	SUE	Y	F	64	95
8	106	SMITH	PETE	M	M	73	100
9	107	DOE	JANE	O	F	66	100
10	108	BRADY	PETE	O	M	73	100
11	109	BRADY	JOE	O	M	71	120
12	110	HEEBER	SALLY	M	F	59	125
13	111	DOLTZ	HAL	O	M	76	130
14	112	PEEBLES	AL	Y	M	76	63

Source: Used with permission from Microsoft Corporation

FIGURE E-18   Runner data used to illustrate built-in functions

The data is the same as that shown in Figure E-1, except that Figure E-18 includes a column for the runners' height in inches.

MIN and MAX Functions

The MIN function determines the smallest value in a range of data. The MAX function returns the largest. Say that we want to know the fastest time for all runners, which would be the minimum time in column G. The MIN function computes the smallest value in a set of values. The set of values could be a data range, or it could be a series of cell addresses separated by commas. The syntax of the MIN function is as follows:

- MIN(set of data)

To show the minimum time in cell C16, you would enter the formula shown in the formula bar in Figure E-19:

C16			f_x	=MIN(G2:G14)			
	A	B	C	D	E	F	G
1	RUNNER#	LAST	FIRST	AGE	GENDER	HEIGHT	TIME (MIN)
2	100	HARRIS	JANE	O	F	60	70
3	101	HILL	GLENN	Y	M	65	70
4	102	GARCIA	PEDRO	M	M	76	85
5	103	HILBERT	DORIS	M	F	64	90
6	104	DOAKS	SALLY	Y	F	62	94
7	105	JONES	SUE	Y	F	64	95
8	106	SMITH	PETE	M	M	73	100
9	107	DOE	JANE	O	F	66	100
10	108	BRADY	PETE	O	M	73	100
11	109	BRADY	JOE	O	M	71	120
12	110	HEEBER	SALLY	M	F	59	125
13	111	DOLTZ	HAL	O	M	76	130
14	112	PEEBLES	AL	Y	M	76	63
15							
16	MINIMUM TIME:		63				

Source: Used with permission from Microsoft Corporation

FIGURE E-19   MIN function in cell C16

(Assume that you typed the label "MINIMUM TIME:" into cell A16.) You can see that the fastest time is 63 minutes.

To see the slowest time in cell G16, use the MAX function, whose syntax parallels that of the MIN function, except that the largest value in the set is determined. See Figure E-20.

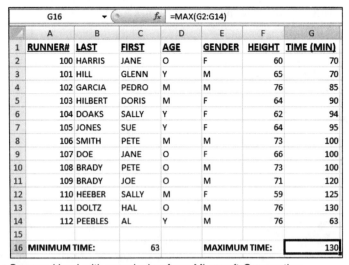

Source: Used with permission from Microsoft Corporation

FIGURE E-20   MAX function in cell G16

AVERAGE, ROUND, and ROUNDUP Functions

The AVERAGE function computes the average of a set of values. Figure E-21 shows the use of the AVERAGE function in cell C17:

	C17			f_x	=AVERAGE(G2:G14)	
	A	B	C	D	E	F	G
1	RUNNER#	LAST	FIRST	AGE	GENDER	HEIGHT	TIME (MIN)
2	100	HARRIS	JANE	O	F	60	70
3	101	HILL	GLENN	Y	M	65	70
4	102	GARCIA	PEDRO	M	M	76	85
5	103	HILBERT	DORIS	M	F	64	90
6	104	DOAKS	SALLY	Y	F	62	94
7	105	JONES	SUE	Y	F	64	95
8	106	SMITH	PETE	M	M	73	100
9	107	DOE	JANE	O	F	66	100
10	108	BRADY	PETE	O	M	73	100
11	109	BRADY	JOE	O	M	71	120
12	110	HEEBER	SALLY	M	F	59	125
13	111	DOLTZ	HAL	O	M	76	130
14	112	PEEBLES	AL	Y	M	76	63
15							
16	MINIMUM TIME:		63		MAXIMUM TIME:		130
17	AVERAGE TIME:		95.53846				

Source: Used with permission from Microsoft Corporation

FIGURE E-21   AVERAGE function in cell C17

Notice that the value shown is a real number with many digits. What if you wanted to have the value rounded to a certain number of digits? Of course, you could format the output cell, but doing that changes only what is shown on the screen. You want the cell's contents actually to *be* the rounded number. Therefore, you need to use the ROUND function. Its syntax is:

- ROUND(number, number of digits)

Figure E-22 shows the rounded average time (with two decimal places) in cell G17.

Source: Used with permission from Microsoft Corporation

FIGURE E-22 ROUND function used in cell G17

To achieve this output, cell C17 was used as the value to be rounded. Recall from Figure E-21 that cell C17 had the formula =AVERAGE(G2:G14). The following ROUND formula would produce the same output in cell G17: =ROUND(AVERAGE(G2:G14),2). In this case, Excel evaluates the formula "inside out." First, the AVERAGE function is evaluated, yielding the average with many digits. That value is then input to the ROUND function and rounded to two decimal places.

The ROUNDUP function works much like the ROUND function. ROUNDUP's output is always rounded up to the next value. For example, the value 4 would appear in a cell that contained the following formula: =ROUNDUP(3.12,0). In Figure E-22, if the formula in cell G17 had been =ROUNDUP(AVERAGE(G2:G14),2), the value 96 would have been the result. In other words, 95.54 rounded up with no decimal places becomes 96.

COUNTIF Function

The COUNTIF function counts the number of values in a range that meet a specified condition. The syntax is:

- COUNTIF(range of data, condition)

The condition is a logical expression such as "=1", ">6", or "=F". The condition is shown with quotation marks, even if a number is involved.

Assume that you want to see the number of female runners in cell C18. Figure E-23 shows the formula used.

Source: Used with permission from Microsoft Corporation

FIGURE E-23 COUNTIF function used in cell C18

The logic of the formula is: Count the number of times that "F" appears in the data range E2:E14.

As another example of using COUNTIF, assume that column H shows the rounded ratio of each runner's height in inches to the runner's time in minutes (see Figure E-24).

	H2	▾	*fx*	=ROUND(G2/F2,2)				
	A	B	C	D	E	F	G	H
1	RUNNER#	LAST	FIRST	AGE	GENDER	HEIGHT	TIME (MIN)	RATIO
2	100	HARRIS	JANE	O	F	60	70	1.17
3	101	HILL	GLENN	Y	M	65	70	1.08
4	102	GARCIA	PEDRO	M	M	76	85	1.12
5	103	HILBERT	DORIS	M	F	64	90	1.41
6	104	DOAKS	SALLY	Y	F	62	94	1.52
7	105	JONES	SUE	Y	F	64	95	1.48
8	106	SMITH	PETE	M	M	73	100	1.37
9	107	DOE	JANE	O	F	66	100	1.52
10	108	BRADY	PETE	O	M	73	100	1.37
11	109	BRADY	JOE	O	M	71	120	1.69
12	110	HEEBER	SALLY	M	F	59	125	2.12
13	111	DOLTZ	HAL	O	M	76	130	1.71
14	112	PEEBLES	AL	Y	M	76	63	0.83
15								
16	MINIMUM TIME:		63		MAXIMUM TIME:		130	
17	AVERAGE TIME:		95.53846		ROUNDED AVERAGE:		95.54	
18	NUMBER OF FEMALES:		6					

Source: Used with permission from Microsoft Corporation

FIGURE E-24   Ratio of height to time in column H

Assume that all runners whose height in inches is less than their time in minutes will get an award. How many awards are needed? If the ratio is less than 1, an award is warranted. The COUNTIF function in cell G18 computes a count of ratios less than 1, as shown in Figure E-25.

	G18	▾	*fx*	=COUNTIF(H2:H14,"<1")				
	A	B	C	D	E	F	G	H
1	RUNNER#	LAST	FIRST	AGE	GENDER	HEIGHT	TIME (MIN)	RATIO
2	100	HARRIS	JANE	O	F	60	70	1.17
3	101	HILL	GLENN	Y	M	65	70	1.08
4	102	GARCIA	PEDRO	M	M	76	85	1.12
5	103	HILBERT	DORIS	M	F	64	90	1.41
6	104	DOAKS	SALLY	Y	F	62	94	1.52
7	105	JONES	SUE	Y	F	64	95	1.48
8	106	SMITH	PETE	M	M	73	100	1.37
9	107	DOE	JANE	O	F	66	100	1.52
10	108	BRADY	PETE	O	M	73	100	1.37
11	109	BRADY	JOE	O	M	71	120	1.69
12	110	HEEBER	SALLY	M	F	59	125	2.12
13	111	DOLTZ	HAL	O	M	76	130	1.71
14	112	PEEBLES	AL	Y	M	76	63	0.83
15								
16	MINIMUM TIME:		63		MAXIMUM TIME:		130	
17	AVERAGE TIME:		95.53846		ROUNDED AVERAGE:		95.54	
18	NUMBER OF FEMALES:		6		RATIOS<1:		1	

Source: Used with permission from Microsoft Corporation

FIGURE E-25   COUNTIF function used in cell G18

RANDBETWEEN Function

If you wanted a cell to contain a randomly generated integer in the range from 1 to 9, you would use the formula =RANDBETWEEN(1,9). Any value between 1 and 9 inclusive would be output by the formula. An example is shown in Figure E-26.

Source: Used with permission from Microsoft Corporation

FIGURE E-26 RANDBETWEEN function used in cell A2

Assume that you copied and pasted the formula to generate a column of 100 numbers between 1 and 9. Every time a value was changed in the spreadsheet, Excel would recalculate the 100 RANDBETWEEN formulas to change the 100 random values. Therefore, you might want to settle on the random values once they are generated. To do this, copy the 100 values, click Paste Special, and then click Values to put the values in the same range. The contents of the cells will change from formulas to literal values.

CREATING EXCEL MACROS

Assume that you have a workbook to record your business sales. A worksheet is devoted to each month's sales. Over the course of a year, therefore, your workbook would have 12 worksheets. Assume further that you want to make the same calculations with the data in each worksheet. For example, in each sheet, you might want to sort the data and then compute certain operating statistics. Performing the same operations in each worksheet would be time consuming and error-prone.

An Excel macro lets you automate these repetitive tasks. You can create the macro in the macro recorder and then invoke the macro in each worksheet. A macro runs as a stored program within Excel. The process is illustrated in this section.

The Developer tab must appear in the main Excel Ribbon so you can select the macro recorder. To display the tab, take the following steps:

1. Select the File menu on the left side of the screen, and then select Options.
2. In the window that appears, select Customize Ribbon.
3. A list box appears on the right side of the window. Click the Developer check box.
4. Click OK.

The Developer tab should appear in the Ribbon, as shown in Figure E-27.

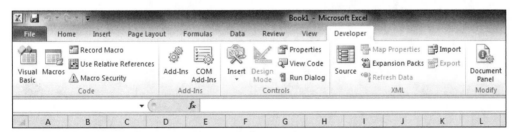

Source: Used with permission from Microsoft Corporation

FIGURE E-27 Developer tab in main Ribbon

When you click the Developer tab, macro options appear in the Code group, as shown on the left side of the figure. Click the Record Macro button to start recording a macro. Click the Macros button to see a menu of your macros.

The following example explains how to create a macro. Let's say you keep data files to record your company's sales and that you have a worksheet devoted to each month. January's data is shown in Figure E-28.

▲	A	B	C
1	**Invoice Number**	**Date of Sale**	**Quantity**
2	1001	1/4/2013	2000
3	1002	1/6/2013	1000
4	1003	1/9/2013	500
5	1004	1/15/2013	6000

Source: Used with permission from Microsoft Corporation

FIGURE E-28　January sales data

Each sale is indicated by an invoice number. The date of each sale is shown, along with the quantity of items sold. Your sales data for February and March are shown in Figures E-29 and E-30, respectively.

▲	A	B	C
1	**Invoice Number**	**Date of Sale**	**Quantity**
2	2001	2/9/2013	4000
3	2002	2/16/2013	2000
4	2003	2/19/2013	1500
5	2004	2/25/2013	700

Source: Used with permission from Microsoft Corporation

FIGURE E-29　February sales data

▲	A	B	C
1	**Invoice Number**	**Date of Sale**	**Quantity**
2	3001	3/9/2013	3000
3	3002	3/16/2013	7000
4	3003	3/19/2013	1000
5	3004	3/25/2013	1400

Source: Used with permission from Microsoft Corporation

FIGURE E-30　March sales data

Assume that you want to perform the following tasks in each sheet: (1) sort the records by quantity sold, from smallest to largest, and (2) delete the records with the two largest quantities sold. Based on the sort, the records with the two largest quantities sold would be the last two records. You can develop a macro that performs both tasks.

To begin building your macro, start with the January data. Select the Developer tab, and then select Record Macro in the Code group. The Record Macro window appears, as shown in Figure E-31.

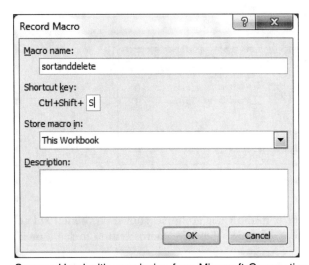

Source: Used with permission from Microsoft Corporation

FIGURE E-31　Record Macro window

Excel enters a default name in the Macro name text box, such as macro1 or macro2, but you should give the macro a more descriptive name. Also, you should enter a shortcut key that you use to invoke the macro.

As shown in Figure E-31, the new macro is named sortanddelete. The shortcut key to invoke the macro is Ctrl+Shift+S. (In other words, hold down the Ctrl key and select an uppercase *S*.) To enter the shortcut, click in the Shortcut key box, and then select Shift and the letter *s*. Click OK to begin the recording. Excel records and encodes each step you take until you stop the recording.

In January's data, you can select cell A1. (Alternatively, you could select the entire range of cells.) Click the Data tab in the Ribbon, and then select Sort in the Sort & Filter group. The Sort window appears, as shown in Figure E-32; you use the window to control how the sort is performed.

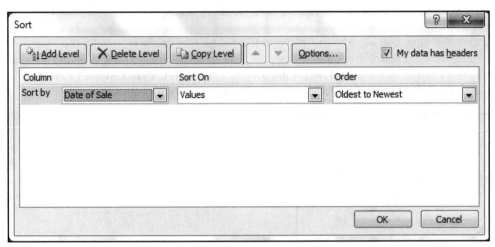

Source: Used with permission from Microsoft Corporation

FIGURE E-32   Sort window

You want to sort by Quantity from smallest to largest, so you should make the choices shown in Figure E-33.

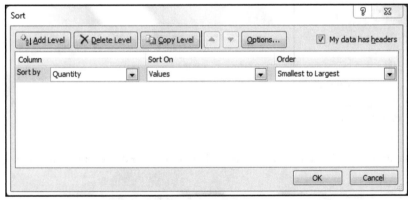

Source: Used with permission from Microsoft Corporation

FIGURE E-33   Sort window selections

Click OK. The January data should look like that in Figure E-34.

	A	B	C
1	**Invoice Number**	**Date of Sale**	**Quantity**
2	1003	1/9/2013	500
3	1002	1/6/2013	1000
4	1001	1/4/2013	2000
5	1004	1/15/2013	6000

Source: Used with permission from Microsoft Corporation

FIGURE E-34   Sorted January data

Note that the macro recorder is still running. The next step is to delete the bottom two records. Click and drag to select the data in A4:C5. Click the Home tab in the Ribbon. In the Editing group, select Clear – Clear All. The data should look like that in Figure E-35.

	A	B	C
1	**Invoice Number**	**Date of Sale**	**Quantity**
2	1003	1/9/2013	500
3	1002	1/6/2013	1000

Source: Used with permission from Microsoft Corporation

FIGURE E-35   January data with two records deleted

You are done, so you should stop recording. Click the Developer tab, and then select Stop Recording in the Code group. This option appears in place of the Record Macro option; the label changes depending on the status of the recorder.

You might assume that the next step would be to move to the February worksheet, invoke the macro, and then do the same thing with the March worksheet. However, you must resolve a problem before you can run the macro properly. You need to look at the macro's programming code to understand the problem. Select Macros in the Developer tab Code group. The Macro window appears, as shown in Figure E-36.

Source: Used with permission from Microsoft Corporation

FIGURE E-36   Macro window

The macro is already selected because it is the only one in the workbook. Click Edit to open the code window shown in Figure E-37.

```
Sub sortanddelete()
'
' sortanddelete Macro
'
' Keyboard Shortcut: Ctrl+Shift+S
'
    Range("A1").Select
    ActiveWorkbook.Worksheets("Jan").Sort.SortFields.Clear
    ActiveWorkbook.Worksheets("Jan").Sort.SortFields.Add Key:=Range("C2:C5"), _
        SortOn:=xlSortOnValues, Order:=xlAscending, DataOption:=xlSortNormal
    With ActiveWorkbook.Worksheets("Jan").Sort
        .SetRange Range("A1:C5")
        .Header = xlYes
        .MatchCase = False
        .Orientation = xlTopToBottom
        .SortMethod = xlPinYin
        .Apply
    End With
    Range("A4:C5").Select
    Selection.Clear
End Sub
```

Source: Used with permission from Microsoft Corporation

FIGURE E-37   Sortanddelete macro code window

Excel transcribes a macro's actions into code using the Visual Basic for Applications programming language (VBA). The code in Figure E-37 is a set of VBA instructions shown in the VBA editor, which is separate from the Excel worksheet window. The macro recorder's code instructs the macro to take the following steps:

1. In the first worksheet, select A1.
2. In the January worksheet, start sorting.
3. In the January worksheet, perform an ascending sort on the cells in C2:C5 (the Quantity cells).
4. For each cell in the range of A1:C5 in the January worksheet, apply the sort routine.
5. Finally, select the range of A4:C5 and clear it.

Perhaps you can see the problem. Say that you move to the February worksheet and run the macro by selecting Ctrl+Shift+S simultaneously. Nothing happens with the February data because the VBA code refers only to the January worksheet. Thus, no matter what sheet you are working in when you invoke the macro, only the January data is affected.

Clearly, you need a more general macro that works regardless of which month's data you are using. VBA can refer to the currently active worksheet by using the general term *ActiveSheet*. Thus, in the VBA code editor, you can change the three references to the January worksheet to ActiveSheet, as shown in Figure E-38.

```
macro1.xlsm - Module1 (Code)

(General)                                              sortanddelete

Sub sortanddelete()

' sortanddelete Macro

' Keyboard Shortcut: Ctrl+Shift+S

    Range("A1").Select
    ActiveWorkbook.ActiveSheet.Sort.SortFields.Clear
    ActiveWorkbook.ActiveSheet.Sort.SortFields.Add Key:=Range("C2:C5"), _
        SortOn:=xlSortOnValues, Order:=xlAscending, DataOption:=xlSortNormal
    With ActiveWorkbook.ActiveSheet.Sort
        .SetRange Range("A1:C5")
        .Header = xlYes
        .MatchCase = False
        .Orientation = xlTopToBottom
        .SortMethod = xlPinYin
        .Apply
    End With
    Range("A4:C5").Select
    Selection.Clear
End Sub
```

Source: Used with permission from Microsoft Corporation

FIGURE E-38   Code changed to refer to ActiveSheet

Before testing the macro, you should save it. Click the Save icon in the editor's main menu. Excel asks if you want to use the macro-enabled file type, which has an .xlsm extension instead of .xlsx. Click No at the prompt, as shown in Figure E-39.

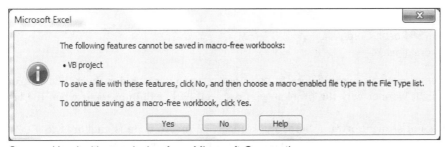

Source: Used with permission from Microsoft Corporation

FIGURE E-39   Saving the macro-enabled workbook

The next window allows you to save the file as a macro-enabled workbook, as shown in Figure E-40. Enter the desired filename, and then click the Save button.

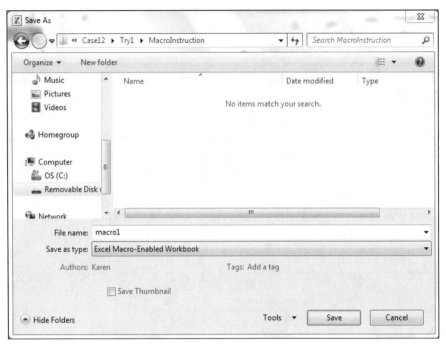

Source: Used with permission from Microsoft Corporation

FIGURE E-40   Saving the file as a macro-enabled workbook

To exit the editor and return to the worksheet, click the X button in the upper-right corner of the window.

Go back to the sheet with your February sales data, select a cell, and invoke the macro by pressing Ctrl+Shift+S. The macro now works, as you can see in Figure E-41.

	A	B	C
1	**Invoice Number**	**Date of Sale**	**Quantity**
2	2004	2/25/2013	700
3	2003	2/19/2013	1500
4			
5			

Source: Used with permission from Microsoft Corporation

FIGURE E-41   February data after invoking macro

Next, run the macro with the cursor in the March worksheet. The results are shown in Figure E-42.

	A	B	C
1	**Invoice Number**	**Date of Sale**	**Quantity**
2	3003	3/19/2013	1000
3	3004	3/25/2013	1400
4			
5			

Source: Used with permission from Microsoft Corporation

FIGURE E-42   March data after invoking macro

Notice that the last two lines of macro code in Figure E-38 do not refer to a particular sheet, but will run properly in any location. You might wonder why the problem you resolved earlier did not recur when the last two lines ran.

Unfortunately, the answer is not straightforward. Sometimes, recorded macro code is anchored to a specific worksheet, and sometimes it is not. If a recorded macro seems to be behaving strangely, you should examine the VBA code with the assistance of a programmer.

PART 7

PRESENTATION SKILLS

GIVING AN ORAL PRESENTATION

Giving an oral presentation in class lets you practice the presentation skills you will need in the workplace. The presentations you create for the cases in this textbook will be similar to professional business presentations. You will be expected to present objective, technical results to your organization's stakeholders, and you will have to support your presentation with visual aids commonly used in the business world. During your presentation, your instructor might assign your classmates to role-play an audience of business managers, bankers, or employees. They might also provide feedback on your presentation.

Follow these four steps to create an effective presentation:

1. Plan your presentation.
2. Draft your presentation.
3. Create graphics and other visual aids.
4. Practice delivering your presentation.

PLANNING YOUR PRESENTATION

When planning an oral presentation, you need to know your time limits, establish your purpose, analyze your audience, and gather information. This section explores each of these elements.

Knowing Your Time Limits

You need to consider your time limits on two levels. First, consider how much time you will have to deliver your presentation. For example, what are the key points in your material that can be covered in 10 minutes? The element of time is the primary constraint of any presentation. It limits the breadth and depth of your talk, and the number of visual aids that you can use. Second, consider how much time you will need for the process of preparing your presentation—drafting your presentation, creating graphics, and practicing your delivery.

Establishing Your Purpose

After considering your time limits, you must define your purpose: what you need to say and to whom you will say it. For the Access cases in this book, your purpose will be to inform and explain. For instance, a business's owners, managers, and employees may need to know how the company's database is organized and how they can use it to fill in forms and create reports. In contrast, for the Excel cases, your purpose will be to recommend a course of action based on the results of your business model. You will make the recommendations to business owners, managers, and bankers based on the results of inputting and running various scenarios.

Analyzing Your Audience

Once you have established the purpose of your presentation, you should analyze your audience. Ask yourself: What does my audience already know about the subject? What do the audience members want to know? What do they need to know? Do they have any biases or personal agendas that I should consider? What level of technical detail is best suited to their level of knowledge and interest?

In some Access cases, you will make a presentation to an audience that might not be familiar with Access or with databases in general. In other cases, you might be giving your presentation to a business owner who started to work on a database but was not able to finish it. Tailor the presentation to suit your audience.

For the Excel cases, you are most often interpreting results for an audience of bankers or business managers. In those instances, the audience will not need to know the detailed technical aspects of how you generated your results. But what if your audience consists of engineers or scientists? They will certainly be more interested in the structure and rationale of your decision models. Regardless of the audience, your listeners need to know what assumptions you made prior to developing your spreadsheets because those assumptions might affect their opinion of your results.

Gathering Information

Because you will have just completed a case as you begin preparing your oral presentation, you will already have the basic information you need. For the Access cases, you should review the main points of the case and your goals. Make sure you include all of the points you think are important for the audience to understand. In addition, you might want to go beyond the requirements and explain additional ways in which the database could be used to benefit the organization, now or in the future.

For the Excel cases, you can refer to the tutorials for assistance in interpreting the results from your spreadsheet analysis. For some cases, you might want to use the Internet or the library to research business trends or background information that can support your presentation.

DRAFTING YOUR REPORT AND PRESENTATION

When you have completed the planning stage, you are ready to begin drafting the presentation. At this point, you might be tempted to write your presentation and then memorize it word for word. Even if you could memorize your presentation verbatim, however, your delivery would sound unnatural because people use a simpler vocabulary and shorter sentences when they speak than when they write. For example, read the previous paragraph out loud as if you were presenting it to an audience.

In many business situations, you will be required both to submit a written report of your work and give a PowerPoint presentation. First, write your report, and then design your PowerPoint slides as a "brief" of that report to discuss its main points. When drafting your report and the accompanying PowerPoint slides, follow this sequence:

1. Write the main body of your report.
2. Write the introduction to your report.
3. Write the conclusion to your report.
4. Prepare your presentation (the PowerPoint slides) using your report's main points.

Writing the Main Body

When you draft your report, write the body first. If you try to write the opening paragraph first, you might spend an inordinate amount of time attempting to craft your words perfectly, only to revise the introduction after you write the body of the report.

Keeping Your Audience in Mind

To write the main body, review your purpose and your audience profile. What are the main points you need to make? What are your audience's needs, interests, and technical expertise? It is important to include some technical details in your report and presentation, but keep in mind the technical expertise of your audience.

Remember that the people reading your report or listening to your presentation have their own agendas—put yourself in their places and ask, "What do I need to get out of this presentation?" For example, in the Access cases, an employee might want to know how to enter information on a form, but the business owner might be more interested in generating queries and reports. You need to address their different needs in your presentation. For example, you might say, "And now, let's look at how data entry associates can input data into this form."

Similarly, in the Excel cases, your audience will consist of business owners, managers, bankers, and perhaps some technical professionals. The owners and managers will be concerned with profitability, growth, and customer service. In contrast, the bankers' main concern will be repayment of a loan. Technical professionals will be more concerned with how well your decision model is designed, along with the credibility of the results. You need to address the interests of each group.

Using Transitions and Repetition in your Presentation

During your presentation, remember that the audience is not reading the text of your report, so you need to include transitions to compensate. Words such as *next, first, second*, and *finally* will help the audience follow the sequence of your ideas. Words such as *however, in contrast, on the other hand*, and *similarly* will help the audience follow shifts in thought. You can use your voice to convey emphasis.

Also consider using hand gestures to emphasize what you say. For instance, if you list three items, you can use your fingers to tick off each item as you discuss it. Similarly, if you state that profits will be flat, you can make a level motion with your hand for emphasis.

You may be speaking behind a podium or standing beside a projection screen, or both. If you feel uncomfortable standing in one place and you can walk without blocking the audience's view of the screen, feel free to move around. You can emphasize a transition by changing your position. If you tend to fidget, shift, or rock from one foot to the other, try to anchor yourself. A favorite technique of some speakers is to come from behind the podium and place one hand on it while speaking. They get the anchoring effect of the podium while removing the barrier it places between them and the audience. Use the stance or technique that makes you feel most comfortable, as long as your posture or actions do not distract the audience.

As you draft your presentation, repeat key points to emphasize them. For example, suppose your main point is that outsourcing labor will provide the greatest gains in net income. Begin by previewing that concept, and state that you will demonstrate how outsourcing labor will yield the biggest profits. Then provide statistics that support your claim, and show visual aids that graphically illustrate your point. Summarize by repeating your point: "As you can see, outsourcing labor does yield the biggest profits."

Relying on Graphics to Support Your Talk

As you write the main body, think of how to integrate graphics into your presentation. Do not waste words with a long description if a graphic can bring instant comprehension. For instance, instead of describing how information from a query can be turned into a report, show the query and a completed report. Figures F-1 and F-2 illustrate an Access query and the resulting report.

Order Query 1					
Customer Name	City	Product Name	Qty	Price per Unit	Total
Applewood Restaurant	Martinsburg	Frozen Alligator on a Stick	20	$27.99	$559.80
Applewood Restaurant	Martinsburg	Nogales Chipotle Sauce	15	$11.49	$172.35
Applewood Restaurant	Martinsburg	Mom's Deep Dish Apple Pie	12	$12.49	$149.88
Fresh Catch Fishery	Salem	Brumley's Seafood Cocktail Sauce	24	$4.79	$114.96
Fresh Catch Fishery	Salem	NY Smoked Salmon	21	$21.99	$461.79
Fresh Catch Fishery	Salem	Mama Mia's Tiramisu	15	$17.99	$269.85
Jimmy's Crab House	Elkton	Frozen Alligator on a Stick	12	$27.99	$335.88
Jimmy's Crab House	Elkton	Brumley's Seafood Cocktail Sauce	24	$4.79	$114.96
Jimmy's Crab House	Elkton	Mama Mia's Tiramisu	18	$17.99	$323.82
Jimmy's Crab House	Elkton	Mom's Deep Dish Apple Pie	36	$12.49	$449.64

Source: Used with permission from Microsoft Corporation

FIGURE F-1 Access query

| May 2012 Orders--Fine Foods, Inc. | | | | Monday, June 18, 2012 | |
| | | | | 3:48:30 PM | |
Customer Name	City	Product Name	Qty	Price per Unit	Total
Applewood Restaurant	Martinsburg	Frozen Alligator on a Stick	20	$27.99	$559.80
Applewood Restaurant	Martinsburg	Nogales Chipotle Sauce	15	$11.49	$172.35
Applewood Restaurant	Martinsburg	Mom's Deep Dish Apple Pie	12	$12.49	$149.88
Fresh Catch Fishery	Salem	Brumley's Seafood Cocktail Sauce	24	$4.79	$114.96
Fresh Catch Fishery	Salem	NY Smoked Salmon	21	$21.99	$461.79
Fresh Catch Fishery	Salem	Mama Mia's Tiramisu	15	$17.99	$269.85
Jimmy's Crab House	Elkton	Frozen Alligator on a Stick	12	$27.99	$335.88
Jimmy's Crab House	Elkton	Brumley's Seafood Cocktail Sauce	24	$4.79	$114.96
Jimmy's Crab House	Elkton	Mama Mia's Tiramisu	18	$17.99	$323.82
Jimmy's Crab House	Elkton	Mom's Deep Dish Apple Pie	36	$12.49	$449.64
				Total Orders	$2,952.93
					Page 1 of 1

Source: Used with permission from Microsoft Corporation

FIGURE F-2   Access report

Also consider what kinds of graphic media are available and how well you can use them. Your employer will expect you to be able to use Microsoft PowerPoint to prepare your presentation as a slide show. Luckily, many college freshmen are required to take an introductory course that covers Microsoft Office and PowerPoint. If you are not familiar with PowerPoint, several excellent tutorials on the Web can help you learn the basics.

Anticipating the Unexpected

Even though you are only drafting your report and presentation at this stage, eventually you will answer questions from the audience. Being able to handle questions smoothly is the mark of a business professional. The first steps to addressing audience questions are being able to anticipate them and preparing your answers.

You will not use all the facts you gather for your report or presentation. However, as you draft your report, you might want to jot down those facts and keep them handy, in case you need them to answer questions from the audience. PowerPoint has a Notes section where you can include notes for each slide and print them to help you answer questions that arise during your presentation. You will learn how to print notes for your slides later in the tutorial.

The questions you receive depend on the nature of your presentation. For example, during a presentation of an Excel decision model, you might be asked why you are not recommending a certain course of action, or why you left it out of your report. If you have already prepared notes that anticipate such questions, you will probably remember your answers without even having to refer to the notes.

Another potential problem is determining how much technical detail you should display in your slides. In one sense, writing your report will be easier because you can include any graphics, tables, or data you want. Because you have a time limit for your presentation, the question of what to include or leave out becomes more challenging. One approach to this problem is to create more slides than you think you need, and then use the Hide Slide option in PowerPoint to "hide" the extra slides. For example, you might create slides that contain technical details you do not think you will have time to present. However, if you are asked for more details on a particular technical point, you can "unhide" a slide and display the detailed information needed to answer the question. You will learn more about the Hide Slide and Unhide Slide options later in the tutorial.

Writing the Introduction

After you have written the main body of your report and presentation, you can develop the introduction. The introduction should be only a paragraph or two, and it should preview the main points you will cover.

For some of the Access cases, you might want to include general information about databases: what they can do, why they are used, and how they can help a company become more efficient and profitable. You will not need to say much about the business operation because the audience already works for the company.

For the Excel cases, you might want to include an introduction of the general business scenario and describe any assumptions you used to create and run your decision support models. Excel is used for decision support, so you should describe the decision criteria you selected for the model.

Writing the Conclusion

Every good report or presentation needs a good ending. Do not leave the audience hanging. Your conclusion should be brief—only a paragraph or two—and it should give your presentation a sense of closure. Use the conclusion to repeat your main points or, for the Excel cases, to recap your findings and recommendations.

On many occasions, information learned during a business project reveals new opportunities for other projects. Your conclusion should provide closure for the immediate project, but if the project reveals possibilities for future improvements, include them in a "path forward" statement.

CREATING GRAPHICS

Visual aids are a powerful means of getting your point across and making it understandable to your audience. Visual aids come in a variety of forms, some of which are more effective than others. The integrated graphics tools in Microsoft Office can help you prepare a presentation with powerful impact.

Choosing Presentation Media

The media you use will depend on the situation and the media you have available, but remember: *You must maintain control of the media or you will lose the attention of your audience.*

The following list highlights the most common media used in a classroom or business conference room, along with their strengths and weaknesses:

- **PowerPoint slides and a projection system**—These are the predominant presentation media for academic and business use. You can use a portable screen and a simple projector hooked up to a PC, or you can use a full multimedia center. Also, although they are not yet universal in business, touch-sensitive projection screens (for example, Smart Board™ technology) are gaining popularity in college classrooms. The ability to project and display slides, video and sound clips, and live Web pages makes the projection system a powerful presentation tool. *Negatives:* Depending on the complexity of the equipment, you might have difficulties setting it up and getting it to work properly. Also, you often must darken the room to use the projector, and it may be difficult to refer to written notes during your presentation. When using presentation media, you must be able to access and load your PowerPoint file easily. Make sure your file is available from at least two sources that the equipment can access, such as a thumb drive, CD, DVD, or online folder. If your presentation has active links to Web pages, make sure that the presentation computer has Internet access.
- **Handouts**—You can create handouts of your presentation for the audience, which once was the norm for many business meetings. Handouts allow the audience to take notes on applicable slides. If the numbers on a screen are hard to read from the back of the room, your audience can refer to their handouts. With the growing emergence of "green" business practices, however, unnecessary paper use is being discouraged. Many businesses now require reports and presentation slides to be posted at a common site where the audience can access them later. Often, this site is a "public" drive on a business network. *Negatives:* Giving your audience reading material may distract their attention from your presentation. They could read your slides and possibly draw wrong conclusions from them before you have a chance to explain them.

- **Overhead transparencies**—Transparencies are rarely used anymore in business, but some academics prefer them, particularly if they have to write numbers, equations, or formulas on a display large enough for students to see from the back row in a lecture hall. *Negatives:* Transparencies require an overhead projector, and frequently their edges are visually distorted due to the design of the projector lens. You have to use special transparency sheets in a photocopier to create your slides. For both reasons, it is best to avoid using overheads.
- **Whiteboards**—Whiteboards are common in both the business conference room and the classroom. They are useful for posting questions or brainstorming, but you should not use one in your presentation. *Negatives:* You have to face away from your audience to use a whiteboard, and if you are not used to writing on one, it can be difficult to write text that is large enough and legible. Use whiteboards only to jot down questions or ideas that you will check on after the presentation is finished.
- **Flip charts**—Flip charts (also known as easel boards) are large pads of paper on a portable stand. They are used like whiteboards, except that you do not erase your work when the page is full— you flip over to a fresh sheet. Like whiteboards, flip charts are useful for capturing questions or ideas that you want to research after the presentation is finished. Flip charts have the same negatives as whiteboards. Their one advantage is that you can tear off the paper and take it with you when you leave.

Creating Graphs and Charts

Strictly speaking, charts and graphs are not the same thing, although many graphs are referred to as charts. Usually charts show relationships and graphs show change. However, Excel makes no distinction and calls both entities *charts*.

Charts are easy to create in Excel. Unfortunately, the process is so easy that people frequently create graphics that are meaningless, misleading, or inaccurate. This section explains how to select the most appropriate graphics.

You should use pie charts to display data that is related to a whole. For example, you might use a pie chart when breaking down manufacturing costs into Direct Materials, Direct Labor, and Manufacturing Overhead, as shown in Figure F-3. (Note that when you create a pie chart, Excel will convert the numbers you want to graph into percentages of 100.)

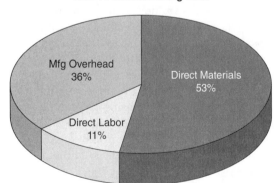

LCD TV Manufacturing Cost

Source: Used with permission from Microsoft Corporation
FIGURE F-3   3D Pie chart: appropriate use

You would *not*, however, use a pie chart to display a company's sales over a three-year period. For example, the pie chart in Figure F-4 is meaningless because it is not useful to think of the period "as a whole" or the years as its "parts."

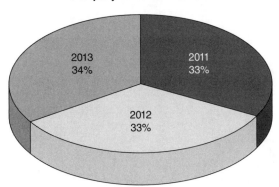

Source: Used with permission from Microsoft Corporation
FIGURE F-4   3D Pie chart: inappropriate use

You should use vertical bar charts (also called column charts) to compare several amounts at the same time, or to compare the same data collected for successive periods of time. The same type of company sales data shown incorrectly in Figure F-4 can be compared correctly using a vertical bar chart (see Figure F-5).

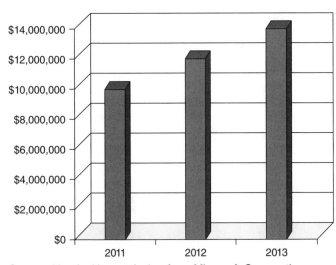

Source: Used with permission from Microsoft Corporation
FIGURE F-5   3D Column chart: appropriate use

As another example, you might want to compare the sales revenues from several different products. You can use a clustered bar chart to show changes in each product's sales over time, as in Figure F-6. This type of bar chart is called a "clustered column" chart in Excel.

When building a chart, include labels that explain the graphics. For instance, when using a graph with an x- and y-axis, you should show what each axis represents so your audience does not puzzle over the graphic while you are speaking. Figures F-6 and F-7 illustrate the necessity of good labels.

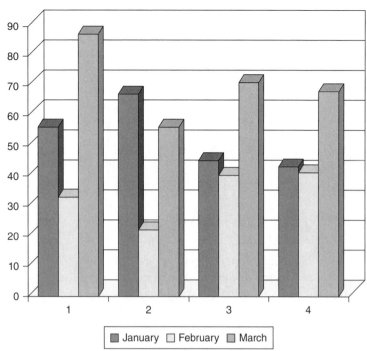

Source: Used with permission from Microsoft Corporation
FIGURE F-6   3-D clustered column graph without labels

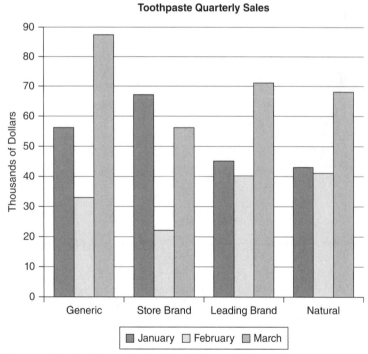

Source: Used with permission from Microsoft Corporation
FIGURE F-7   Graph with labels

In Figure F-6, the graph has no title and neither axis is labeled. Are the amounts in units or dollars? What elements are represented by each cluster of bars? In contrast, Figure F-7 provides a comprehensive snapshot of product sales, which would support a talk rather than create confusion.

Another common pitfall of visual aids is charts that have a misleading premise. For example, suppose you want to show how sales are distributed among your inventory, and their contribution to net income. If you simply take the number of items sold in a given month, as displayed in Figure F-8, the visual fails to give your audience a sense of the actual dollar value of those sales. It is far more appropriate and informative to graph the net income for the items sold instead of the number of items sold. The graph in Figure F-9 provides a more accurate picture of which items contribute the most to net income.

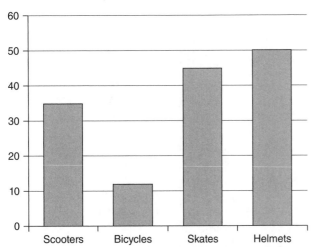

Source: Used with permission from Microsoft Corporation

FIGURE F-8   Graph of number of items sold that does not reflect generated income

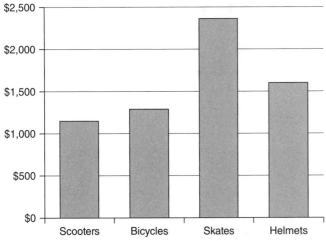

Source: Used with permission from Microsoft Corporation

FIGURE F-9   Graph of net income by item sold

You should also avoid putting too much data in a single comparative chart. For example, assume that you want to compare monthly mortgage payments for two loan amounts with different interest rates and time frames. You have a spreadsheet that computes the payment data, as shown in Figure F-10.

	A	B	C	D	E	F	G
1	**Calculation of Monthly Payment**						
2	Rate	6.00%	6.10%	6.20%	6.30%	6.40%	6.50%
3	Amount	$ 100,000	$ 100,000	$ 100,000	$ 100,000	$ 100,000	$ 100,000
4	Payment (360 Payments)	$ 599	$ 605	$ 612	$ 618	$ 625	$ 632
5	Payment (180 Payments)	$ 843	$ 849	$ 854	$ 860	$ 865	$ 871
6	Amount	$ 150,000	$ 150,000	$ 150,000	$ 150,000	$ 150,000	$ 150,000
7	Payment (360 Payments)	$ 899	$ 908	$ 918	$ 928	$ 938	$ 948
8	Payment (180 Payments)	$ 1,265	$ 1,273	$ 1,282	$ 1,290	$ 1,298	$ 1,306

Source: Used with permission from Microsoft Corporation

FIGURE F-10   Calculation of monthly payment

In Excel, it is possible (but not advisable) to capture all of the information in a single clustered column chart, as shown in Figure F-11.

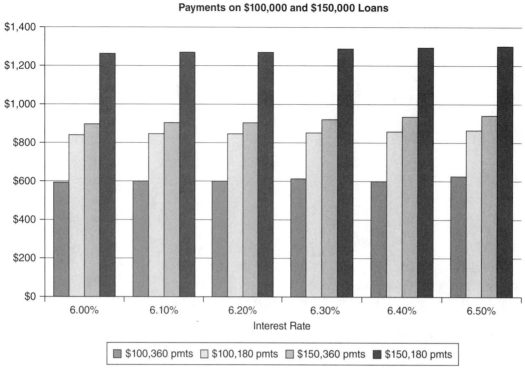

Source: Used with permission from Microsoft Corporation

FIGURE F-11   Too much information in one chart

The chart contains a great deal of information. Putting the $100,000 and $150,000 loan payments in the same "cluster" may confuse the readers. They would probably find it easier to understand one chart that summarizes the $100,000 loan (see Figure F-12) and a second chart that covers the $150,000 loan.

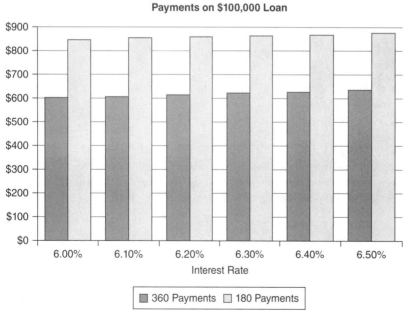

Source: Used with permission from Microsoft Corporation

FIGURE F-12   Good balance of information

You could then augment the charts with text that summarizes the main differences between the payments for each loan amount. In that fashion, the reader is led step by step through the analysis.

Excel no longer has a Chart Wizard; instead, the Insert tab includes a Charts group. Once you create a chart and click it, three chart-specific tabs appear under a Chart Tools heading on the Ribbon to assist you with chart design, layout, and formatting. If you are unfamiliar with the charting tools in Excel, ask your instructor for guidance or refer to the many Excel tutorials on the Web.

Creating PowerPoint Presentations

PowerPoint presentations are easy to create. When you open PowerPoint, it starts a new presentation for you. You can select from many different themes, styles, and slide layouts by clicking the Design tab. If none of PowerPoint's default themes suit you, you can download theme "templates" from Microsoft Office Online. When choosing a theme and style for your slides, such as background colors or graphics, fonts, and fills, keep the following guidelines in mind:

- In older versions of PowerPoint, users were advised to avoid pastel backgrounds or theme colors and to keep their slide backgrounds dark. Because of the increasing quality of graphics in both computer hardware and projection systems, most of the default themes in PowerPoint will project well and be easy to read.
- If your projection screen is small or your presentation room is large, consider using boldface type for all of your text to make it readable from the back of the room. If you have time to visit the presentation site beforehand, bring your PowerPoint file, project a slide on the screen, and look at it from the back row of the audience area. If you can read the text, the font is large enough.
- Use transitions and animations to keep your presentation lively, but do not go overboard with them. Swirling letters and pinwheeling words can distract the audience from your presentation.
- It is an excellent idea to animate the text on your slides with entrance effects so that only one bullet point appears at a time when you click the mouse (or when you tap the screen using a touch-sensitive board). This approach prevents your audience from reading ahead of the bullet point being discussed and keeps their attention on you. Entrance effects can be incorporated and managed using the Add Animation button in PowerPoint 2010, as shown in Figures F-13 and F-14.

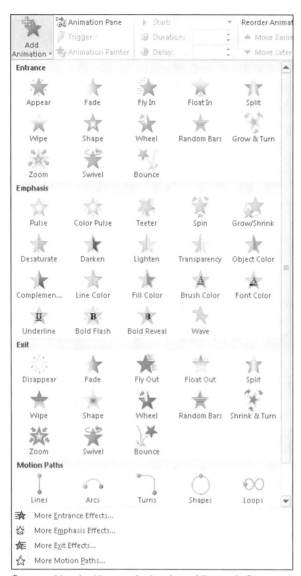

Source: Used with permission from Microsoft Corporation

FIGURE F-13 The Add Animation button on the Ribbon in PowerPoint

Source: Used with permission from Microsoft Corporation

FIGURE F-14 Add Entrance Effect window

NOTE—DIFFERENCES IN POWERPOINT ANIMATION TOOLS—2010 VS. 2007

The structure of the animation tools has changed considerably from PowerPoint 2007 to the 2010 version. The Custom Animation button and pane are both gone. Most of the custom animation tools are now incorporated using the Add Animation button in PowerPoint 2010. The look and feel is different, but the interface is more intuitive and easier to use. You can still use an animation pane to organize and edit your animations within a slide.

- Consider creating PowerPoint slides that have a section for your notes. You can print the notes from the Print dialog box by choosing Notes Pages from the Print menu, as shown in Figure F-15. Each slide will be printed at half its normal size, and your notes will appear beneath each slide, as shown in Figure F-16.

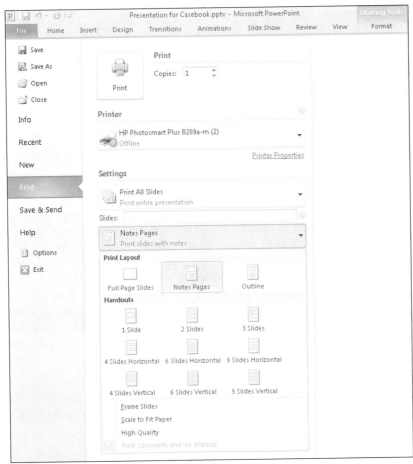

Source: Used with permission from Microsoft Corporation

FIGURE F-15 Printing notes page

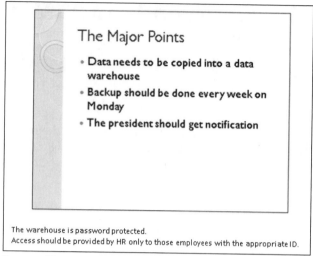

Source: Used with permission from Microsoft Corporation

FIGURE F-16   Sample notes page

- Finally, you should check your PowerPoint slides on a projection screen before your presentation. Information that looks good on a computer display may not be readable on the projection screen.

Using Visual Aids Effectively

Make sure you choose the visual aids that will work most effectively, and that you have enough without using too many. How many is too many? The amount of time you have to speak will determine the number of visual aids you should use, as will your target audience. A good rule of thumb is to allow at least one minute to present each PowerPoint slide. Leave a minimum of two minutes for audience questions after a 10-minute presentation, and allow up to 25 percent of your total presentation time to address questions after longer presentations. (For example, for a 20-minute presentation, figure on taking five minutes for questions.) For a 10-minute talk, try to keep the body of your presentation to eight slides or less. Your target audience will also influence your selection of visual aids. For instance, your slides will need more graphics and animation if you are addressing a group of teenagers than if you are presenting to a board of directors. Remember to use visual aids to emphasize your main points, not to detract from them.

Review each of your slides and visual aids to make sure it meets the following criteria:

- The font size of the text is large enough to read from the back of the presentation area.
- The slide or visual aid does not contain misleading graphics, typographical errors, or misspelled words—the quality of your work is a direct reflection on you.
- The content of your visual aid is relevant to the key points of your presentation.
- The slide or visual aid does not detract from your message. Your animations, pictures, and sound effects should support the text. Your visuals should look professional.
- A visual aid should look good in the presentation environment. If possible, rehearse your PowerPoint slides beforehand in the room where you will give the presentation. Make sure you can read your slides easily from the back row of seats in the room. If you have a friend who can sit in, ask her or him to listen to your voice from the back row of seats. If you have trouble projecting your voice clearly, consider using a microphone for your presentation.
- All numbers should be rounded unless decimals or pennies are crucial. For example, your company might only pay fractions of a cent per Web hit, but this cost may become significant after millions of Web hits.
- Slides should not look too busy or crowded. Many PowerPoint experts have a "6 by 6" rule for bullet points on a slide, which means you should include no more than six bullet points per slide and no more than six words per bullet point. Also avoid putting too many labels or pictures on a slide. Clip art can be "cutesy" and therefore has no place in a professional business presentation. A well-selected picture or two can add emphasis to the theme of a slide. For examples of a slide that is too busy versus one that conveys its points succinctly, see Figures F-17 and F-18.

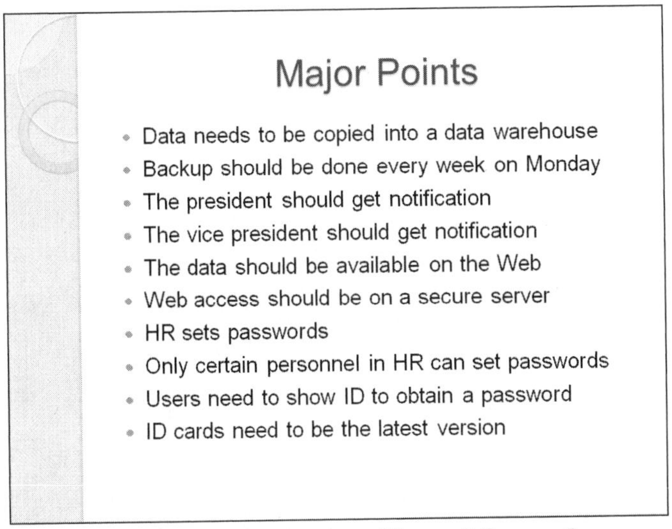

Major Points

- Data needs to be copied into a data warehouse
- Backup should be done every week on Monday
- The president should get notification
- The vice president should get notification
- The data should be available on the Web
- Web access should be on a secure server
- HR sets passwords
- Only certain personnel in HR can set passwords
- Users need to show ID to obtain a password
- ID cards need to be the latest version

Source: Used with permission from Microsoft Corporation
FIGURE F-17   Busy slide

The Major Points

- Data needs to be copied into a data warehouse
- Backup should be done every week on Monday
- The president should get notification

Source: Used with permission from Microsoft Corporation
FIGURE F-18   Slide with appropriate number of bullet points and a supporting photo

You may find that you have created more slides than you have time to present, and you are unsure of which slides you should delete. Some may have data that an audience member might ask about. Fortunately, PowerPoint lets you "hide" slides; these hidden slides will not be displayed in Slide Show view unless you "unhide" them in Normal view. Hiding slides is an excellent way to keep detailed data handy in case your audience asks to see it. Figure F-19 shows how to hide a slide in a PowerPoint presentation. Right-click the slide you want to hide, and then click Hide Slide from the menu to mark the slide as hidden in the presentation. To unhide the slide, right-click it and then click Unhide Slide from the menu. Click the slide to display it in Slide Show view.

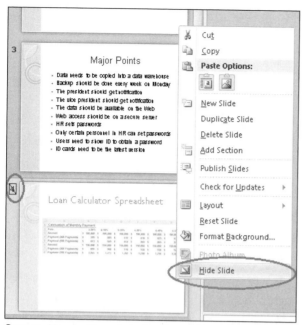

Source: Used with permission from Microsoft Corporation

FIGURE F-19 Hiding a slide in PowerPoint

PRACTICING YOUR DELIVERY

Surveys indicate that public speaking is the greatest fear of many people. However, fear or nervousness can be channeled into positive energy to do a good job. Remember that an audience is not likely to think you are nervous unless you fidget or your voice cracks. Audience members want to hear what you have to say, so think about them and their interests—not about how you feel.

Your presentations for the cases in this textbook will occur in a classroom setting with 20 to 40 students. Ask yourself: Am I afraid when I talk to just one or two of my classmates? The answer is probably no. In addition, they will all have to give presentations as well. Think of your presentation as an extended conversation with several classmates. Let your gaze move from person to person, making brief eye contact with each of them randomly as you speak. As your focus moves from one person to another, think to yourself: I am speaking to one person at a time. As you become more proficient in speaking before a group, your gaze will move naturally among audience members.

Tips for Practicing Your Delivery

Giving an effective presentation is not the same as reading a report to an audience. You should rehearse your message well enough so that you can present it naturally and confidently, with your slides or other visual aids smoothly intermingled with your speaking. The following tips will help you hone the effectiveness of your delivery:

- Practice your presentation several times, and use your visual aids when you practice.
- Show your slides at the right time. Luckily, PowerPoint makes this easy; you can click the slide when you are ready to talk about it. Use cues as necessary in your speaker's notes.
- Maintain eye and voice contact with the audience when using the visual aid. Do not turn your back on your audience. It is acceptable to turn sideways to glance at your slide. A popular trick of experienced speakers is to walk around and steal a glance at the slide while moving to a new position.
- Refer to your visual aids in your talk and use hand gestures where appropriate. Do not ignore your own visual aid, but do not read it to your audience—they can read for themselves.
- Keep in mind that your slides or visual aids should support your presentation, not *be* the presentation. Do not try to crowd the slide with everything you plan to say. Use the slides to illustrate key points and statistics, and fill in the rest of the content with your talk.
- Check your time, especially when practicing. If you stay within the time limit when practicing, you will probably finish a minute or two early when you actually give the presentation. You will be a little nervous and will talk a little faster to a live audience.

- Use numbers effectively. When speaking, use rounded numbers; otherwise, you will sound like a computer. Also make numbers as meaningful as possible. For example, instead of saying "in 83 percent of cases," say "in five out of six cases."
- Do not extrapolate, speculate, or otherwise "reach" to interpret the output of statistical models. For example, suppose your Excel model has many input variables. You might be able to point out a trend, but often you cannot say with mathematical certainty that if a company employs the inputs in the same combination, it will get the same results.
- Some people prefer recording their presentation and playing it back to evaluate themselves. It is amazing how many people are shocked when they hear their recorded voice—and usually they are not pleased with it. In addition, you will hear every *um, uh, well, you know*, throat-clearing noise, and other verbal distraction in your speech. If you want feedback on your presentation, have a friend listen to it.
- If you use a pointer, be careful where you wave it. It is not a light saber, and you are not Luke Skywalker. Unless you absolutely have to use one to point at crucial data on a slide, leave the pointer home.

Handling Questions

Fielding questions from an audience can be tricky because you cannot anticipate all of the questions you might be asked. When answering questions from an audience, *treat everyone with courtesy and respect.* Use the following strategies to handle questions:

- Try to anticipate as many questions as possible, and prepare answers in advance. Remember that you can gather much of the information to prepare those answers while drafting your presentation. The Notes section under each slide in PowerPoint is a good place to write anticipated questions and your answers. Hidden slides can also contain the data you need to answer questions about important details.
- Mention at the beginning of your talk that you will take questions at the end of the presentation, which helps prevent questions from interrupting the flow and timing of your talk. In fact, many PowerPoint presentations end with a Questions slide. If someone tries to interrupt, say that you will be happy to answer the question when you are finished, or that the next graphic answers the question. Of course, this point does not apply to the company CEO—you *always* stop to answer the CEO's questions.
- When answering a question, a good practice is to repeat the question if you have any doubt that the entire audience heard it. Then deliver your answer to the whole audience, but make sure you close by looking directly at the person who asked the question.
- Strive to be informative, not persuasive. In other words, use facts to answer questions. For instance, if someone asks your opinion about a given outcome, you might show an Excel slide that displays the Solver's output; then you can use the data as the basis for answering the question. In that light, it is probably a good idea to have computer access to your Excel model or Access database if your presentation venue permits it, but avoid using either unless you absolutely need it.
- If you do not know the answer to a question, it is acceptable to say so, and it is certainly better than trying to fake the answer. For instance, if someone asks you the difference between the Simplex LP and GRG solving methods in Excel Solver, you might say, "That is an excellent question, but I really don't know the answer—let me research it and get back to you." Then follow up after the presentation by researching the answer and contacting the person who asked the question.
- Signal when you are finished. You might say that you have time for one more question. Wrap up the talk yourself and thank your audience for their attention.

Handling a "Problem" Audience

A "problem" audience or a heckler is every speaker's nightmare. Fortunately, this experience is rare in the classroom: Your audience will consist of classmates who also have to give presentations, and your instructor will be present to intervene in case of problems.

Heckling can be a common occurrence in the political arena, but it does not happen often in the business world. Most senior managers will not tolerate unprofessional conduct in a business meeting. However, fellow business associates might challenge you in what you perceive as a hostile manner. If so, remain calm, be professional, and rely on facts. The rest of the audience will watch to see how you react—if you behave professionally, you make the heckler appear unprofessional by comparison and you'll gain the empathy of the audience.

A more common problem is a question from an audience member who lacks technical expertise. For instance, suppose you explained how to enter data into an Access form, but someone did not understand your explanation. Ask the questioner what part of the explanation was confusing. If you can answer the question briefly and clearly, do so. If your answer turns into a time-consuming dialogue, offer to give the person a one-on-one explanation after the presentation.

Another common problem is receiving a question that you have already answered. The best solution is to give the answer again, as briefly as possible, using different words in case your original answer confused the person. If someone persists in asking questions that have obvious answers, you might ask the audience, "Who would like to answer that question?" The questioner should get the hint.

PRESENTATION TOOLKIT

You can use the form in Figure F-20 for preparation, the form in Figure F-21 for evaluation of Access presentations, and the form in Figure F-22 for evaluation of Excel presentations.

Preparation Checklist

Facilities and Equipment

☐ The room contains the equipment that I need.
☐ The equipment works and I've tested it with my visual aids.
☐ Outlets and electrical cords are available and sufficient.
☐ All the chairs are aligned so that everyone can see me and hear me.
☐ Everyone will be able to see my visual aids.
☐ The lights can be dimmed when/if needed.
☐ Sufficient light will be available so I can read my notes when the lights are dimmed.

Presentation Materials

☐ My notes are available, and I can read them while standing up.
☐ My visual aids are assembled in the order that I'll use them.
☐ A laser pointer or a wand will be available if needed.

Self

☐ I've practiced my delivery.
☐ I am comfortable with my presentation and visual aids.
☐ I am prepared to answer questions.
☐ I can dress appropriately for the situation.

Source: Used with permission from Microsoft Corporation

FIGURE F-20   Preparation checklist

Evaluating Access Presentations

Course: _____   Speaker: _____   Date: _____

Rate the presentation by these criteria:
4=Outstanding   3=Good   2=Adequate   1=Needs Improvement
N/A=Not Applicable

Content

_____ The presentation contained a brief and effective introduction.

_____ Main ideas were easy to follow and understand.

_____ Explanation of database design was clear and logical.

_____ Explanation of using the form was easy to understand.

_____ Explanation of running the queries and their output was clear.

_____ Explanation of the report was clear, logical, and useful.

_____ Additional recommendations for database use were helpful.

_____ Visuals were appropriate for the audience and the task.

_____ Visuals were understandable, visible, and correct.

_____ The conclusion was satisfying and gave a sense of closure.

Delivery

_____ Was poised, confident, and in control of the audience

_____ Made eye contact

_____ Spoke clearly, distinctly, and naturally

_____ Avoided using slang and poor grammar

_____ Avoided distracting mannerisms

_____ Employed natural gestures

_____ Used visual aids with ease

_____ Was courteous and professional when answering questions

_____ Did not exceed time limit

Submitted by: _____

Source: Used with permission from Microsoft Corporation

FIGURE F-21   Form for evaluation of Access presentations

Evaluating Excel Presentations

Course: _____ Speaker: _____ Date: _____

Rate the presentation by these criteria:
4=Outstanding   3=Good   2=Adequate   1=Needs Improvement
N/A=Not Applicable

Content

_____ The presentation contained a brief and effective introduction.

_____ The explanation of assumptions and goals was clear and logical.

_____ The explanation of software output was logically organized.

_____ The explanation of software output was thorough.

_____ Effective transitions linked main ideas.

_____ Solid facts supported final recommendations.

_____ Visuals were appropriate for the audience and the task.

_____ Visuals were understandable, visible, and correct.

_____ The conclusion was satisfying and gave a sense of closure.

Delivery

_____ Was poised, confident, and in control of the audience

_____ Made eye contact

_____ Spoke clearly, distinctly, and naturally

_____ Avoided using slang and poor grammar

_____ Avoided distracting mannerisms

_____ Employed natural gestures

_____ Used visual aids with ease

_____ Was courteous and professional when answering questions

_____ Did not exceed time limit

Submitted by: _____

Source: Used with permission from Microsoft Corporation

FIGURE F-22   Form for evaluation of Excel presentations

E